Roadside Geology of IDAHO

Second Edition

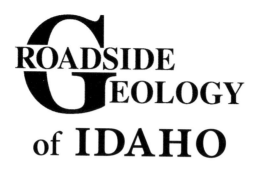

ROADSIDE GEOLOGY of IDAHO

Second Edition

Paul K. Link
Shawn Willsey
Keegan Schmidt

ILLUSTRATED BY
Chelsea M. Feeney

2021
Mountain Press Publishing Company
Missoula, Montana

Text © 2021 by Paul K. Link, Shawn Willsey, and Keegan Schmidt

First printing, May 2021
All rights reserved

Geologic maps constructed by Chelsea M. Feeney (www.cmcfeeney.com) using geologic data from the Idaho Geological Survey.

Photos by authors unless otherwise credited.
Back cover photo:
City of Rocks National Reserve. —Photo by Wallace Keck

Roadside Geology is a registered trademark of Mountain Press Publishing Company.

Library of Congress Cataloging-in-Publication Data

Names: Link, P. K., author. | Willsey, Shawn, author. | Schmidt, Keegan L., author.
Title: Roadside geology of Idaho / Paul K. Link, Shawn Willsey, Keegan Schmidt ; illustrated by Chelsea M. Feeney.
Description: Second edition. | Missoula, Montana : Mountain Press Publishing Company, 2021. | Series: Roadside geology series | Revised edition of: Roadside geology of Idaho / David Alt, Donald W. Hyndman. 1989. | Includes bibliographical references and index. | Summary: "Learn about the remarkable geologic diversity of the Gem State with the completely revised, full-color edition of *Roadside Geology of Idaho*. Excellent graphics, spectacular photographs, and straightforward writing describe and interpret the rocks and landscapes visible outside your car window. With this book as their travel companion, residents and visitors alike are sure to understand and appreciate Idaho's sprawling plains, forested hills, and deep canyons in a completely new way"—Provided by publisher.
Identifiers: LCCN 2021008899 | ISBN 9780878427024 (trade paperback)
Subjects: LCSH: Geology—Idaho—Guidebooks. | Idaho—Guidebooks.
Classification: LCC QE103 .A37 2021 | DDC 557.96—dc23
LC record available at https://lccn.loc.gov/2021008899

PRINTED IN THE UNITED STATES

P.O. Box 2399 • Missoula, MT 59806 • 406-728-1900
800-234-5308 • info@mtnpress.com
www.mountain-press.com

To those exploring the amazing landscapes and geology of Idaho.
Enjoy your adventure!

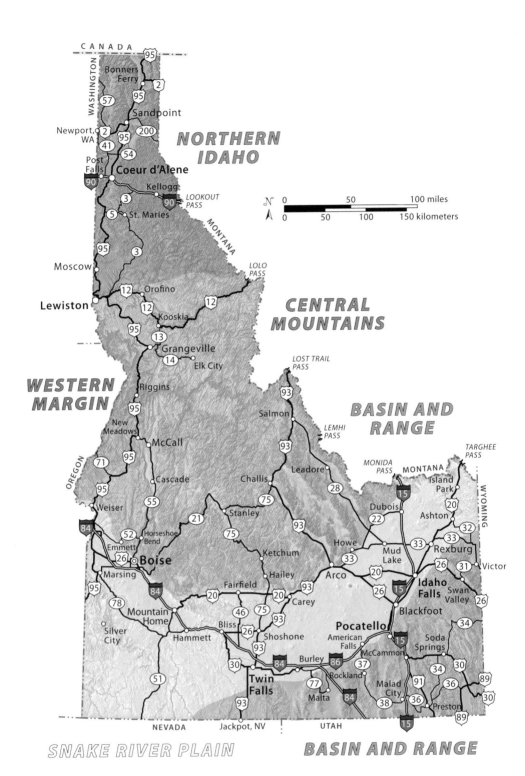

Roads and sections of *Roadside Geology of Idaho*.

CONTENTS

IDAHO GEOLOGY MAP   x

PREFACE   xii

INTRODUCTION   1
   Geology Basics   2
   Plate Tectonics   5
   Idaho's Geologic History   6

SNAKE RIVER PLAIN   15
   Yellowstone Hot Spot   16
   Basaltic Volcanism   19
   Lake Idaho   22
   Owyhee Mountains   24
   Bonneville Flood   24
   <u>Guides to the Snake River Plain</u>
   I-15: Pocatello—Idaho Falls—Monida Pass at Montana Border   26
   I-84: Oregon Border—Boise—Twin Falls—I-86 Junction   31
      Malad Gorge of Thousand Springs State Park   37
   I-86: I-84 Junction—Pocatello   40
   US 20: Mountain Home—Carey   43
   US 20: Arco—Idaho Falls   47
   US 20: Idaho Falls—Montana Border   50
      1976 Teton Dam Disaster   52
      Mesa Falls Scenic Byway   53
   US 26: US 20 Junction—Blackfoot   56
   US 30: Bliss—Twin Falls   58
      Bliss Slide   61
      Hagerman Fossil Beds National Monument   63
      Ritter Island and Box Canyon   63
   US 93: Jackpot, Nevada—Shoshone—Arco   66
      Snake River Canyon   70
      Craters of the Moon National Monument and Preserve   74
   US 95: Oregon Border—Weiser   78
   ID 32 (Teton Scenic Byway): Tetonia—Ashton   81
   ID 33: US 26 Junction—Howe—Rexburg   83
      ID 22 to Dubois   86
      Menan Buttes   87
   ID 46: Wendell—Fairfield   88
      Little City of Rocks   89
   ID 51: Nevada Border—Mountain Home   90

ID 75: Shoshone—Timmerman Junction   94
   Black Magic Canyon   96
ID 78: Marsing—Hammett   97
   Lizard Butte   99
   Upper Reynolds Creek Road   100
   Celebration Park   101
   Bruneau Dunes State Park   102

BASIN AND RANGE   105

Neoproterozoic to Mesozoic Sedimentary Rocks   108
   Phosphate Mining   110
Sevier Fold-and-Thrust Belt   111
Metamorphic Core Complexes   113
Lake Bonneville   114

Guides to the Basin and Range

I-15: Utah Border—Pocatello   114
I-84: I-86 Junction—Utah Border   122
US 26: Idaho Falls—Swan Valley—Wyoming Border   125
US 30: McCammon—Soda Springs—Wyoming Border   129
US 89: Utah Border—Montpelier—Wyoming Border   135
   Minnetonka Cave   136
US 91: Utah Border—Preston—I-15 Junction   138
US 93: Arco—Challis   142
ID 28: Mud Lake—Leadore—Salmon   148
ID 31: Swan Valley—Victor   156
ID 34: Preston—Soda Springs—Wyoming Border   159
ID 36: Malad City—Preston—Montpelier   165
ID 37 and ID 38: American Falls—Malad City   172
ID 77: Malta—Declo   176
   City of Rocks National Reserve   178

WESTERN MARGIN   181

A Pacific Ocean Crustal Collage   182
Terrane Collision and the Salmon River Suture Zone   184
Miocene Flood Basalts   188
Canyon Cutting, Uplift, and Ice-Age Floods   192

Guides to the Western Margin

US 12: Lewiston—Kooskia   194
US 95: Weiser—New Meadows   202
   Hells Canyon Road   205
US 95: New Meadows—Grangeville   209
   Salmon River Road from Riggins   214
   Hells Canyon at Pittsburg Landing   222
US 95: Grangeville—Lewiston Hill   227
ID 13: Grangeville—Kooskia   233
ID 14: ID 13 Junction—Elk City   235

CENTRAL MOUNTAINS   244
  Lemhi Subbasin   245
  Rodinia Rifting   247
  Antler Orogeny   248
  Big Collisions and the Idaho Batholith   249
  Eocene Extension and the Challis Magmatic Event   251
  Miocene, Pliocene, and Pleistocene Time   254

  <u>Guides to the Central Mountains</u>
  US 12: Kooskia—Lolo Pass at Montana Border   255
    Coolwater Culmination along Selway Road   258
  US 93: Challis—Salmon—Lost Trail Pass at Montana Border   261
  ID 21: Boise—Stanley   265
  ID 55: Boise—New Meadows   271
    Snowbank Mountain   276
  ID 75: Timmerman Junction—Ketchum—Stanley   278
    Trail Creek Road   284
  ID 75: Stanley—Challis   288

NORTHERN IDAHO   293
  Ancient Continental Basement Rocks   293
  Belt Supergroup of a Long-Lived Continental Basin   294
  Sevier Fold-and-Thrust Belt and the Kaniksu and Idaho Batholiths   299
  Eocene Magmatism and the Priest River Core Complex   300
  Miocene Flood Basalts on the Edge of the Columbia Plateau   302
  Purcell Ice Lobe and the Rathdrum Prairie   302

  <u>Guides to Northern Idaho</u>
  I-90: Post Falls—Lookout Pass at Montana Border   305
    Tubbs Hill   308
  US 2 and ID 200: Newport—Sandpoint—Montana Border   316
    Priest Lake   318
  US 95: Lewiston—Moscow—Coeur d'Alene   326
  US 95: Coeur d'Alene—Sandpoint—Canadian Border   332
    Farragut State Park   334
    Round Mountain and the Priest River Core Complex   335
    Flood Megaripples and Spirit Lake   336
  ID 3: US 12 Junction—St. Maries—I-90 Junction   342

GLOSSARY   352

REFERENCES   360

INDEX   367

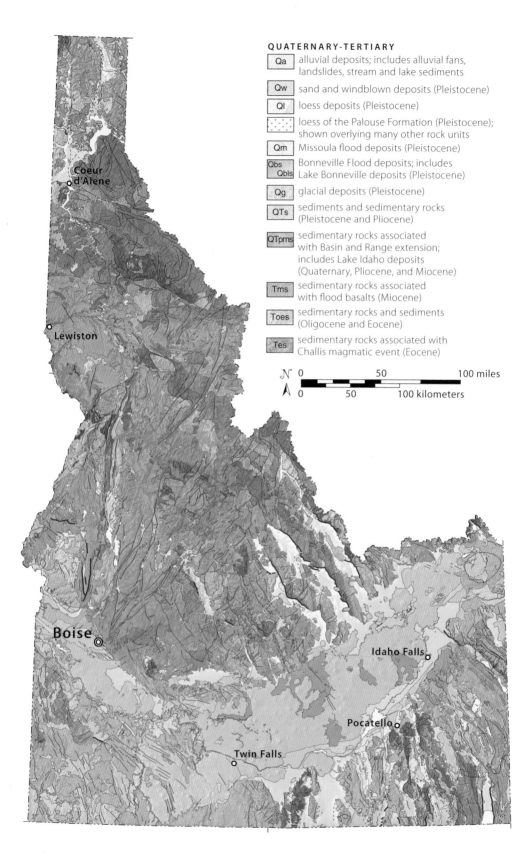

VOLCANIC ROCKS

- Qb — basalt (Quaternary)
- Qr — rhyolite (Pleistocene)
- QTb — basalt (Pleistocene and Pliocene)
- Tpmr — rhyolite (Pliocene and Miocene)
- Tpmb — basalt (Pliocene and Miocene)
- Tmr — rhyolite (Miocene)
- Tmfo — older rhyolite, latite, and andesite (Miocene)
- Tcr — Columbia River Basalt Group (Miocene)
- Tov — volcanic rocks (Oligocene)
- Tcv — Challis Volcanic Group (Eocene)

INTRUSIVE ROCKS

- Toi — granite (Oligocene)
- Tei — Challis intrusive rocks (Eocene)

IDAHO BATHOLITH
- TKg, Kg — granodiorite, granite, and two-mica granite (Paleocene and Cretaceous)
- Ktg — tonalite, granodiorite, and quartz diorite; includes deformed and metamorphosed units (Cretaceous)

BLUE MOUNTAIN PROVINCE
- Ktt — tonalite and trondhjemite (Cretaceous)
- KJqd — quartz diorite (Cretaceous and Jurassic)
- TRPi — intrusive rocks (Triassic and Permian)

- KJp — plutonic rocks along the Western Idaho shear zone (Cretaceous and Jurassic)
- Kis — syenite and related rocks (Cretaceous)
- Ji — tonalite, hornblendite, and gabbro (Jurassic)
- OCi — syenitic intrusive rocks (Ordovician and Cambrian)
- Zi — granitic gneiss (Neoproterozoic)
- Yag — augen gneiss (Mesoproterozoic)
- Yam — amphibolite (Mesoproterozoic)

FAULT SYMBOLS ON ROAD GUIDE MAPS

- recent normal fault; active within last 15,000 years
- fault; dashed where concealed
- normal fault; ball and bar on down-dropped side
- detachment fault; hachures on down-dropped side
- thrust fault; teeth on upper plate
- reactivated thrust fault; teeth on upper plate; bar and ball on down-dropped side on reactivated fault segments
- strike-slip fault; arrows indicate motion

ROCKS OF THE BLUE MOUNTAINS PROVINCE

- KPro — Riggins Group, Orofino Metamorphic Suite, and related rocks (Permian to Cretaceous)
- Jcw — Coon Hollow and Weatherby Formations; includes sedimentary rocks deposited across the combined Olds Ferry and Wallowa terranes (Jurassic)
- JTRsv — sedimentary and volcanic rocks (Jurassic and Triassic)
- JTRof — Olds Ferry terrane (Jurassic and Triassic)
- TRPsd — Seven Devils Group of the Wallowa terrane (Triassic and Permian)
- MzPzb — Baker terrane (Mesozoic and Paleozoic)

SEDIMENTARY BEDROCK

- TKs, Ks — Paleocene-Cretaceous and Cretaceous rocks
- TRJs, MzS — Mesozoic rocks (Triassic to Jurassic)
- Pzls — late Paleozoic rocks (Mississippian to Permian)
- Pzes — early Paleozoic rocks (Cambrian to Devonian)

Windermere Supergroup
- €Zs — sedimentary and metasedimentary rocks (Cambrian and Neoproterozoic)
- Zp — Pocatello Formation (Neoproterozoic)

MESOPROTEROZOIC TO PALEOZOIC

- PzYs — metasedimentary rocks of uncertain age
- Ydt — metasedimentary rocks of the Deer Trail Group (Mesoproterozoic)

MESOPROTEROZOIC BELT SUPERGROUP

- Ymi — Missoula Group
- Ypi — Piegan Group
- Yra — Ravalli Group
- Yp — Prichard Formation
- Ygs — gneiss, schist, and quartzite

MESOPROTEROZOIC METASEDIMENTARY ROCKS OF LEMHI SUBBASIN OF BELT BASIN

- Yal — Apple Creek and Lawson Creek Formations
- Ys, Yha — Swauger Formation; includes Hoodoo Quartzite and argillaceous quartzite
- Yl, Yy — Lemhi Group; includes Yellowjacket Formation
- Yq — quartzitic metamorphic rocks
- Ym — gneissic and schistose metasedimentary rocks

BASEMENT ROCKS

- Yagl — Laclede augen gneiss (Mesoproterozoic)
- Xan — anorthosite (Paleoproterozoic)
- Xog — orthogneiss (Paleoproterozoic)
- XAm — metamorphic rocks (Paleoproterozoic and Archean)

PREFACE

The first edition of *Roadside Geology of Idaho* was published in 1989 by David Alt and Donald Hyndman, a dynamic geologic duo who authored several early books in the iconic Roadside Geology series. Alt and Hyndman were not afraid to advance new theories about geologic history. Sometimes they were right. Since that first edition, our collective understanding and knowledge about Idaho's rocks and landscapes have grown and evolved due to advances in technology, dating techniques, and good old-fashioned, boots-on-the-ground fieldwork by hundreds of geologists and students. With so much more known about Idaho's geology, we present an entirely new version of *Roadside Geology of Idaho*, complete with new maps, photographs, and illustrations. This new edition also includes many highways and roadways not covered by the first edition. We are fiercely proud of this new edition but remain indebted to the original authors of the first edition for establishing the book's foundation.

Exploring, understanding, and describing the outstanding but sometimes complicated geology of an immense state like Idaho is a multiperson job. We joined forces in 2017, with each of us working on the region we know best. In general, Paul Link and Shawn Willsey wrote the Snake River Plain and Basin and Range sections, Keegan Schmidt authored the Western Margin and Northern Idaho sections, and all three authors contributed to the Central Mountains section. By blending our collective expertise, we document and describe the incredible geologic story of Idaho.

It is our hope and desire that this updated edition will inspire lifelong Idahoans, recent transplants, and interested visitors to appreciate and love Idaho in a fun, new way. Idaho's diverse and stunning scenery is even more impressive when you understand its origins and history.

We have attempted to make this edition easy to use and helpful to you, the reader. Road guides are written from either south to north or west to east, but these guides can easily be followed when traveling in the opposite direction. Vivid and colorful geologic maps produced by Chelsea M. Feeney accompany each road guide and provide overall context and points of interest. Most road guides reference mileposts or other prominent markers for locating key geologic features. Photo captions include GPS coordinates of where the outcrop is or where the photo was taken of a distant view. You can type these coordinates directly into the search box in Google Maps.

Beyond the authors, two other key individuals were invaluable in putting this edition together and deserve special recognition. Chelsea M. Feeney masterfully drafted all the illustrations and maps in this book, and those maps are one of its most important assets. Chelsea's expertise as both a designer and geologist was priceless. Her contributions to this book cannot be overstated. Jennifer Carey, our editor at Mountain Press, was the captain of this project. She smoothly worked with the three authors and blended our individual ideas into one cohesive entity. On numerous occasions, Jenn provided focus when one or more of us was lost on some tangential path. Her attention to detail, patience, and perspective are deeply appreciated.

This unnamed lake high above Hell Roaring Lake in the Sawtooth Range of central Idaho lies in a small glacial cirque carved by ice during the Pleistocene ice ages. Most of the Sawtooths are sculpted from Eocene-age granite of the Sawtooth pluton.

A number of colleagues also provided much needed assistance with helping us sort out some of the fine details of Idaho's geology. We wish to thank Mark Anders (Columbia University), Diana Boyack, Dave Pearson, Dave Rodgers and Glenn Thackray (Idaho State University), Andy Buddington (Spokane Community College), Jim Coogan (Western Colorado University), Reed Lewis and Dennis Feeney (Idaho Geological Survey), Terry Maley (retired BLM), Mark McFaddan (retired Northern Idaho College), David Miller (USGS), and Dan Moore (BYU-Idaho). Thank you to Kristy Hill and Nancy Linscott for reviewing early drafts of parts of this book.

GEOLOGIC TIME SCALE

ERA or EON	PERIOD	EPOCH	AGE (mya)*	IMPORTANT GEOLOGIC EVENTS IN IDAHO
CENOZOIC	QUATERNARY	HOLOCENE	0.01	ash from eruption of Mt. Mazama, 7,700 years ago
		PLEISTOCENE		Bonneville Flood along Snake River, 17,500 years ago
				Missoula floods of northern Idaho, 19,000–14,000 years ago
				alpine glaciers in Idaho's high country; Canadian ice sheet advances into Idaho's panhandle region, 200,000–10,000 years ago
				Lake Idaho is captured by a new drainage system through Hells Canyon to create modern Snake River drainage, 3–2 mya
	NEOGENE	PLIOCENE	2.6	Lake Idaho intermittently fills western Snake River Plain supporting rich ecosystem, 10–3 mya
		MIOCENE	5.3	
				eruption of most Columbia River Basalt Group, 17–16 mya
	PALEOGENE (TERTIARY)	OLIGOCENE	23	intrusion of magma and volcanic eruptions coupled with extension in core complexes: 51–40 mya in northern and central Idaho (Challis magmatism), Oligocene in southern Idaho
		EOCENE	33.9	
		PALEOCENE	56	
MESOZOIC	CRETACEOUS		66	northern Bitterroot lobe of Idaho batholith forms, 80–53 mya
				southern Atlanta lobe of Idaho batholith, 98–70 mya
				northern portion of Idaho batholith (Kaniksu batholith) forms in northern Idaho, 120–90 mya
				Sevier orogeny creates thrust faults and folds in eastern Idaho while post-accretion convergence continues in western Idaho, 140–70 mya
	JURASSIC		145	accretion of Blue Mountains terranes in western Idaho along Salmon River suture zone, 150–140 mya
	TRIASSIC		201	
PALEOZOIC	PERMIAN		252	formation of supercontinent Pangea, 280–230 mya
	PENNSYLVANIAN		299	
	MISSISSIPPIAN		323	Antler orogeny and deposition of conglomeratic Copper Basin Group east of Ketchum, 380–320 mya
	DEVONIAN		359	
	SILURIAN		419	
	ORDOVICIAN		444	
	CAMBRIAN		485	Beaverhead plutons intrude Lemhi arch in east-central Idaho, 500–480 mya
PROTEROZOIC	NEOPROTEROZOIC		541	breakup of supercontinent Rodinia, 700–550 mya; deposition of Windermere Supergroup
				Snowball Earth—Sturtian glacial rocks deposited in Idaho, 710–660 mya
	MESOPROTEROZOIC		1,000	deposition of Belt basin sedimentary rocks, 1,470–1,370 mya
	PALEOPROTEROZOIC		1,600	island arcs formed along Great Falls tectonic zone, running northeast from Boise to Salmon, 1,800–1,700 mya
				intrusions in northern Idaho, 1,860 mya
ARCHEAN			2,500	Idaho's oldest basement rocks intruded by granitic plutons, 2,650 mya

Right-side annotations:
- eruptions of rhyolite and basalt create Snake River Plain as Idaho passes over Yellowstone hot spot; Basin and Range Province forms due to extensional faulting, 17 mya to present
- subduction along western edge of North America, 280 mya to present
- deposition of limestone, shale, and sandstone in shallow marine setting along passive margin on west side of Laurentia, 640–260 mya

*mya = millions of years ago

INTRODUCTION

Idaho is a state of striking contrasts. Green forested hills and deep blue lakes in the north, airy mountain peaks and cascading rivers in the center, and stark lava fields amidst sprawling desert in the south. Spanning nearly 500 miles in length from north to south and with an east-west width varying from 45 to 300 miles, this immense state records a vast geologic history, including massive volcanic eruptions, catastrophic floods, scouring glaciers, and crashing tectonic plates.

Idaho's diverse topography presents a challenge to the intrepid roadside geologist. The rugged central portion of the state is largely remote wilderness with few roads—an effective barrier between northern and southern Idaho. North-south travel is limited to a few winding highways squeezed into narrow, deep canyons mainly along the state's western fringe. Idaho's interstate freeways are few and far between, running through relatively flat ground and around the wilderness island in the middle. They follow railroads built in the late 1800s, which in turn followed Native American and trapper trails. While some of Idaho's highways are engineering marvels, plastered onto steep mountainsides or threaded through sinuous canyons, others are nearly straight and level. Collectively, these roadways offer spectacular access and exposure to Idaho's fantastic geology and landscapes.

For convenience, we divide Idaho into five major geologic regions: Northern Idaho, the Western Margin, the Central Mountains, the Basin and Range, and the Snake River Plain. Northern Idaho's lakes lie among forested hills of Cretaceous granitic rocks and a thick sequence of Precambrian sedimentary rock. The Western Margin's thick stack of Miocene basalt of the Columbia Plateau overlie Paleozoic and Mesozoic rocks that were part of tropical islands smeared onto the ancient edge of North America and are now mainly exposed at the bottoms of deep canyons. The Central Mountains in Idaho's rugged interior are dominated by vast expanses of Cretaceous granitic rocks of the Idaho batholith. The Basin and Range of southeast Idaho is defined by north- or northwest-trending mountains divided by intervening valleys, the product of extensional faulting that began in the Miocene Epoch and continues today. The once-continuous Basin and Range region is cleaved in two by the Snake River Plain, a broad swath of mostly low-lying terrain across the southern portion of the state that is dominated by Miocene to Holocene volcanic and sedimentary rocks.

We begin our exploration of Idaho by first explaining some key geologic concepts before turning our attention to the incredibly diverse and rich geologic history of the Gem State.

GEOLOGY BASICS

Everything in the universe is made of elements, the basic building blocks of all matter, such as oxygen, iron, and carbon. Minerals are solid, inorganic compounds of specific elements possessing a crystalline structure. Many of the Earth's minerals are silicates, which utilize a basic pyramidal building block composed of one atom of silicon and four oxygen atoms. Quartz, one of Earth's most common minerals, is entirely silicon dioxide, also known as silica.

Combinations of minerals are called rocks. Geologists classify the incredible diversity of rocks into three categories based on how or where they form: igneous, sedimentary, and metamorphic. Idaho has more than its fair share of all three types. Rocks are primarily identified and described based on the type, shape, size, and arrangement of their minerals, which vary considerably.

Igneous Rocks

Igneous rocks form due to the cooling and crystallization of molten rock, either above the surface (where molten rock is called lava) or below the surface (called magma). The two main variables used to differentiate one igneous rock from another are the cooling rate and the chemical makeup of the magma or lava. Cooling rates are largely inferred by observing the size of the rock's crystals. Volcanic (or extrusive) rocks erupt on the surface of the Earth where lava cools rapidly, leaving little time for crystals to grow. Extrusive rocks generally contain only microscopic crystals or even volcanic glass when cooling is so rapid that the lava does not have time to form even microscopic crystals. In contrast, subsurface magma cools much more slowly into intrusive rock, allowing crystals to grow large enough to be easily visible. Of course, nature is much more complex and messy than these ideal conditions. For example, sometimes extrusive rocks have large crystals if the magma began cooling slowly for a while before finally erupting at the surface.

A body of underground magma is called a magma chamber, and when it cools underground, the body of intrusive rock is called a pluton. A collection of plutons forms a batholith. All of the rocks in a batholith are plutonic.

Batches of magma may form only plutonic rocks if they never reach the surface, or both plutonic and volcanic rocks if they do erupt, but not all magmas are made of the same material. Magmas containing high amounts of silica are called felsic magmas and form granite if they cool slowly underground or rhyolite if they erupt from a volcano. Mafic magmas contain relatively low amounts of silica, forming gabbro where they cool underground or basalt where they erupt from volcanoes. The terms *mafic* and *felsic* are also applied to igneous rocks, reflecting their original magma compositions. Plutonic rocks with silica compositions that are intermediate between felsic and mafic are called diorite, whereas volcanic rocks with intermediate compositions are called andesite.

With so much variability in magma chemistry, cooling rates, and other minor factors, professional geologists employ literally hundreds of names to differentiate igneous rocks. For the roadside geologist, however, it is much easier to use broader terms. For example, we frequently use the term *granitic rocks* to describe light-colored, felsic plutonic rocks. However, in places, it is important to use more specific rock names. Granitic rocks fall into three main categories: granite, granodiorite, and tonalite. All three of these rocks have similar amounts of silica in the mineral quartz

The distribution of extrusive and intrusive igneous rocks in Idaho.

The Copeland granodiorite of Northern Idaho features large pink rectangular crystals of potassium feldspar surrounded by quartz and plagioclase feldspar.

but vary in the proportion of other constituents, such as potassium in the mineral potassium feldspar, which decreases dramatically from granite to tonalite.

Igneous rocks are quite prolific in Idaho because much of the state has been forged by molten rock. A vast expanse of granitic rock from the Cretaceous Period forms the Kaniksu batholith of northern Idaho. Farther south in the Central Mountains lie the granitic rocks of the even larger Idaho batholith from the Cretaceous and Paleogene Periods, with its northern Bitterroot lobe and southern Atlanta lobe. The Central Mountains also contain volcanic and plutonic rocks from the Eocene Epoch, part of the Challis magmatic event. The pinkish granitic rocks of the Sawtooth Range are an example of part of a Challis batholith. In the Western Margin, andesite and basalt are associated with Paleozoic and Mesozoic volcanic island arcs that were accreted onto the ancient western edge of North America. About 17 million years ago, basaltic lavas that erupted from volcanic fissures in eastern Oregon and Washington and western Idaho poured out to form the Columbia Plateau. The Snake River Plain of southern Idaho contains volcanic rocks, mainly rhyolite and basalt, erupted 17 million years ago to present from both explosive and quiescent volcanic vents.

Sedimentary Rocks

Sedimentary rocks form where sediment (sand, silt, mud, or pebble rock fragments) is compacted and cemented together. Sandstone, siltstone, mudstone (shale) and conglomerate are examples. Sedimentary rocks also form by the accumulation of organic material or the precipitation of chemical compounds. For example, coal forms from compacted peat, and chert is silica that precipitates from water. Because sedimentary rocks accumulate at the Earth's surface, they usually contain horizontal layers, called beds, varying in thickness from less than an inch to thousands of feet. Sedimentary rocks are instructive because careful analysis of their mineral composition, sediment sizes and shapes, and the fossils they contain allows geologists to determine the type of environment in which the particles were deposited. Sand, for example, is deposited in a variety of environments including alluvial fans, river deltas, desert sand dunes, and beaches. Determining the specific environment helps geologists reconstruct the geologic history of an area.

Sedimentary rocks dominate several large chunks of Idaho. Northern Idaho contains an impressively thick stack of sedimentary rocks deposited 1.5 to 1.4 billion years ago in the huge Belt basin of Proterozoic time. In the Western Margin, sedimentary rocks were deposited with volcanic rocks from the Permian to Jurassic Periods on and around tropical islands that were later accreted onto the edge of North America. In the Basin and Range and Central Mountains, more than 10,000 feet of late Precambrian, Paleozoic, and Mesozoic sedimentary rocks were deposited along the coastal margin of North America. These rocks were then shoved eastward in late Mesozoic time and subsequently stretched and cut by normal faults of the Basin and Range. Sedimentary rocks in the Snake River Plain mainly include Neogene to Holocene lake and stream deposits.

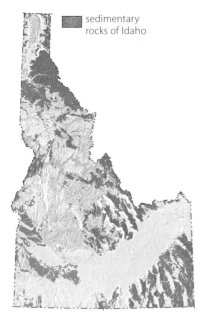

The distribution of sedimentary rocks in Idaho.

Metamorphic Rocks

Rocks changed by heat and pressure are called metamorphic rocks. An increase in heat or pressure causes minerals to recrystallize, often forming new minerals that are more stable. Metamorphism is a solid-state change; the rock is not melted. Metamorphic changes can be slight, such as shale being metamorphosed into slate, or dramatic, such as mudstone turning into schist. When unequal pressure is involved, minerals in the metamorphic rock become aligned perpendicular to the applied pressure to produce layering, or foliation. The heat and pressure for metamorphism typically comes from deep burial. Some rocks are metamorphosed only by the heat of adjacent bodies of magma. Most metamorphic rocks are closely associated with the rock that existed prior to metamorphism (also called the parent rock). For example, quartzite is metamorphosed quartz-rich sandstone, and marble is metamorphosed limestone.

Because metamorphic rocks form deep in the Earth's crust, many of Idaho's metamorphic rocks are the oldest rocks in the state and lie below younger unmetamorphosed rocks. We call these older, lower-lying rock packages the basement rocks.

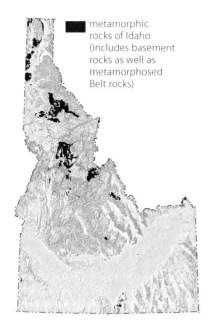

metamorphic rocks of Idaho (includes basement rocks as well as metamorphosed Belt rocks)

The distribution of metamorphic rocks in Idaho.

The state's oldest rocks are Archean metamorphic rocks found in scattered areas of the Basin and Range, Central Mountains, and Northern Idaho. In Northern Idaho and the Central Mountains, the margins of many igneous intrusions contain Proterozoic metamorphic rocks. Some of the rocks accreted onto North America in the Western Margin region were highly metamorphosed when they were buried 15 to 20 miles deep during the collision.

PLATE TECTONICS

A foundational knowledge of plate tectonics is an essential part of understanding the geologic events of any landscape. The Earth's rigid outer layer on which we dwell, the lithosphere, is broken into discrete slabs called tectonic plates. The lithosphere includes both Earth's crust and upper mantle. Lying in the mantle beneath the lithosphere, about 50 to 120 miles below the surface, is a region of soft, weak rock called the asthenosphere. The low-density lithospheric plates literally float on denser asthenosphere, moving around at speeds on the order of inches per year. The edges of these plates interact with each other near Earth's surface. Some of the Earth's dynamic processes, such as earthquakes and volcanic eruptions, occur as the plate margins bump, grind, separate, and slide.

The jostling of tectonic plates along their margins form three basic types of plate boundaries: divergent, convergent, and transform. Divergent boundaries form where two plates move away from each other, most commonly in the oceans where they are called mid-ocean ridges. The stretching motion, or extension, of the lithosphere at divergent boundaries thins the lithosphere, reducing the pressure on the underlying asthenosphere. This process allows some of the asthenosphere's soft rock to melt, forming magma, which ascends into the intervening area between the plates, erupting as lava onto the seafloor.

Convergent boundaries form where two plates collide with each other, commonly as a long, drawn-out process at rates of 1 to 2 inches per year. Where two continental plates converge, the lithosphere is thickened, creating high mountains but no volcanoes. The Alps and Himalayan Mountains are great modern examples. A tectonic collision between an oceanic and continental plate results in the denser oceanic plate sliding beneath the continent, a process called subduction. As the descending oceanic plate plunges beneath the continent in a subduction zone, water trapped in

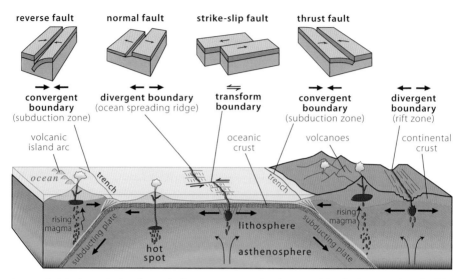

The types of plate boundaries that exist between tectonic plates and the faults associated with each.

the oceanic plate reacts with the hot mantle and crustal rocks above, melting some of these rocks to form magma. The less-dense magma rises through and into the overlying continental plate, creating magma chambers and, at the surface, volcanoes. The powerful compressional forces at subduction zones and colliding continents produce folded rock layers and reverse, or thrust, faults.

Where the motion of two plates forces them to slide past each other laterally, a transform plate boundary exists. Quite simply, transform boundaries connect sections of divergent and convergent plate boundaries. The San Andreas fault in California is a premier example of such a boundary, allowing the Pacific Plate to move northwesterly relative to the North American Plate and connecting the divergent boundary in the Gulf of California to the subduction zone along the northern California coast.

IDAHO'S GEOLOGIC HISTORY

PRECAMBRIAN ERA
(4.6 billion to 541 million years ago)

Idaho's oldest rocks are part of the continental core of ancestral North America (Laurentia) and are exposed in several discrete locations: the Albion Mountains of south-central Idaho, the Pioneer Mountains of central Idaho, and the Boehls Butte–Clearwater and Priest River complexes of northern Idaho. These Archean and early Proterozoic metamorphic rocks (mainly gneiss and schist) were originally granitic rocks that were subsequently buried to middle-crustal depths and metamorphosed. Geophysical studies suggest that these old rocks are part of the North American craton, the tectonically stable nucleus of the continent. In Idaho, these rocks were brought to Earth's surface during regional uplift in the last 50 million years.

About 1.5 to 1.4 billion years ago, extensional or stretching stresses within the proto–North America continent created a large basin that filled with an extremely thick sequence of mostly mud and sand. This immense package of sedimentary rock, known as the Belt Supergroup, is over 10 miles thick and covers much of the northern half of Idaho along with sections of western Montana, easternmost Washington, and southeastern British Columbia. Many outcrops of the Belt Supergroup show exquisite preservation of minute sedimentary details that offer important clues regarding deposition at a time when only basic lifeforms, such as algae and bacteria, existed.

Earth's continents joined together into a supercontinent called Rodinia between 1.1 and 0.9 billion years ago. Supercontinents are unstable on Earth, and the breakup of Rodinia began about 700 million years ago, disconnecting western North America from Australia, South China, and Antarctica. The opening rift eventually created a widening ocean basin, the paleo–Pacific Ocean, between the separated continents and split the Belt basin deposits in two. Rifting occurred in a stuttering process of fits and starts, lasting as much as 200 million years. During the rifting, an extensive package of sedimentary and volcanic rocks, the Windermere Supergroup, was deposited on the continental margin, which extended from Utah to Canada, running the length of Idaho. After continental separation began, most of Idaho resided along the shallow coastal margin of North America, an ideal place for sand and mud to accumulate and harden into sandstone and shale.

Around the time Rodinia was breaking apart, several massive glacial episodes entombed much or perhaps all of the Earth's surface in ice. Evidence for these episodes, known as Snowball Earth, are found in the Proterozoic Pocatello

Diamictite of the Scout Mountain Member of the Pocatello Formation, exposed north of Portneuf Gap, just north of I-15. The deposit of angular boulders floating in a shaly matrix is interpreted as reworked glacial till, some 680 million years old, from an episode of Snowball Earth. Photographed in 1981 with dog named Blue. (42.7965, -112.3553)

Formation of southeastern Idaho and near Edwardsburg in the Central Mountains. The Pocatello Formation includes diamictites, oceanic mudstones containing rocks dropped from icebergs.

PALEOZOIC ERA
(541 to 252 million years ago)

In the wake of Rodinia's breakup, most of Idaho entered a period of relative geologic quiescence on the newly formed continental shelf of North America. With the majority of the state submerged beneath shallow tropical waters, Idaho became the dumping ground for thousands of feet of mud, sand, and newly evolved marine organisms. Collectively, up to 50,000 feet of Paleozoic rock accumulated, some of which is preserved in the eastern part of the state. Many of the Paleozoic sedimentary rocks in Idaho are carbonates (limestone and dolostone) derived from the accumulation of tiny exoskeletons of marine organisms on the ocean floor. Today, some of these carbonates lie atop the summits of Idaho's tallest mountains, providing compelling evidence of uplift as a dynamic Earth process.

In the latter half of the Paleozoic Era, the next supercontinent on Earth, Pangea, was beginning to assemble. Most of the action took place in eastern North America with multiple continental collisions and massive mountain building episodes. By the late Paleozoic, however, subduction began along the western margin of North America, creating several notable mountain building events that impacted Idaho and the inland Pacific Northwest. During the late Devonian and Mississippian Periods, a chain of islands collided with the coast to the southwest of Idaho and generated mountains in Nevada as part of the Antler orogeny. Deep basins formed immediately east of these highlands, and the basin in what is now Idaho was filled by thick sedimentary deposits of the Mississippian Copper Basin Group. By the Pennsylvanian and Permian, the Antler highlands were largely eroded, and basins such as those hosting the Sun Valley Group near Ketchum were forming at the same time the Ancestral Rocky Mountains were rising to the southeast. In the Late Permian, the phosphate-rich Phosphoria Formation was deposited in a calm marine basin centered near Soda Springs.

MESOZOIC ERA
(252 to 66 million years ago)

The supercontinent Pangea began to break up in the Triassic Period, about 200 million years ago, with rifting eventually forming the Atlantic Ocean. As the Atlantic opened, the North American continent moved westward and initiated a new round of subduction along the western edge of North America. Island chains that previously resided thousands of miles away in the Pacific Ocean were carried eastward atop oceanic plates, ultimately colliding with the edge of North America. In Idaho, these volcanic island chains and their associated marine sedimentary rocks were sutured to the continent between 150 and 140 million years ago to become what is now Idaho's Western Margin, forming a complexly deformed suite of accreted rocks known as the Blue Mountains Province. Ongoing subduction along the newly expanded continental margin during this period led to magma generation and the intrusion of plutons along, and to either side of, the suture zone.

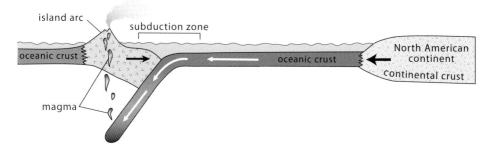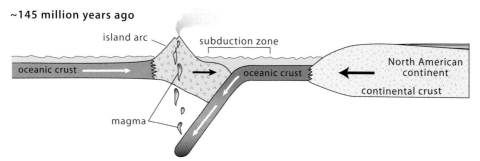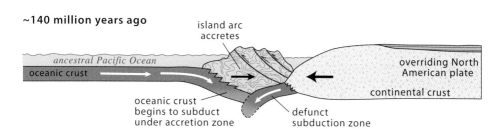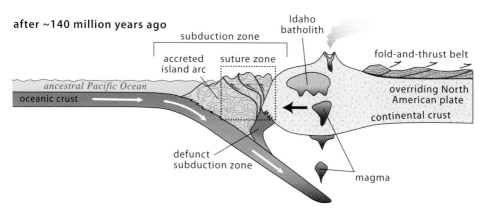

At the beginning of Mesozoic time, an island arc, produced by west-dipping subduction, approached North America from the west. As the island arc accreted to the continent about 140 million years ago, a new subduction zone formed to its east, dipping east under the continent. The collision generated the magma of the Idaho batholith and thrusted Paleozoic sedimentary rocks eastward.

As exotic terranes were accreting to the edge of North America, farther inland, in eastern Idaho, the upper crust buckled and broke into a series of massive slabs separated by thrust faults, or low-angle reverse faults. Episodic movement on these faults shoved these immense slabs eastward, stacking one package of rock atop another, thickening the Earth's crust and creating mountains during an event known as the Sevier orogeny. The package of stacked slabs lies near, and west of, the Idaho-Wyoming border and is called the Sevier fold-and-thrust belt.

From 120 to 53 million years ago, huge volumes of subduction-generated, granitic magma cooled and crystallized slowly to form the Kaniksu batholith in northern Idaho and the two lobes of the Idaho batholith in central Idaho. Chemical reactions between the magma and surrounding rocks generated fluids rich in metals such as gold, silver, lead, and zinc. As the fluids circulated through pores and fractures in the rock, they cooled to concentrate valuable ore deposits later exploited by miners.

To the east, thrust faulting associated with the Sevier orogeny changed. Rather than breaking the crust into large thrust sheets bounded by low-angle faults, a new style of deformation ensued which involved the uplift of basement rocks along steep reverse faults. This episode of uplift, known as the Laramide orogeny, continued into the Paleogene Period and is responsible for much of the highlands just east of Idaho in the Rocky Mountain region of Wyoming and Montana.

CENOZOIC ERA
(66 million years ago to present)

The compressive tectonic regime that dominated the Mesozoic Era in western North America began changing in Cenozoic time, and the rocks in Idaho finally relaxed. Thick welts of crust produced during Sevier thrusting underwent a very distinctive style of normal faulting in which Earth's crust was stretched and pulled apart along low-angle normal faults due to northwest-southeast extension. This process moved the ancient basement rocks upward from deep within the crust while sliding the younger upper-crustal rocks to the side and downward. These formerly deep rocks, known as metamorphic core complexes, are now exposed in the Albion Mountains of southern Idaho, the Pioneer Mountains of central Idaho, and the Priest River and the Boehls Butte–Clearwater complexes of northern Idaho.

At about the same time, an enigmatic episode of magmatism occurred in central and northern Idaho between 51 and 40 million years ago—the Challis magmatic event. Explosive and nonexplosive volcanic eruptions produced ash deposits and lava flows that blanketed Idaho's landscapes. Some of the magma generated at this time never erupted, pooling in vast magma chambers where it cooled and crystallized slowly. Uplifting along normal faults later brought these rocks to the surface in places like the Sawtooth Range and Boulder Mountains of central Idaho.

More than 250 million years of subduction along the western margin of North America eventually consumed an entire oceanic plate beneath it, ending subduction along what is now the southern California coast by about 25 million years ago as a divergent plate boundary was swallowed beneath the continent. A transform boundary, the San Andreas fault, developed in California in place of subduction there. Subduction of the small Juan de Fuca Plate off the coasts of northernmost California, Oregon, and Washington continues today but does not have a pronounced effect on Idaho. By about 17 million years ago, the response to movement along the San Andreas fault in much of western North America, including Idaho, was to spread, or extend, in a mostly east-west direction. The stretched crust broke along north-trending normal faults, moving the block of rock on one side of the fault upward to form a mountain range while dropping the other side down to form a sedimentary basin. Movement along these faults was episodic, building stress to the point of failure and then releasing that stress along the fault, causing movement and earthquakes over intervals of thousands of years.

This episode of stretching in Idaho and much of the interior western United States is called Basin and Range extension. East-west stretching of the crust along both low- and high-angle normal faults produced a landscape dominated by alternating north- to northwest-trending mountains and valleys. Many of Idaho's loftiest mountains owe their elevated position to Basin and Range extension. Historic and recent earthquakes such as the 6.9-magnitude Borah Peak earthquake of 1983 near Mackay and the 6.5-magnitude earthquake northwest of Stanley in 2020 demonstrate that the process continues today.

Roughly concurrent with the onset of Basin and Range extension, a massive wave of volcanism struck the Pacific Northwest. The cause of this volcanic episode is still

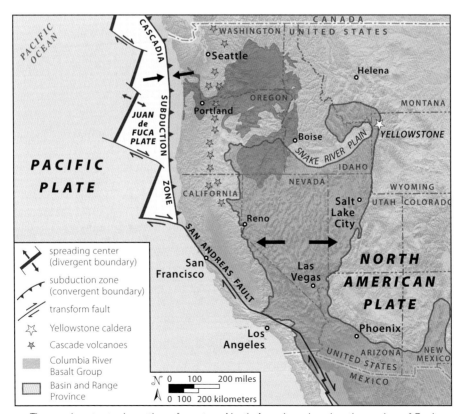

The modern tectonic setting of western North America, showing the region of Basin and Range extension and area covered with flood basalts.

actively researched and debated by geologists, but the leading model postulates that a stationary plume of rising hot mantle, called a hot spot, created copious volumes of magma that rose through the lithosphere. The arrival of the initial plume generated huge outpourings of basaltic lava in westernmost Idaho, eastern Oregon, and eastern Washington. The lava cooled and crystallized into a thick sequence of rock known as the Columbia River Basalt Group. The lava erupted from north- to northwest-trending fissure vents, mainly during a short interval of time from 16.7 to 15.9 million years ago. The sheer volume of lava erupted during this relatively short period is staggering. Flows of this magnitude are called flood basalts because they inundate the landscape, covering large swaths of existing topography. The eruption of these flood basalts created the Columbia Plateau of eastern Oregon and Washington.

Following the initial flood basalt phase of volcanism that produced the Columbia River Basalt Group, volcanism became more focused across a more limited area. As the North American Plate gradually drifted to the southwest, the location of volcanoes above the hot spot shifted to the northeast, forming a successive string of volcanic centers along a track stretching from southwestern Idaho to present-day Yellowstone National Park. The ages of Miocene and Pliocene rhyolitic volcanic rocks across southern Idaho depict the position of the hot spot over time. The volcanoes initially erupted explosively, ejecting massive volumes of rhyolitic ash that caused some volcanoes to collapse on their partially vacated magma chamber to create huge calderas as much as 50 miles in diameter.

As areas of southern Idaho moved off the hot spot due to plate movement, the crust cooled and subsided, ultimately creating the band of low topography known as the Snake River Plain. The westward drainage of the Snake River developed on this subsiding area, eventually linking with the Columbia River and the Pacific Ocean. Later, more benign basaltic lava erupted, covering the calderas and forming the primary rock found atop much of the Snake River Plain. The eruption of basaltic lava continued into more recent times, with the youngest eruptions at Craters of the Moon about 2,000 years ago.

The low-lying volcanic landscape formed by the rise and fall of the Yellowstone hot spot coupled with the cool, wet climate of the Late Miocene and Pliocene Epochs created ideal conditions for the formation of lakes in the Snake River Plain. The largest of these was Lake Idaho, a freshwater lake covering the western Snake River Plain and comparable in size to today's Lake Ontario. The lake formed intermittently, waxing and waning in response to shifts in climate, but existed during two main periods, one from 10 to 6 million years ago, and another from about 4 to 3 million years ago.

During cooler and wetter periods of the Pleistocene Epoch, many parts of Idaho were profoundly affected by extensive erosion and deposition by glaciers. In the high mountains of central Idaho, alpine glaciers filled stream valleys, carving dramatic U-shaped canyons and forming rugged ridgelines. These glaciers carried sediment of all sizes downslope to their terminus, where their load of sediment was deposited on valley floors and along valley margins. Once the glaciers retreated and melted, some of these valleys filled with water to form scenic lakes.

In northern Idaho, the Pleistocene glaciation was manifested differently. The immense Canadian ice sheet repeatedly advanced down major valleys into the Idaho Panhandle. Alpine glaciers in this area fed ice directly onto lobes of the ice sheet that

An outcrop of Eocene-age Sawtooth granite in Redfish Canyon in the Sawtooth Range displays excellent glacial polish and striations. These impressive features formed during the Pleistocene ice age as sediment trapped in the moving glacier's ice was dragged across the rock surface.

filled the large north-oriented valleys between mountain ranges. Priest Lake and Lake Pend Oreille fill depressions left by these large valley-filling ice lobes.

As the last of the Pleistocene ice ages drew to a close, two massive flood events tore across Idaho, one in the north and another in the south. The northerly event actually consisted of multiple floods that occurred between about 19,000 and 14,000 years ago and are known as the Missoula floods. A lobe of the continental ice sheet repeatedly dammed the Clark Fork River in northern Idaho and western Montana, impounding water behind the ice. The ice dams failed, unleashing colossal torrents of water westward across the Rathdrum Prairie of northern Idaho and into eastern Washington, where it stripped off sediments and gouged canyons into the underlying basalt to produce the Channeled Scablands. Once the floodwaters subsided, the ice crept back across the river valley to block the river once again, setting the stage for the next flooding event.

The second massive flood affected parts of southeastern Idaho and the path of the Snake River. Known as the Bonneville Flood, this megaflood of about 17,500 years ago partially drained Lake Bonneville, an immense freshwater lake that occupied much of western Utah and bits of Nevada and southern Idaho. Unlike the Missoula floods, the Bonneville Flood was a singular event, but its magnitude was truly enormous, causing substantial erosion of the Snake River Canyon and forming some of the Snake River Plain's iconic landscapes such as Shoshone Falls.

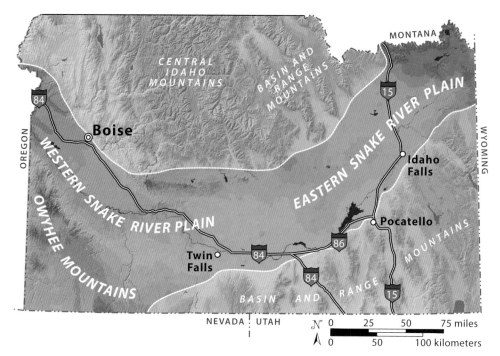

The Snake River Plain region is divided into a western and eastern plain and also includes the Owyhee Mountains.

View to the south across the vast Snake River Plain from the southern end of the Lost River Range. Pleistocene basalt from nearby shield volcanoes forms the low middle ground. East Butte (left) and Middle Butte (center) are uplifts produced by rhyolite volcanism. In the far distance on the right is the Portneuf Range near Pocatello.

SNAKE RIVER PLAIN

Idaho's Snake River Plain is one of the most unique and interesting geologic regions, not just in western North America but in the entire world. Defined by its mostly low topography and home to its large namesake river, the plain cuts a swath across the southern belly of Idaho, interrupting the rugged, mountainous regions that surround it. Many of Idaho's major population centers, prime agricultural regions, and interstate freeway systems lie within the Snake River Plain, all a consequence of the region's fascinating geologic history—a story that includes eastward migration of the Continental Divide, two distinct styles of volcanism, a period of large lakes rich in fossils, and a catastrophic deluge.

Forming an arcuate "smile" across the entire span of southern Idaho, the Snake River Plain stretches from the Oregon border on the west to the Yellowstone Plateau on the east, a distance of about 370 miles. Across this vast lowland flows the Snake River, one of the largest rivers in the western United States and the primary destination for most of southern Idaho's rivers. The Snake River originates at the Continental Divide in Yellowstone National Park, flows southward along the base of the Grand Tetons before turning westward as it enters Idaho and the Snake River Plain. In western Idaho, the river heads northward, forming the border with Oregon through the scenic and deep Hells Canyon before joining the Columbia River near Kennewick, Washington.

Geologically, the Snake River Plain is a volcanic wonderland, forged by two very different styles of volcanic activity. Most of the region is covered with young (Pliocene to Holocene) basalt, which flowed out of localized volcanic vents as runny lava. Basalt's characteristically low silica content, as compared to other magma types, accounts for its fluidity. Multiple eruptions and overlapping lavas created stacked layers of basalt, ranging from tens to thousands of feet in thickness.

Lying mostly beneath the low-silica basalt in the eastern Snake River Plain are several thousand feet of late Miocene to Pleistocene (12 million to 600,000 years old) high-silica rhyolite, a volcanic rock produced by explosive eruptions and sticky outpourings of viscous lava. In most places, the rhyolite is buried beneath the basalt and not exposed. Thus, the geology of the Snake River Plain is somewhat like a frosted cake with the much thicker rhyolite lying beneath the thin veneer of basalt. In general, the rhyolite is only exposed in the deepest canyons that penetrate through the basalt frosting or along the margins of the Snake River Plain, where the basalt thins and laps up against the older rhyolite. So much volcanic activity has ensued within the Snake River Plain that any of the older subsurface rocks that existed here prior to the eruptions have likely been completely disrupted.

Although it outwardly appears to be one continuous entity, the Snake River Plain is divided into eastern and western regions based on several geologic differences. A crude boundary between the two lies near Hagerman in south-central Idaho. The western Snake River Plain trends northwest-southeast and averages about 40 miles wide by 180 miles long. It is a structural basin, a graben, bounded by normal faults on its northeast and southwest margins. These faults became active about 12 million years ago as part of Basin and Range extension, resulting in up to 9,000 feet of vertical displacement. The end product is this northwest-trending basin, bounded by the Owyhee Mountains to the south and the Danskin and Boise Mountains to the north. The basin-bounding faults created pathways for hot groundwater to move upward toward the surface, resulting in hot springs along the fault zones and geothermal wells, such as those near downtown Boise.

The eastern Snake River Plain trends northeast, nearly perpendicular to the western Snake River Plain, and measures about 60 miles wide by 240 miles long. Unlike its western counterpart, the eastern Snake River Plain is a downwarped basin, not a fault-bounded basin. This downwarped region marks the passage of the continent over the Yellowstone hot spot.

YELLOWSTONE HOT SPOT

The origin of much of the Snake River Plain is best understood by examining the age pattern of rhyolite across the region. Rhyolites of the Snake River Plain are oldest near the southwest corner of Idaho, where they are about 17 to 15 million years old. Moving northeast across the plain, the rhyolite progressively gets younger, at the same rate and in the opposite direction that the North American tectonic plate has moved during the past 17 million years. The volcanic track today occupies the Yellowstone region, where the youngest rocks are about 2.1 million years old or less. Throughout the Snake River Plain, this rhyolite is nearly identical in its composition and outward appearance, suggesting most of it originated from melting of the deep crustal rocks below.

While the details regarding the origin of the Snake River Plain are still being unraveled and scrutinized by geologists, the current model suggests that a large,

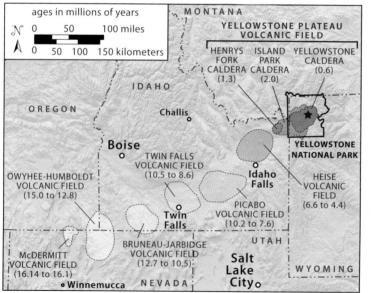

Rhyolite volcanic fields of the Snake River Plain get younger to the northeast. Ages of each volcanic field are shown in parentheses as millions of years ago.

stationary plume of ascending magma, called the Yellowstone hot spot, formed about 17 million years ago near the southwest corner of Idaho. The authors of this book's previous edition advocated a meteor impact as the origin of the Yellowstone hot spot and Snake River Plain. While once considered a possible explanation, geologists have found no evidence for the impact or its debris. Hot spots exist at several places on Earth, and their ultimate origin lies deep in the Earth's mantle, a difficult region to gather evidence.

One recent and promising model postulates that the hot spot is much older than previously thought and produced a thick, vast pile of oceanic basalt off the coast of Oregon about 56 to 49 million years ago during the Eocene Epoch. This large basaltic region was then accreted onto the edge of the continent beginning around 50 million years ago. Magma production from the hot spot was temporarily terminated during Oligocene time as the ancestral Cascade subduction zone and the west-moving North American Plate passed over it. The hot spot then resurfaced east of the Cascade Range in the Miocene Epoch when it produced lavas of the Columbia River Basalt Group (discussed in the Western Margin chapter), mainly from 16.6 to 15.9 million years ago. The reemergence of the hot spot also initiated eruptions near the Oregon, Nevada, and Idaho border around 17 million years ago and tracked northeastward over time to the present location of Yellowstone National Park.

In southwestern Idaho, the rising basaltic magma from the hot spot partially melted some of Idaho's ancient crust, material rich in silica and similar to granite in composition to form a batch of felsic magma. This rhyolitic magma, enriched in silica, is a less dense and stickier, or more viscous, type of magma as compared to basalt. The buoyant magma rose toward the surface, causing the land above to dome upward, forming a broad, elevated region. Like cracks in rising bread, fractures and faults formed at the surface above the inflated magma chambers. Unable to escape the pasty lava, gases accumulated in the magma chambers, increasing the underlying pressure, similar to an over-carbonated can of soda. Eventually, the pressure within the magma exceeded the strength of the rocks above, triggering a large volcanic eruption.

On the Snake River Plain, many of these rhyolitic eruptions were extremely violent and explosive. The rapid release in pressure blasted the rhyolitic magma into tiny glass shards of ash, forming billowing ash columns above the volcanoes and sending surging pyroclastic flows barreling outward across the landscape and well beyond the boundaries of the plain. Fine ash particles were blown upward into the stratosphere and drifted eastward with the prevailing westerly winds into the Great Plains and as far as the Gulf of Mexico. Some ash clouds were blown northward to deposit ash within sedimentary interbeds of the Columbia River basalts of western and northern Idaho. These colossal eruptions lasted for weeks or even months, emptying so much of the underlying magma chambers that they collapsed downward upon themselves, forming large depressions as much as 50 miles wide known as calderas. The sheer size of these caldera-forming eruptions in the Snake River Plain nearly defy comprehension, erupting as much as 600 cubic miles of ash, about 2,500 times larger than the 1980 Mt. St. Helens eruption. Collectively, the last three caldera-forming eruptions in Yellowstone produced enough ash to completely fill the Grand Canyon.

While some rhyolitic eruptions were large enough to create a caldera, others did not. Smaller explosive eruptions and the extrusion of pasty, gas-poor rhyolitic lava were also common. Collectively, the region above the hot spot accumulated thick

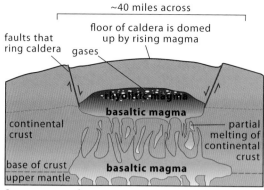

A. Basaltic magma from the hot spot partially melts the base of the continental crust to form a deep rhyolitic magma chamber. This new magma buoyantly rises into the crust.

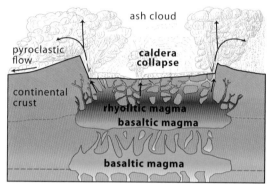

B. Pressure from gases in the magma finally exceeds the weight of rocks above, triggering a huge eruption where magma is blasted into tiny shards of ash that blankets the landscape and hardens into tuff. The magma chamber is partially emptied, causing the ground to collapse and form a caldera.

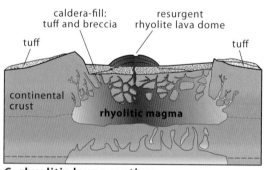

C. Gas-poor rhyolitic lava erupts within the caldera, forming steep lava domes.

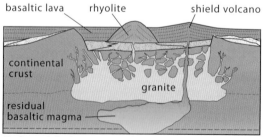

D. Rhyolite magma solidifies into granite. Basaltic magma rises to erupt fluid lava that builds small shield volcanoes and covers the land.

Simplified eruptive sequence above the Yellowstone hot spot. —Modified from Willsey, 2017

sequences of erupted rhyolitic tuffs and lavas, produced by both explosive and nonexplosive volcanic events. Regions of exposed and buried rhyolite lava and ash that erupted during a particular period of time define named volcanic fields across the Snake River Plain. In many cases, the only evidence for their existence is derived from deposits that are preserved at the margins or outside of the plain.

Eruptions at individual volcanic fields on the Snake River Plain lasted for a few million years on average. Eventually, the westward movement of the North American Plate severed the active volcanic field's connection with the underlying hot spot. As the formerly active volcanic field was dragged westward off the hot spot, a new section of fresh continental crust moved in over the hot spot from the east. This initiated anew the process of partial melting and started the whole sequence over again. The result is a series of large volcanic fields that form a chain along the length of the Snake River Plain, older in the southwest and younger to the northeast. The youngest of the batch now lies in and around Yellowstone National Park, where three major caldera-forming eruptions and several smaller eruptions have rocked the region over the past 2.1 million years. Today, the Yellowstone hot spot lies beneath its namesake national park. The unrelenting movement of the North American Plate means future hot spot eruptions will eventually exit the park's northeast boundary and enter south-central Montana.

Before migrating over the hot spot, the eastern Snake River Plain was pulled apart and the crust thinned by Basin and Range extension. North- and northwest-trending mountain ranges, bounded by normal faults and continuous with today's existing ranges, probably existed where the plain now lies. As the North American Plate passed over the hot spot, the mountain ranges were disrupted by the series of large rhyolitic volcanic fields that become younger to the northeast. Passage of the hot spot beneath southeast Idaho effectively ironed out its previous topography.

As the Snake River Plain formed over the past 17 million years, the position of the underlying hot spot exerted a strong influence on the drainage patterns of rivers and streams. Above the hot spot, ascending magma and heating of crustal rocks caused the land to rise, forming an elevated plateau hundreds of miles across, with streams flowing outward from it. As plate motion continued, the original drainage divide, parts of which were the Continental Divide, moved northeastward with the underlying hot spot and the location of the active volcanic field. Thus, the Continental Divide has migrated northeast across Idaho to its current location in the Yellowstone region. Downwarping and subsidence of the eastern Snake River Plain is associated with loading of the crust by dense mafic plutons in the lower crust (the magmas that melted the upper crust to produce the rhyolite) and cooling as each active volcanic area moved off the Yellowstone hot spot.

BASALTIC VOLCANISM

The dark volcanic rock, basalt, is by far the preeminent rock type exposed within the Snake River Plain. Eruptions of basaltic lava were the final chapter for many of the volcanic fields initiated by the Yellowstone hot spot. Once much of the rhyolitic magma chamber had cooled and crystallized, basaltic magma that melted from the lower crust or the underlying mantle was able to ascend as dikes along well-established magma plumbing systems and erupt as fluid lava at the surface. In some areas throughout the Snake River Plain, basaltic volcanism persisted intermittently long after the rhyolitic volcanism ceased.

Other eruptions of basalt in the Snake River Plain have been linked to Basin and Range extension. As the crust was thinned by faulting over the past 12 million years, the pressure on the underlying asthenosphere was reduced, causing some of this material to melt and form basaltic magma. Less dense than the rock surrounding it, the magma rises upward, traveling as dikes along fault and fracture zones to reach the surface and form volcanic vents.

Despite the apparent monotony of basalt across the Snake River Plain, this singular rock type creates a surprising variety of landscapes, rock textures, and other features of interest to the passing traveler. Basaltic vents range from linear fissures to steep-sided cinder cones to broad shield volcanoes, depending on gas content of the magma and geometry of the magma plumbing system.

Where flowing basaltic lava pours from a vent, it forms a smooth, ropey surface known as pahoehoe. Sometimes lava lakes form, with a pahoehoe crust overlying still molten basalt. As the lava travels farther from its vent, it cools, becomes pasty, and runs over itself, tearing and shredding the lava to form sharp, spiny lava called aa. Lava tubes form where lava crusts over at the surface but continues to flow beneath

A. As basaltic lava cools, it runs over itself and forms a jagged surface known as aa. The low profile of a shield volcano rises in the distance. **B.** Hot basaltic lava forms a smooth, ropy surface called pahoehoe. **C.** Columns or columnar joints form in thick lava flows where uniform cooling creates equally spaced fractures or joints. **D.** Lava tubes begin as lava flows whose upper surface has cooled and crusted over, keeping the lava insulated as it travels. When the eruption wanes, the enclosed flow becomes a hollow cave.

as molten material. Once the eruption ends, a hollow cavern can be left behind. Thick flows of lava that cool uniformly at the surface develop regular spaced fractures called columnar joints, an attractive geometric landform seen in many basalt cliffs.

Recent basalt eruptions form some of the most interesting volcanic fields on the Snake River Plain. The largest young lava field in the contiguous United States is found at Craters of the Moon National Monument and Preserve, where 618 square miles are covered in basalt. The lava field was formed by eight major eruptions between 15,000 and 2,000 years ago. The dry climate of Idaho and the slow rate of soil development allow the basalt to look quite fresh, as if it had erupted yesterday.

The eruptions at Craters of the Moon are part of a larger, impressive volcanic feature of the eastern Snake River Plain called the Great Rift. Stretching nearly 60 miles across the plain, the Great Rift is a northwest alignment of fissures and other volcanic vents. The southern end of the Great Rift lies just west of American Falls at the Wapi and Kings Bowl lava fields and extends across the plain to the vents at the northern end of Craters of the Moon. Based on its alignment with normal faults north and south of the plain, the Great Rift was likely associated with Basin and Range extension, which thinned and broke the crust, allowing an easy conduit for basaltic lava to erupt.

While the raw lava landscapes of the eastern Snake River Plain were a serious impediment to early human migration and settlement, the extensive outpouring of basaltic lava proved to be a boon for modern civilization. The adjacent high mountains capture large amounts of snow during winter. In the spring, snowmelt feeds

Aerial view of the linear fissure of Kings Bowl and its adjacent lava field. Kings Bowl lies at the southern end of the Great Rift. —Photo by Scott Hughes and Susan Sakimoto

rivers and streams flowing into the eastern Snake River Plain, where the water encounters the permeable basaltic bedrock. The water seeps downward into the fractures to become groundwater stored in the largest fractured-rock aquifer in the western United States. Covering about 10,800 square miles and with a volume comparable to that of Lake Erie, the eastern Snake River Plain aquifer is a critical water resource, supplying irrigation water to large agricultural regions as well as drinking water to many of southern Idaho's cities and towns.

LAKE IDAHO

As the western Snake River Plain subsided due to faulting in the last 10 million years, a large freshwater lake filled its basin several times. Called Lake Idaho, this massive water body at its greatest extent measured about 160 miles long by 40 miles wide, nearly the size of Lake Ontario. Lake Idaho existed at a time when the climate and life forms of the western Snake River Plain were markedly different from those of today. With a much wetter climate and lush vegetation, the area attracted a host of interesting animals such as sloths, camels, and horses. A noteworthy denizen of ancient Lake Idaho was the spike-toothed salmon, a huge ancestor to modern salmon weighing up to 400 pounds and measuring as much as 9 feet in length. The abundance of vertebrate fossils and the contrast between an ancient lush lake and the modern southwest Idaho desert attracted the interest of the US Geological Survey as early as the 1870s.

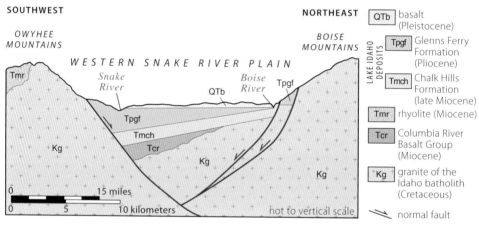

Sediments of Lake Idaho fill the down-dropped basin of the western Snake River Plain.
—Modified from Wood and Clemens, 2002; Beranek et al., 2006

Two stratigraphic formations, consisting of sediment deposited within the lake and along its margin, record the lake's existence. The Chalk Hills Formation was deposited from 10 to 6 million years ago and consists of lakebed and stream deposits. Following deposition of the Chalk Hills Formation, Lake Idaho largely dried up or separated into smaller lakes connected by a system of rivers and streams. The lake reformed and rose again around 4 million years ago as evidenced by the Glenns Ferry Formation. Like the Chalk Hills, sediments of this formation were deposited in floodplains, marshes, streams, and lakes. Lava flows and ash beds sandwiched in

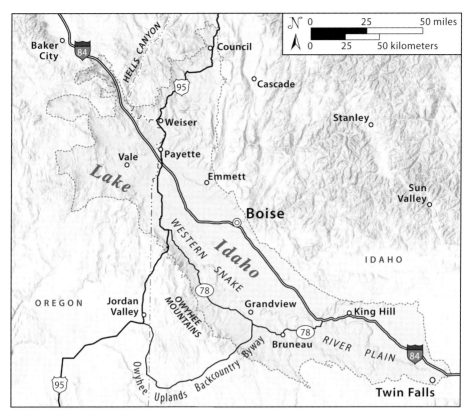

Lake Idaho, a large freshwater lake, covered the western Snake River Plain during the late Miocene and Pliocene.

and among the sedimentary layers provide age constraints for the formation and the myriad of fossils it contains.

Some of the organic-rich material encased in Lake Idaho sediments eventually was transformed into oil and gas deposits that have been recently exploited in the western Snake River Plain. The Chalk Hills Formation is the best producer thus far, but exploration continues on this and other Lake Idaho deposits.

The volcanic activity that accompanied the formation of the western Snake River Plain coexisted with Lake Idaho and the formation of its basin. As faulting thinned the crust, magma rose to create vents near and along the faults that form the plain's margins. Most eruptions involved basaltic lava, which underlies much of the area, but some of the earliest eruptions were rhyolitic in composition. Several of the volcanic eruptions occurred beneath Lake Idaho, where the interaction between hot lava and cool water produced pillow lavas and locally explosive steam-driven eruptions called hydrovolcanic eruptions.

What happened to Lake Idaho? Gradual erosion by a northward-flowing tributary to the ancestral Salmon River eventually eroded southward to the northern divide of the Lake Idaho drainage basin, causing the lake to spill northward into Hells Canyon and to drain the western Snake River Plain. The draining of the lake integrated the Salmon and ancestral Snake River systems, resulting in the much longer and larger modern Snake River that is linked to streams from the east that once fed into Lake Idaho.

The development of a through-flowing Snake River linked to the Columbia River lowered the base level of all streams and rivers in the region. Rather than flowing into a lake at an elevation of about 3,600 feet, streams cut downward to much lower elevations, carving many of the deep canyons of southwestern Idaho such as those of the Bruneau and Jarbidge Rivers.

OWYHEE MOUNTAINS

A notable exception to the low-lying region of the Snake River Plain is the Owyhee Mountains, an elevated region in the southwest corner of Idaho. This area has been included as part of the Snake River Plain chapter due to its similar rock types and common geologic origin. The Owyhee Mountains consist of mostly rhyolite associated with the passage of the Yellowstone hot spot. Faulting along the southwestern margin of the western Snake River Plain elevated these highlands. The highest peaks reach over 8,000 feet and hosted glaciers during the Pleistocene Epoch. The Owyhee Mountains also expose older Miocene basalts of the Columbia River Basalt Group and Cretaceous granites. Extensional faulting of the western Snake River Plain separated these granites from their brethren in the Idaho batholith of central Idaho.

BONNEVILLE FLOOD

In addition to its exceptional volcanic history, the Snake River Plain is also home to one of the largest catastrophic floods in Earth's history. The power and magnitude of the calamitous Bonneville Flood drastically transformed the landscape along the Snake River corridor. Dramatic evidence of the flood extends from the Pocatello region westward through Twin Falls and continues to the Oregon border and northward through Hells Canyon.

The story of the Bonneville Flood begins with Lake Bonneville, an immense freshwater lake occupying much of western Utah, eastern Nevada, and small portions of southern Idaho during the Pleistocene and Pliocene Epochs. At its maximum extent just before the flood, the lake was similar in size to today's Lake Michigan. Covering more than 20,000 square miles, it measured about 300 miles long by 135 miles wide and, with a shoreline at 5,200 feet elevation, was over 1,000 feet deep.

Lying in an enclosed, interior basin with no outlet, the lake only lost water through evaporation. As such, the lake's size fluctuated markedly in response to climatic changes. Cool, wet periods enlarged the lake, while hot, dry episodes caused the water level to drop. These fluctuations were particularly pronounced during ice ages and interglacial periods. The present Great Salt Lake of northern Utah is the latest in a long succession of lakes that have occupied the basin over the past 15 million years.

Lake Bonneville rose to its highest level about 20,000 years ago during the last ice age. In addition to the cool climate, the lake level was enhanced by the addition of the Bear River to its watershed 55,000 years ago when a lava flow changed the river's course. Around 17,500 years ago, the lake catastrophically overspilled the lowest point of its divide, an alluvial fan near Red Rock Pass in southeastern Idaho. An earthquake and the resulting seismic water wave on the lake may have initiated the spillover. Regardless of the cause, lake water poured over the divide, cutting a chasm into the alluvial fan sediments and the underlying Paleozoic rocks. As the divide was breached, more lake water emptied northward, increasing the flood's power. Overall, the immense lake dropped about 400 feet in a massive flood whose peak flow lasted perhaps a few months, with the entire process taking about one year.

The raging floodwaters raced down the Portneuf River drainage, overwhelming its narrow valley and plowing through the location where Pocatello sits today. Here, the surging maelstrom joined the westward-flowing Snake River. On its course across southern Idaho to the Columbia River, the flood's velocity and energy increased dramatically wherever the floodwaters passed through narrow canyon sections. The deluge ripped huge basalt boulders as large as cars from the canyon walls and flung them downstream. In the roiling torrent, boulders smashed into each other, eventually changing their shape from angular to rounded. Where the canyon sections widened, the velocity slowed, allowing the large rounded boulders to drop onto the canyon floor. In places, the boulders look like a field of giant petrified watermelons and are known as Melon Gravels. The Bonneville Flood ranks as one of the great flood events on Earth.

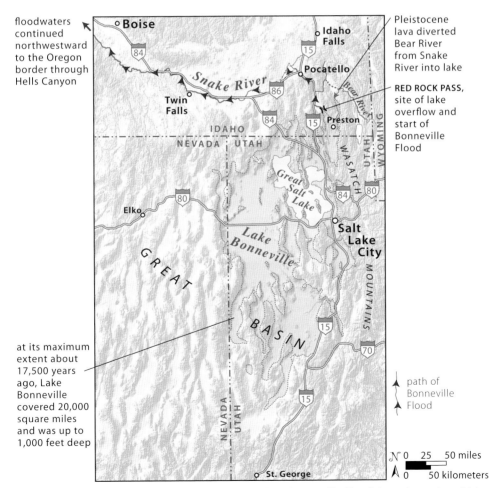

Lake Bonneville about 17,500 years ago and the path of the Bonneville Flood through southern Idaho.

Cliffs of the 52,000-year-old McKinney Basalt on the left (east) side of the Snake River with Melon Gravels from the Bonneville Flood near milepost 128 of I-84.

I-15
Pocatello—Idaho Falls—Monida Pass
124 miles

North of Pocatello, I-15 passes through the land of famous potatoes, fertile sandy and silty soils carried by wind and the Snake River system. The soil is irrigated with water from the Snake River, which lies west of the freeway. Fort Hall Bottoms, a wetland fed by springs, benefits wildlife and fish in the Snake River. Big Southern Butte, a 300,000-year-old, huge dome made of rhyolite, is prominent in the distance to the northwest.

Ferry Butte forms the low mound on the skyline north of the town of Fort Hall at milepost 78. Ferry Butte is a mound of basalt thought to be uplifted by a buried intrusion of rhyolite. In the distance farther northwest are Middle Butte, composed of several basalt flows that lie above a rhyolite plug, and East Butte, a 600,000-year-old rhyolite dome. Between and surrounding the volcanic buttes is an ocean of basalt lava flows, younger than 1 million years, and in some cases as young as 2,000 years.

At Fort Hall, near milepost 80, I-15 crosses the Oregon Trail, which followed Ross Fork Creek as it drained west toward the Snake River. Fort Hall was an important stop on the Oregon Trail, which was active from about 1840 to about 1860, with wagon trains of westward migrants seeking a new life in agricultural lands of the Willamette Valley in Oregon. About 15 miles east of Fort Hall, out of sight beyond the nearest

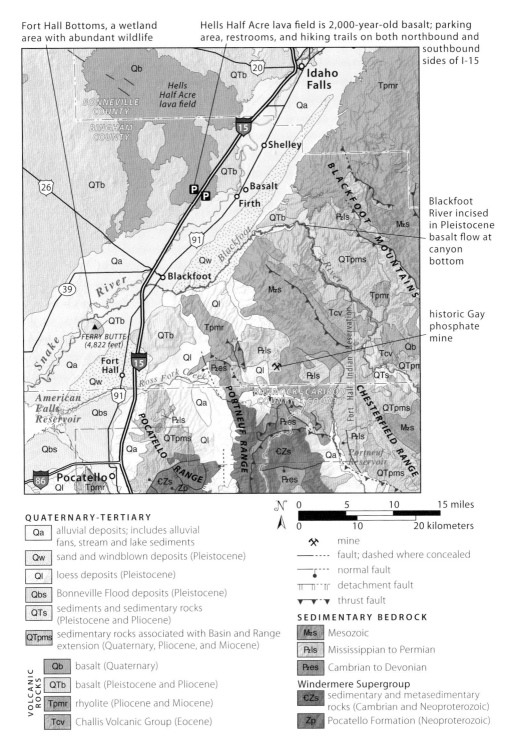

Geology along I-15 between Pocatello and Idaho Falls.

ridges, is the historic Gay Mine, where phosphate deposits were mined for fertilizer during the middle part of the twentieth century by the J. R. Simplot Company. The phosphate, which originated in the shells and bones of marine life, lies in the Phosphoria Formation, deposited in a deep, anoxic marine basin in Permian time.

I-15 crosses Hells Half Acre lava field north of milepost 100. A geological site at the rest area at milepost 101 affords access to 2,000-year-old basalt on a paved

Kinport Peak south of Pocatello, viewed looking south from milepost 75. The peak (center) pokes up above a flat surface that has been interpreted as the regional land surface when the rhyolitic Picabo caldera occupied the Snake River Plain where the freeway is located. There has been about 3,000 feet of subsidence of the Snake River Plain since the caldera erupted about 8 million years ago. (42.9861 -112.4183)

Telephoto view of Big Southern Butte, a 300,000-year-old rhyolite dome looming above the Snake River riparian zone on the Fort Hall Indian Reservation near milepost 76. The Lost River Range is in the distance. (42.9901 -112.4168)

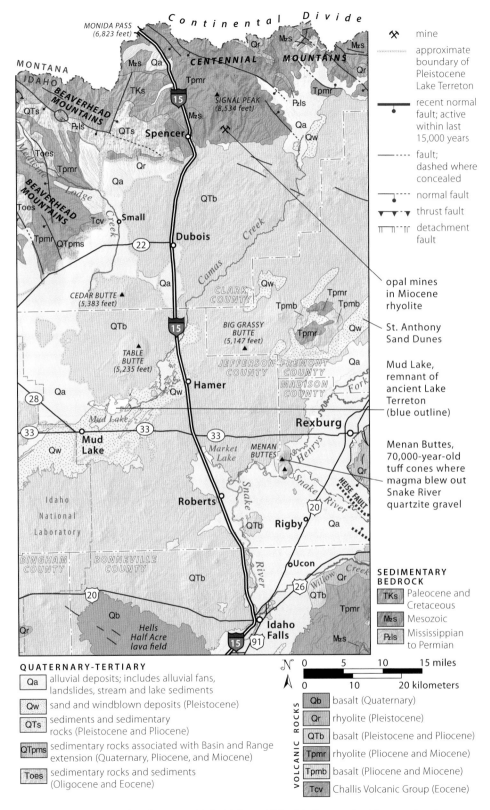

Geology along I-15 between Idaho Falls and Monida Pass at the Montana border.

walking trail. The young lava at Hells Half Acre flowed southeast from vents about 16 miles to the northwest, close to the axis of the Snake River Plain. The name is fitting because walking on uneven, jagged young lava is hellish. To the east, the Blackfoot River drains westward from the southern edge of the Blackfoot Mountains where it emerges from a narrow canyon in Pleistocene basalt.

In Idaho Falls, the Snake River flows over the edges of Pleistocene basalt. The original rapids here was converted to a falls for hydropower. To see the falls, take exit 118 for Broadway Street and head east to the river, where a series of parks form the Idaho Falls Greenbelt through the middle of downtown. The Idaho Falls Temple is immediately on the east side of the river upstream from the falls.

North of Idaho Falls, I-15 follows the Snake River for about 15 miles and then heads straight north across young basalt to Montana. Near Market Lake, just north of milepost 134, the highway crosses what was formerly the course of the Henrys Fork of the Snake River. The river was diverted eastward and away from the Market Lake area by basalt lava in the last 100,000 years. The highway crosses these basalts north of Roberts.

Mountains visible northwest of Market Lake include, from west to east, the Lost River Range, Lemhi Range, and Beaverhead Mountains, all formed from Paleozoic sedimentary rocks. A few miles west of the freeway, near the junction with ID 33, lies the location of Pleistocene Lake Terreton, which is described in the ID 33 road guide.

Opal in Spencer is quarried from nearby exposures of Miocene rhyolite and sold by locals.

On a clear day, you can see to the east the peaks of the Teton Range where Archean metamorphic rock is uplifted along a steep fault on the east side of the range. These old rocks are part of the Wyoming Province of the North American craton.

I-15 passes Spencer, self-proclaimed Opal Capital of America, at milepost 180. The prized opal occupies irregular voids and pockets in Miocene rhyolite from the Heise volcanic field. Hot groundwater circulated through the rhyolite and dissolved silica as it passed through. Later, the groundwater cooled, forcing the silica out of solution to form the spectacular opal deposits. Small shops in Spencer sell opal quarried from the nearby hills.

North of Spencer, the freeway ascends the east-trending Centennial Mountains on a smooth grade that follows a branch of the Union Pacific Railroad, first built as a narrow-gauge route in 1878 to access the Montana gold mines. North-flowing basalt lava filled the canyon east of the highway north of milepost 188. The Centennial Mountains have been uplifted in the past 6 million years, so prior to that, lava and streams flowed northeast from the Snake River Plain into Montana across what is now the Continental Divide. The divide and Montana border, at an elevation of 6,870 feet, is reached at Monida Pass at milepost 196.

I-84
Oregon Border—Boise—Twin Falls—I-86 Junction
222 miles

I-84 is the principal east-west route through southern Idaho. For many travelers blitzing across Idaho on their way elsewhere, it may be their only impression of the state. At first glance, the geology and scenery along this corridor appear bland and uninspiring. However, subtle hints exposed along the freeway reveal evidence of dramatic geologic events. Much of the freeway passes young volcanoes and Lake Idaho deposits within the low-lying terrain of the Snake River Plain.

The Snake River largely delineates the border between Oregon and Idaho as well as the western entrance of I-84 into Idaho. The fertile soils of the river's floodplain and the mild climate of this relatively low-elevation region converge to create ideal conditions for agriculture. Urban and agricultural development from the Oregon border eastward toward Boise define an area called the Treasure Valley, so named based on the abundant agricultural and natural resources of the region.

Near milepost 5, the freeway begins climbing out of the Snake River floodplain and into undulating, beige hills of silt and clay that accumulated at the bottom of Lake Idaho. The freeway eventually descends into the Boise River drainage, crossing the river between mileposts 26 and 27. One of the principal tributaries of the Snake River in southern Idaho, the Boise River begins in the rugged highlands of central Idaho.

Around milepost 53, the lower foothills immediately north of the freeway and the city of Boise expose primarily Lake Idaho sediments, while the mountains to the north consist of granitic rocks of the Idaho batholith. The abrupt elevation change from the city into the foothills is due to northwest-trending normal faults along the base of the foothills. Cumulatively, the total vertical displacement along these faults exceeds 9,000 feet, which partly accounts for the approximately 4,000 feet of elevation difference between the city and the mountains about 10 miles to the northeast. These

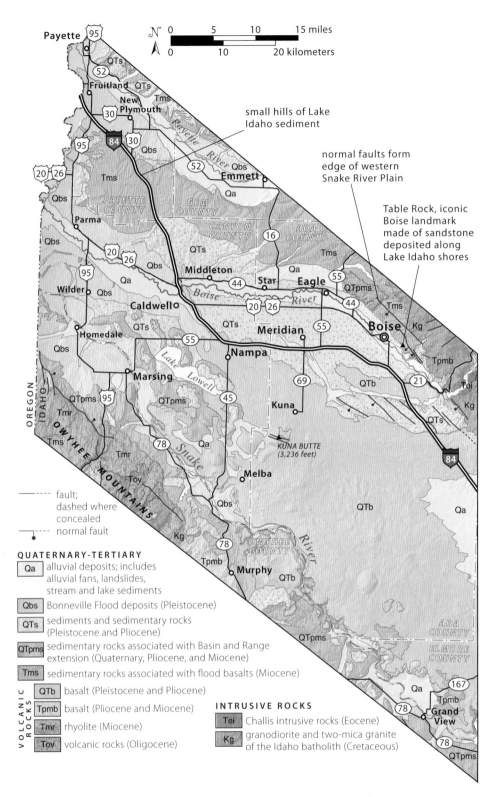

Geology along I-84 between the Oregon border and Boise.

faults and others define the boundaries of the western Snake River Plain. Similar faults lie along the southern margin of the Snake River Plain where the Owyhee Mountains rise to the south. Collectively, these faults and the intervening valley define a graben, a down-dropped area between normal faults.

The area southwest of Boise displays eight well-developed Pleistocene terraces produced by the Boise River. Each terrace consists of sand and gravel deposits that were later dissected during a period of downcutting. The terraces get progressively younger as they drop toward the modern Boise River. Many homes and other developments have been constructed on these terraces, taking advantage of their level profiles.

After exiting the Boise urban area, I-84 leaves the Boise River basin and crawls onto the volcanic tablelands to the southeast. Small roadcuts around milepost 68 expose Pleistocene basalt, which flowed southward from volcanic vents a few miles to the north. Volcanoes along this stretch of freeway include the typical broad shield volcanoes so ubiquitous in the Snake River Plain, along with small, steep-sided cinder cones. Good examples of cinder cones include the small hill about 7 miles south of the freeway near milepost 80 and a pair of heavily quarried hills about 1 mile south of the freeway near milepost 90. Cinders extracted from these locations are used in Boise and other neighboring towns for landscaping. Crater Rings, a set of large pit craters at the summit of a shield volcano, lies about 5 miles east of the quarried cinder cones but is not visible from the freeway. Shield volcanoes include Lockman Butte, north of the freeway near milepost 90, and Rocky Butte, topped with communication towers north of milepost 95, the main exit for Mountain Home.

Collectively, the volcanoes in this region are part of the Kuna–Mountain Home rift zone, a nearly 60-mile-long, nearly east-west alignment of volcanic vents that cuts obliquely across the western Snake River Plain. Lava and other material erupted from these volcanoes have forced the Snake River to the southern margin of the plain.

An unnamed Pleistocene shield volcano near Mountain Home. (43.1320, -115.6577)

34 SNAKE RIVER PLAIN (84)

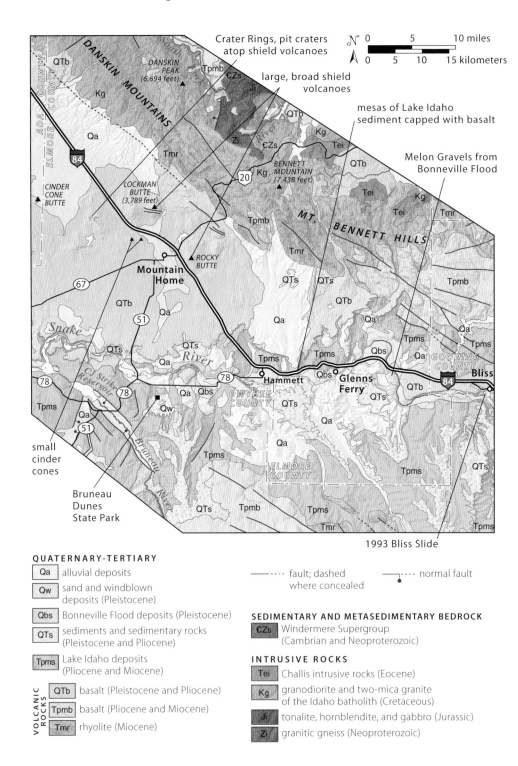

Geology along I-84 between Mountain Home and Bliss.

The highlands north of Mountain Home are the Mt. Bennett Hills, a west-trending range primarily composed of Miocene rhyolite erupted as sluggish lava and also as super-hot ash deposits that remelted before cooling. Much of the rhyolite erupted along volcanic vents trending northwest, parallel to the faults that bound the northern side of the western Snake River Plain. The rhyolite of the Mt. Bennett Hills near Mountain Home ranges from about 11 to 9 million years in age, corresponding to the time when the region was above the Yellowstone hot spot.

Near milepost 108, the freeway descends into the Snake River Canyon. Here the canyon is about 3 to 4 miles wide but narrows to less than 1 mile just east of Hammett. The basalt that caps the mesas and forms the canyon rimrock lies above Lake Idaho deposits of the Pliocene Glenns Ferry Formation, a geologic unit well known for the fossils it holds at nearby Hagerman Fossil Beds National Monument. Outcrops of these beige lake, shoreline, and stream deposits are visible on the slopes north of the freeway near mileposts 114 and 117.

The freeway crosses the Snake River near milepost 122 and the town of Glenns Ferry, historically used as a river crossing for westbound travelers on the Oregon Trail. East of Glenns Ferry, the river makes a wide, sweeping meander a few miles to the north, and I-84 crosses the Snake River again around milepost 128. Large, rounded boulders deposited by the Bonneville Flood lie scattered near the freeway east of the river. East of the river crossing, I-84 ascends basalt erupted from McKinney Butte, a shield volcano located about 18 miles to the northeast. Basaltic lava erupted here about 52,000 years ago and poured into the Snake River, filling the existing canyon with a lava dam several miles long and impounding a temporary lake upstream of the dam. Between mileposts 134 and 137, the irregular lava surface is readily apparent.

The beige, cross-bedded sandstone of the Pliocene-age Glenns Ferry Formation lies beneath Pleistocene basalt capping the mesa near milepost 117. (42.9580, -115.6577)

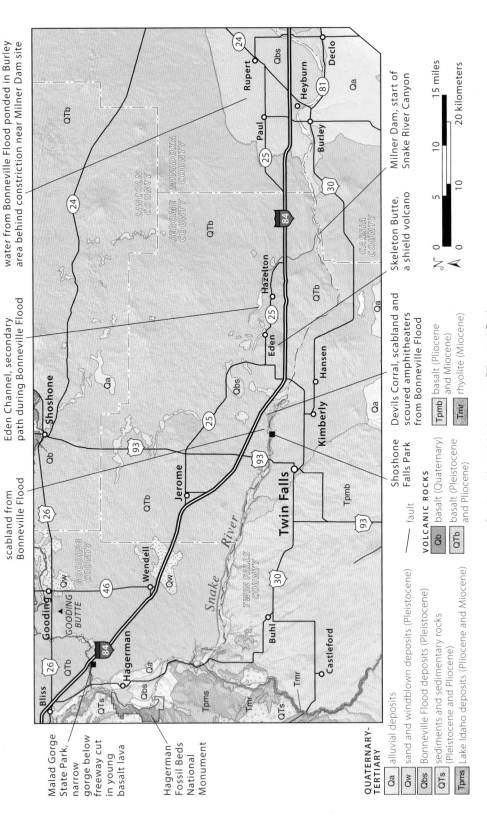

Geology along I-84 between Bliss and Burley.

MALAD GORGE OF THOUSAND SPRINGS STATE PARK

I-84 crosses the impressive Malad Gorge at milepost 146, but it's hard to see much of this narrow, deep canyon as you hurtle along at freeway speeds. For a more satisfying view and a fine picnic spot, take exit 147 at Tuttle and head west for about 0.3 mile. Turn north on Ritchie Road for 0.7 mile as it bends west to an intersection. Turn north to visit Malad Gorge or turn south for the picnic area. A small parking fee is required.

The Malad River may be one of the shortest named rivers anywhere. Not to be confused with the Malad River in southeast Idaho near the city of Malad, the river here is the combination of the Big Wood and Little Wood Rivers, which drain the mountains of central Idaho, flowing southward into and across the Snake River Plain. The two rivers merge near Gooding, where the conjoined river's name changes to the Malad River. This Malad travels a mere 12 miles before merging with the larger Snake River.

Devils Washbowl waterfall on the Malad River at Malad Gorge State Park near milepost 146. (42.8679, -114.8548)

Aerial view to south of Malad Gorge with I-84 in foreground. —Photo by Gage Willsey

The narrow gorge containing the Malad River begins just north of the freeway where the river drops precipitously into a 250-foot-deep chasm. The walls of the gorge expose stacked basalt lavas erupted from nearby Gooding Butte about 373,000 years ago. The Malad River was diverted to its present location by the McKinney Butte eruption about 52,000 years ago, and its impressive gorge was carved sometime thereafter by huge floods. The details are still being investigated but may involve catastrophic outburst floods wrought by collapse of glacial dams impounding drainages in the mountains to the north. The subsequent floods released huge volumes of water that quickly eroded through the underlying basalt.

Agricultural fields, dairy farms, and a handful of towns dominate the margins of the freeway between the Malad River and Twin Falls. Between mileposts 172 and 179, the freeway crosses a swath of exposed basalt, largely stripped clean of its overlying topsoil by the catastrophic Bonneville Flood about 17,500 years ago. As this massive deluge backed up at a narrow constriction in the Snake River's path just east of Burley, the rising waters began to flow westward along another path, the Eden Channel, north of the river through the present towns of Eden and Hazelton. The Eden Channel parallels the river for about 30 miles before merging back into the Snake River north of Twin Falls. This overland flood stripped away all available topsoil, exposing the underlying basalt and creating the barren, raw landscape, called scabland, visible along both sides of the freeway. As the overland channel merged

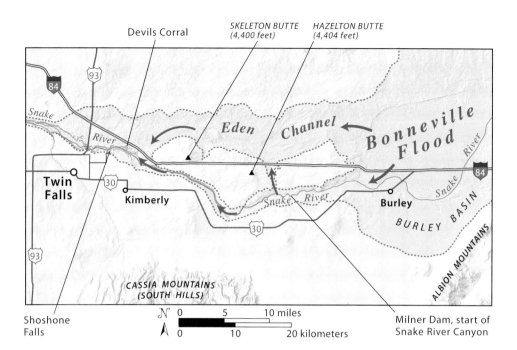

The path of the Bonneville Flood in south-central Idaho. Floodwater backed up behind the narrow constriction that defines the start of the Snake River Canyon (near present-day Milner Dam), flooding the Burley basin. The water spilled over to the north, flowing overland to form the Eden Channel. The two paths merged near Twin Falls, with the floodwater eroding several spectacular landforms.

with the Snake River Canyon near Shoshone Falls, it cut downward through the layers of basalt, creating secondary canyons now left dry. A spectacular example, known as Devils Corral, lies south of the interstate near milepost 178.

A splendid shield volcano, Skeleton Butte, lies just north of the freeway near milepost 185. Erupting 1.76 million years ago, much of its lower slopes has been buried by lava from younger, neighboring volcanoes, making it appear smaller than its original size. A large and conspicuous cut on the south side of the volcano served as a quarry for road base material during construction of I-84.

The freeway crosses the Snake River near milepost 216 east of Burley. The river level lies just beneath the freeway bridge where it flows through a broad valley, a stark contrast to the nearly 500-foot-deep confines of the Snake River Canyon near Twin Falls.

In general, the river's course across southern Idaho is dictated by the region's past volcanic activity. Basaltic lava erupted from shield volcanoes is hot and fluid enough to travel tens of miles before cooling and crystallizing into solid rock. Over the past 5 or so million years, eruptions of lava have flowed into the Snake River. Large eruptions overwhelmed the river with lava, disrupting the river's course and forcing it to flow elsewhere, where it established a new channel.

I-86
I-84 Junction—Pocatello
63 miles

This short freeway connects I-84 to I-15 in southeast Idaho. The Basin and Range mountains of southeast Idaho are lined up south of the road. The highway closely follows the path of the Snake River, which has been forced to the southern edge of the Snake River Plain by volcanism. To the north are young volcanoes of the Snake River Plain, ranging from low basaltic shield volcanoes to steep rhyolite domes.

At the junction with I-84, the small mountains directly south of I-86 are the Cotterel Mountains, a west-dipping package of Miocene rhyolite tuff and lava bounded by faults. A steep normal fault exists on the east side and merges at depth with the shallowly east-dipping Raft River detachment fault, which is at the surface west of the Cotterel Mountains. The loftier highlands behind and west of the Cotterel Mountains are the Albion Mountains, part of the Raft River–Albion metamorphic core complex that exposes Archean rocks. See the ID 77 road guide in the Basin and Range chapter for more information.

At milepost 2, a textbook-worthy shield volcano, adorned with radio towers, lies south of the freeway. Near milepost 5, a clear view southward showcases a typical Basin and Range landscape of alternating mountains and valleys. The Albion and Cotterel Mountains lie to the west. The Black Pine Mountains, the solitary range to the south best seen from I-84, are composed of late Paleozoic sedimentary rocks. The dome-shaped range west of the Black Pine Mountains is the Raft River Mountains of Utah, composed of Proterozoic and Archean rocks. The Sublett Range, composed of mostly Paleozoic sedimentary rocks, lies to the east. To the north are the jagged peaks of the Pioneer Mountains, often mistakenly called "the Sawtooths." The real Sawtooth Range lies much farther north and out of view.

Near milepost 10, a subtle rise on the north horizon defines Pillar Butte, the summit vent for the vast Wapi lava field, which erupted about 2,200 years ago, roughly concurrent with the youngest eruptions of the Craters of the Moon volcanic field to the north. Pillar Butte is the southernmost vent along the Great Rift, a nearly 60-mile-long, northwest-trending zone of fissures and other volcanic vents extending across the Snake River Plain into the Craters of the Moon region. Near milepost 14, the freeway crosses the Raft River, which originates in the extreme northwest corner of Utah, the only part of Utah that drains northward into the Snake River.

At milepost 19, the freeway descends toward the Snake River. The rest area on the south side of the freeway at the top of the hill marks the route of the Oregon Trail, where it ascended from the Snake River and headed south and west across the Raft River Valley.

Eastward, the freeway cuts through the late Miocene Massacre Volcanic Complex, a cluster of basaltic volcanoes that interacted with shallow groundwater, producing explosive eruptions and building cones of ash and larger particles around 6 million years ago. While erosion has removed as much as 1,000 feet from the tops of these interesting volcanoes, exposures along the Snake River and its tributaries provide clues and evidence of their existence. A long, terraced roadcut on the south side of I-86 near milepost 27 exposes flat-lying Pleistocene basalt. This flat basalt flowed onto a hill of tuff and breccia from the Massacre Volcanic Complex.

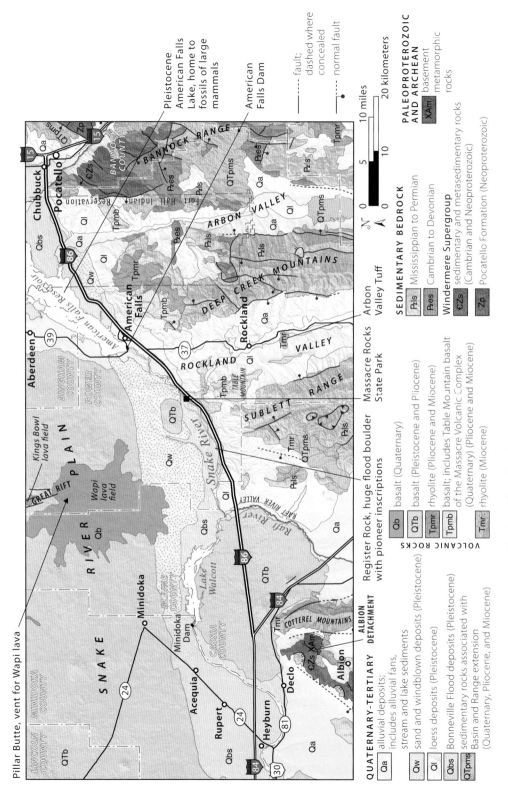

Geology along I-86 between the I-84 junction and Pocatello.

Pleistocene basalt laps onto west-dipping tan basaltic tuff and breccia of the late Miocene Massacre Volcanic Complex, where groundwater mixed with lava, creating explosive conditions. (42.6549, -113.0081)

The exit at milepost 28 provides access to two sites of historic and geologic note, Massacre Rocks State Park and Register Rock. To reach Register Rock, turn south from the off-ramp and follow Register Road westward to a small turnoff and shaded park. Register Rock is a large basalt boulder transported and tumbled by the Bonneville Flood. Upon it, westward emigrants inscribed their names and dates of passage.

Massacre Rocks State Park, marking the site of an 1862 skirmish between westward settlers and Native Americans, can be reached by turning north from the off-ramp. The park provides riverfront access and many other amenities including a campground, visitor center, and hiking trails. Geologically, the park contains abundant outcrops of basaltic tuff and breccia formed by the hydrovolcanic eruptions of the Massacre Volcanic Complex. Large boulder bars scattered throughout the park mark the path of the Bonneville Flood, which dramatically shaped the surrounding landscape.

At milepost 30, the Oregon Trail is south of the freeway. A paved, marked trail begins at the rest area accessed by westbound motorists at milepost 31. The Deep Creek Mountains, the next in the progression of Basin and Range mountains, are south of the freeway near milepost 36.

The largest reservoir on the Snake River, American Falls, lies north of the freeway near milepost 39. This water body and the surrounding lowlands were part of ancestral American Falls Lake about 72,000 years ago. Stream and lake deposits from this period include fossils of Pleistocene mammals such as camels, bison, sloths, and mammoths.

Near milepost 50, eastbound travelers see the Bannock Range to the east with the Portneuf Range farther eastward. The craggy hill behind the Simplot phosphate plant near milepost 57 is 8-million-year-old rhyolite. To the north lie the rhyolite domes of Big Southern Butte, Middle Butte, and East Butte described in the US 26 road log.

US 20
MOUNTAIN HOME—CAREY
101 miles

US 20 (also known as the Sun Valley Highway) is the preferred route between Boise and the Sun Valley area. Northeast of Mountain Home, the highway slices through the western end of the Mt. Bennett Hills, plows through the rich fields of the Camas Prairie, crosses the Big Wood River near the intersection with ID 75, and ends in Carey, where it merges with US 93 and US 26. Along the way, US 20 intercepts a host of volcanic rocks along with fine exposures of the Idaho batholith and scenic mountain vistas.

The Mountain Home area is covered by Quaternary basalt of the western Snake River Plain. The broad profiles of several large, unnamed shield volcanoes dominate the landscape just north of I-84. US 20 heads northeast from Mountain Home and across basaltic lava from these volcanoes to near milepost 102, where the basalt terminates against the steep base of the Mt. Bennett Hills. Northwest-trending normal faults bound this part of the western Snake River Plain and are responsible for the abrupt transition in topography.

Composed mainly of Miocene rhyolite produced by the Yellowstone hot spot, the Mt. Bennett Hills form an east-west highland bounding the northern edge of the Snake River Plain. Collectively, the rhyolite ranges from about 12 to 5 million years in age and erupted as lava and ash from the Twin Falls volcanic field and other localized vents. Red-brown rhyolite is abundantly exposed in roadcuts and craggy outcrops between mileposts 103 and 111.

Around milepost 122, knobs and pinnacles of light-gray granodiorite of the Idaho batholith emerge on the hillsides among small clusters of aspen trees. The highway crests Cat Creek Summit at an elevation of 5,527 feet between mileposts 124 and 125. A large pullout on the north side of the highway just west of the summit provides access to the granodiorite along with scenic views of the surrounding highlands.

Crags of light-gray granodiorite from the Idaho batholith along US 20 near Cat Creek Summit. (43.2990, -115.3291)

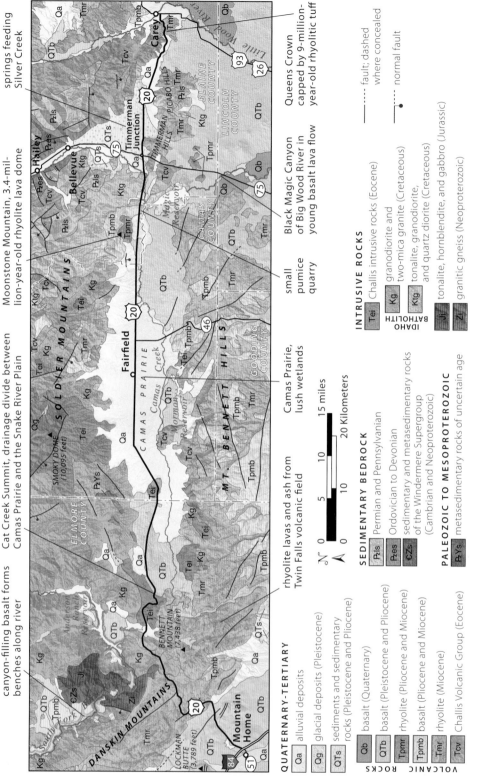

Geology along US 20 between Mountain Home and Carey.

Eastward from Cat Creek Summit, US 20 enters the Camas Prairie, an elongated, east-trending valley straddling the boundary between the Snake River Plain and the central Idaho mountains. The Soldier Mountains form the highlands north of the prairie and expose a mixture of granitic rocks of the Cretaceous Idaho batholith, rocks of the Eocene Challis Volcanic Group, and some Miocene to Pliocene rhyolite, with fine-grained late Paleozoic sedimentary rocks at the mountains' far eastern end. Rocks south of the Camas Prairie are dominated by rhyolite of the Mt. Bennett Hills along with lesser amounts of basalt. The surficial alluvium of Camas Prairie mostly covers Quaternary basalt from local shield volcanoes and vents.

The Camas Prairie's east-west orientation is at odds with the dominant north-south trend of mountains and faults that prevails throughout much of southern Idaho. The most convincing explanation for this orientation is that as the continual westward movement of the North American Plate dragged the central Snake River Plain away from the hot spot, the crust cooled and subsided. Subsidence caused the crust along the margins of the plain to bend downward, stretching the crust in a north-south direction and creating the east-west trending Camas Prairie. With highlands to the north and south, the Camas Prairie collects a large amount of surface water and groundwater, forming abundant wetlands in and among the agricultural fields.

At milepost 168, the highway crosses a tributary to east-flowing Camas Creek, which lies a half mile south of the highway. The shallow draw exposes the 1.45-million-year-old basalt that covers much of the eastern half of the Camas Prairie. The prominent high point to the north is Moonstone Mountain, a volcanic vent that oozed sluggish rhyolitic lava 3.4 million years ago to form a steep-sided dome.

US 20 crests a small summit near milepost 172, exposing a fantastic roadcut through rhyolitic ash erupted about 5 million years ago from a volcanic field located near Magic Reservoir. The north side of the roadcut beautifully displays a gradation from dark, glassy vitrophyre at the bottom of the roadcut upward through pink

The roadcut near milepost 172 of US 20 exposes 5-million-year-old volcanic deposits. The light-colored ash at the top grades downward into pink welded tuff and finally dark-gray vitrophyre along the road. (43.3371, -114.4045)

welded tuff and then light-gray tuff. The weight of the overlying ash and heat upon eruption accounts for the changes in color and hardness of the rock. A small pumice quarry just south of the highway mines pebble-sized pumice used for landscaping and to cover the infield area of baseball fields.

Roadcuts between mileposts 175 and 176 reveal light-gray biotite granodiorite of the Idaho batholith. At milepost 176, the highway crosses the Big Wood River, which drains the Pioneer, Smoky, and Boulder Mountains to the north. The river flows southward through the Wood River Valley for much of its length but is blocked by the east-west Timmerman Hills as it enters the northern edge of the Snake River Plain. Forced westward, the Big Wood River joins Camas Creek at Magic Reservoir, then slices southward on its journey to becoming the Malad River just north of where it joins the Snake west of Hagerman.

Biotite granodiorite of the Idaho batholith along US 20 west of Timmerman Junction. Dark inclusions to the right of the rock hammer are diorite. (43.3278, -114.3273)

Look for tight flow folds within Miocene rhyolite in a roadcut along US 20 west of Carey. Rock hammer for scale. (43.2978, -113.9642)

Eastward from its junction with ID 75, the highlands south of US 20 are the Picabo Hills, mainly 8-million-year-old rhyolite tuff at its western end and a mixture of Idaho batholith rocks and Permian-Pennsylvanian sedimentary rocks of the Sun Valley Group at its eastern end. Red, bold outcrops in the mountains north of the highway at milepost 191 are rocks of the Challis Volcanic Group.

The mesa south of the highway at milepost 193 is Queens Crown, capped by slightly south-dipping, 9-million-year-old rhyolitic tuff. Deposits of white ash are visible on the lower slopes. The top of a roadcut south of the highway between mileposts 193 and 194 exposes the white ash and allows for closer inspection. A terraced roadcut at milepost 195 exposes Miocene rhyolite with intricate flow folding, indicating the viscous nature of the lava as it rolled over itself.

US 20
Arco—Idaho Falls
57 miles

Arco sits at the southern end of the Lost River Range, where the Lost River fault ends at the Snake River Plain. This active normal fault continues to uplift the range. Mississippian limestones of the Scott Peak and Surrett Canyon Formations form the cliffs northeast of Arco, upon which are painted high school graduation years, some more than one hundred years old!

East of Arco, US 20 heads across the wide expanse of the Snake River Plain through the Idaho National Laboratory (INL), a US Department of Energy facility involved with nuclear research. At the Big Lost River rest area at milepost 265, US 20 crosses the Big Lost River. The rest area and the INL buildings north of the highway are in the Big Lost Trough, a Pleistocene and Holocene volcanic-filled basin, in which water of

High school graduation years painted on Mississippian limestone cliffs at the southern end of the Lost River Range, above the town of Arco. (43.6346, -113.2975)

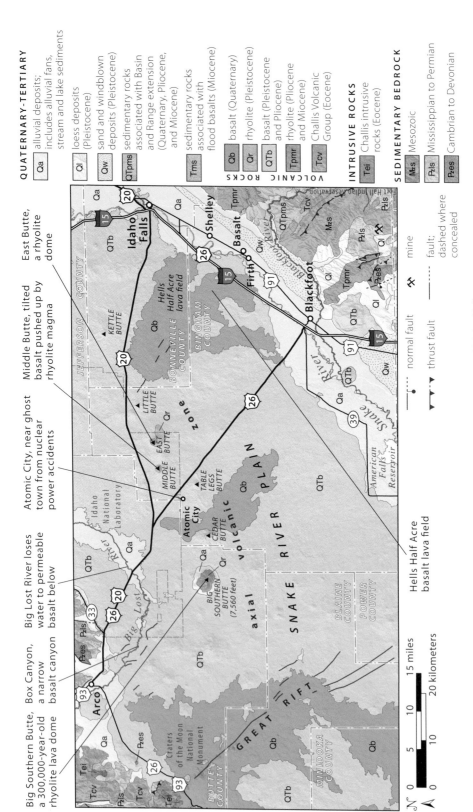

Geology along US 20 between Arco and Idaho Falls.

the Big Lost River sinks into the basalt below. In spring there might be water here, but far less water than where US 93 crosses the river west of Arco. The river's water seeps downward into the permeable basalts of the Snake River Plain, becoming groundwater that flows westward in the Snake River Plain aquifer.

South of the rest area is Big Southern Butte, the tallest structure in the Snake River Plain, rising about 2,400 feet above the plain. Big Southern Butte is a rhyolite dome that erupted about 300,000 years ago when thick pasty lava oozed upward. The silica-rich lava piled upon itself, forming the steep-sided mountain of rhyolite. The north side of the butte is a massive slab of stacked basalt layers that were uplifted and tilted as the rhyolite lava pushed its way to the surface.

US 20 heads east of the rest area across basalt lava fields toward Idaho Falls and crosses the axial volcanic zone, a northeast-trending swath of higher topography and volcanic vents near the center of the eastern Snake River Plain. The highway heads toward East and Middle Buttes before passing them on the north, east of the junction with US 26, which passes south of the buttes. Both of these buttes are rhyolite domes that have erupted in the last half million years along the axial volcanic zone through the cover of Pleistocene basalts. Middle Butte exposes uplifted basalt on its summit because the underlying rhyolite has not broken through.

US 20 crosses a saddle between two shield volcanoes, both topped with communication towers, near milepost 278. Little Butte is the volcano south of the highway; the volcano to the north is unnamed. Yet another shield volcano, Kettle Butte, is visible to

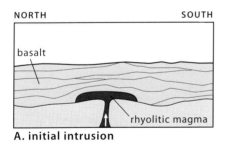
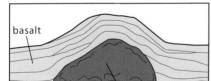
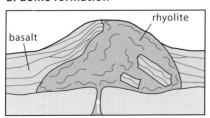

Diagrams depicting evolution of Big Southern Butte. A. Rhyolite magma intrudes beneath the overlying basalt. B. Continued intrusion of rhyolite uplifts the basalt forming an incipient dome. C. Rhyolite lava eventually breaks through to the surface, leaving a large slab of tilted basalt on the north side of Big Southern Butte. Broken slabs of basalt sink into the rhyolite due to higher density.
—From Willsey, 2017

the northeast around milepost 285. South of the highway lies the Hells Half Acre lava field, which erupted about 2,000 years ago. Eastward, the highway intersects farmland irrigated with Snake River water as it continues its gradual descent to Idaho Falls.

US 20
Idaho Falls—Montana Border
100 miles

North of Idaho Falls, US 20 brings tourists to the forests of Island Park and Yellowstone. Most of the rocks are volcanic and very young, geologically, with Pliocene and Pleistocene rhyolites overlain by Pleistocene basalt, all products of the Yellowstone hot spot.

From Idaho Falls, the highway heads north through several irrigated farming communities of the upper Snake River Valley. Near milepost 311, the highway passes a large gravel pit in sediments deposited by the Snake River. The high peaks of the Teton Range are visible to the northeast, above the forested Big Hole Mountains. The Big Holes are on the uplifted side of the Heise fault, that runs northwest, north of the Snake River in the Swan Valley area, and curves northward into the active Rexburg fault. At Rexburg, the distinct rise in ground to the east marks the trace of the fault. Both the BYU–Idaho campus and the Rexburg Temple are located on uplifted land on the east side of the Rexburg fault. The tuff cones of the Menan Buttes are visible to the west near milepost 326 and are discussed in more detail in the ID 33 road guide.

West of St. Anthony near milepost 345 are sand dunes that trend northeast, parallel with the prevailing southwest winds. The sand was blown here from the dry lakebed of Pleistocene Lake Terreton (explained in the ID 33 road log) to the west and dumped in front of the topographic high of Island Park. The lake rose and fell with climatic fluctuations during the Pleistocene ice ages.

North of Ashton, US 20 crosses the Henrys Fork of the Snake River between mileposts 363 and 364, then begins its 1,000-foot climb of the Ashton Grade and the rim of the Henrys Fork caldera. As the highway climbs the grade, the first few roadcuts expose Huckleberry Ridge Tuff, and the upper part of the grade exposes the very similarly looking Mesa Falls Tuff. The crest of the caldera rim is near milepost 368. The caldera floor to the north subsided and filled with younger basalt to form the Island Park region.

West of the highway near milepost 379 is Harriman State Park, where the clear Henrys Fork provides world-class fishing. Also known as Railroad Ranch, this area was the private ranch of the Harriman family who were owners of the Union Pacific railway. A branch line of the Union Pacific ran north from Idaho Falls to West Yellowstone, along the Warm River and through Island Park. This scenic route is now the bicycle trail that can be accessed from the Mesa Falls Scenic Byway.

The highway crosses the Henrys Fork at Osborne Bridge just north of the turnoff to Harriman State Park. Thurmon Ridge forms the western skyline and is the western and northern margin of the Henrys Fork caldera. A few steep hills that lie between the highway and the ridge are rhyolite domes that formed after the caldera-forming explosion 1.3 million years ago but before the basalt eruptions.

Forests of lodgepole pine dominate as the highway heads north through the Island Park area. Lava Creek Tuff, the ash from Yellowstone's most recent big eruption 630,000 years ago, is exposed in small roadcuts north of the town of Island Park. At

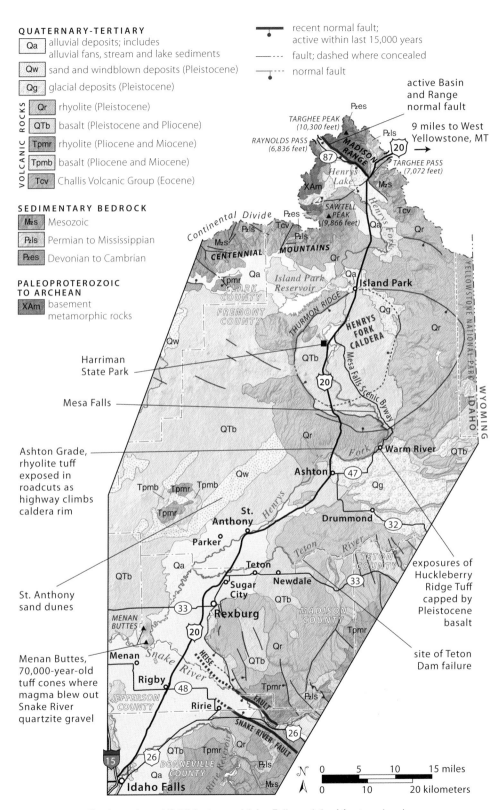

Geology along US 20 between Idaho Falls and the Montana border.

1976 TETON DAM DISASTER

A short diversion from US 20 onto ID 33 near Sugar City leads to the site of the worst disaster in Idaho's Euro-American history. From Sugar City, head east on ID 33 and turn north at 12000 East, near milepost 110. Drive 1.5 miles to a parking area and overlook above the canyon.

Built for hydroelectric power, flood control, and irrigation water, the 300-foot-tall earthen Teton Dam was completed in November 1975. During a wet winter and spring, the reservoir behind the dam began to fill. By the beginning of June 1976, the impounded water neared capacity, about 30 feet below the dam crest and about 3 feet below the spillway crest. Late on June 4, seepage through the cliff face downstream of the north side of the dam was noted. By the morning of June 5, more leaks in the north rock face below the dam were observed as well as an area of seepage on the face of the dam. After several failed efforts to fix the leaks, which continued to release more water, the north face of the dam catastrophically failed just before noon, releasing the contents of the nearly full reservoir over the course of about five hours.

The effects of the flood, reaching as far downstream as Idaho Falls, were devastating. Farms, cities, and towns below the dam had little warning. Eleven lives were lost along with over $400 million in property damage. Flooding in Sugar City and Rexburg was extensive with as much as 9 feet of water inundating homes and businesses.

The overlook provides a commanding location to observe the Teton River, the geology of the canyon, and the remnants of the dam. The bedrock in which the dam was constructed is the 2.1-million-year-old Huckleberry Ridge Tuff, a welded ash deposit created by a massive Yellowstone eruption. As you look across the canyon from the overlook, you can see that the tuff contains many open fractures that formed as columnar joints while the ash cooled and contracted. The ability of water to move through these fractures was a major factor in the dam's collapse. A large boulder bar lies downstream of the dam where blocks of tuff were ripped from the canyon wall and flung downstream during the flood.

View to the north of the site of Teton Dam, which failed in June 1976, causing a large flood downstream and killing eleven people. The long cement channel on the other side of the canyon is the dam's spillway. The pyramid-shaped hill south of the river is what remains of the dam, composed of a mixture of compacted gravel, sand, silt, and clay. (43.9052, -111.5404)

MESA FALLS SCENIC BYWAY
26 MILES

This highly scenic diversion largely parallels US 20 and provides roadside geologists with an alternate route when traveling between Ashton and the Island Park area during nonwinter months. The byway also provides access to two spectacular waterfalls on the Henrys Fork of the Snake River.

After crossing agricultural fields for 7 miles east of Ashton with terrific views of the Tetons to the east, ID 47 drops into the canyon of the Henrys Fork near its confluence with Warm River. Pleistocene basalt forms benches along the canyon rims. The highway crosses Warm River near milepost 9. Light-colored exposures of tuff erupted from Yellowstone lie across the river beneath the basalt. An old railroad grade, now converted to the Yellowstone Branch Trail, allows hikers and bikers to follow Warm River upstream into the Island Park area. ID 47 takes a more difficult route, climbing the

View across Warm River of Pleistocene basalt overlying tuff and rhyolite lava erupted from the Yellowstone volcanic field. (44.1151, -111.3205)

ridge between the canyons of Warm River and Henrys Fork onto the bench above. A pullout on the east side of ID 47 between mileposts 11 and 12 provides one last look into the Warm River drainage. Just north of the pullout are roadcuts exposing tuff erupted from the Yellowstone volcanic field.

A large pullout on the west side of the road and directly across from a short trail that connects to the Yellowstone Branch Trail marks the official end of ID 47. The byway continues north of here as paved Forest Road 294, which lacks milepost markers. This location also marks the approximate southern rim of the huge Island Park caldera, which formed during a violent eruption about 2.1 million years ago.

A parking area and overlook for Lower Mesa Falls lies about 2.5 miles north of the large pullout, and the short road leading to Upper Mesa Falls lies about 0.7 mile farther north. Each Henrys Fork waterfall drops over resistant Mesa Falls Tuff that has been welded by heat and the sheer weight of the overlying ash. Look for outcrops of basalt on the east sides of each waterfall. Lava and the river have been battling over this young landscape for the past million years or so. Eruptions of basaltic lava from area volcanoes have filled the river channel, forcing the river to carve a new canyon. Subsequent eruptions of lava continued to displace the river.

North of Upper and Lower Mesa Falls, the byway wanders through forest and meadows within the Henrys Fork caldera, which formed 1.3 million years ago and is the smallest of the three Yellowstone calderas. Small roadcuts of Pleistocene basalt that partially fills the caldera floor are intermittently exposed along the road. The intersection with US 20 lies near the Henrys Fork of the Snake River, here a slow, meandering stream.

Upper Mesa Falls on the Henrys Fork River drops over Mesa Falls Tuff, which erupted 1.3 million years ago from the Henrys Fork caldera. (44.1874, -111.3294) —Photo by Dan Moore

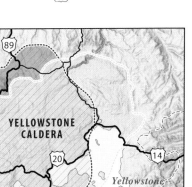

Age	Amount of Ash Erupted	Caldera	Ash Deposit
630,000 years	240 cubic miles	Yellowstone	Lava Creek Tuff
1.3 million years	67 cubic miles	Henrys Fork	Mesa Falls Tuff
2.1 million years	600 cubic miles	Island Park	Huckleberry Ridge Tuff

Map of the three most recent caldera-forming eruptions in the Yellowstone region.

milepost 387, US 20 crosses the Buffalo River, which marks the northern edge of the Island Park caldera. The highway breaks into a large open area around milepost 395, providing rewarding views of the Centennial Mountains to the west. The crest of this east-west trending mountain range forms the Continental Divide and the border with Montana. Sawtell Peak, with a square radio facility on its summit, is the prominent peak nearest the highway and is composed of Eocene volcanic rocks.

The southern tip of the Madison Range, a 70-mile-long, north-trending range extending well into Montana, lies north of US 20 and terminates southward near the junction with ID 87. This mountain range was uplifted during the past 4.5 million years by Basin and Range extension and exposes Archean basement rocks as well as Paleozoic sedimentary rocks. With many peaks over 10,000 feet, its upper reaches have been extensively sculpted and shaped by glacial erosion. US 20 bends eastward as it rises to Targhee Pass at the Continental Divide and the Montana border. West Yellowstone, the western gateway to Yellowstone National Park, lies 8 miles east.

View north from the Snake River at Harriman State Park to the Centennial Mountains, with Sawtell Peak at far right. Thurmon Ridge, the margin of the Henrys Fork caldera, crosses the middle of the photo. (44.32, -111.45)

US 26
US 20 Junction—Blackfoot
34 miles
See map on page 48.

From the US 20 junction at the Idaho National Laboratory, US 26 climbs onto the axial volcanic zone of the Snake River Plain, a northeast-trending zone of elevated topography and volcanic vents in the center of the eastern Snake River Plain. En route to Blackfoot, the highway descends nearly 1,000 feet to the irrigated flats surrounding the Snake River, a journey between the Idaho desert and the agricultural economy of southern Idaho. Starting from no soil and sagebrush on young basalt, the route lowers progressively from lava to alluvium and from irrigation by well water to irrigation by Snake River water in canals built in the late nineteenth century.

South of the US 20 junction, several prominent volcanoes are visible. To the southwest is Big Southern Butte, the tallest volcano in the Snake River Plain, jutting upward more than 2,000 feet above the basalt at its base. Big Southern Butte is a rhyolite dome that erupted about 300,000 years ago when thick, pasty lava oozed upward, piling on itself to form a steep-sided mass of rhyolite. East of Big Southern Butte lie two shield volcanoes: Cedar Butte and Table Legs Butte. East of the US 20 junction lies Middle Butte, a stack of basalt uplifted by underlying rhyolite magma. Farther east is East Butte, another rhyolite dome easily identified by the cluster of towers at its summit. East Butte erupted about 600,000 years ago.

The alignment of these volcanoes collectively defines the axial volcanic zone, which the highway crosses around the turnoff to Atomic City near milepost 279. A series of nuclear meltdowns and explosions at nearby INL facilities in the 1950s and 1960s cast an ominous shadow over the area. Today, Atomic City is a near ghost town of a few dozen residents.

Intermittent piles of basalt boulders, coated with white caliche, a weathering product made of calcium carbonate, form crude snow fences on the southwest side of the freeway. Small roadcuts and low basalt outcrops dot the highway and surrounding landscape.

A particularly interesting roadcut lies at milepost 280. East of the highway are several tumuli, buckled sections of lava where the solidified lava broke and was pushed upward from internal pressure of the lava beneath the surface. West of the highway is a low roadcut exposing basalt from Table Legs Butte, the prominent shield volcano about 1.5 miles to the southwest. Table Legs Butte erupted sometime between 400,000 and 200,000 years ago. A close look at the basalt in this roadcut reveals clusters of large intersecting crystals of plagioclase feldspar with gas bubbles, or vesicles, in the intervening space between crystals. The fancy term used by geologists for this type of texture in basalt is diktytaxitic; it's pronounced phonetically.

Middle Butte, a stack of basalt uplifted by rhyolite magma, from the west. (43.4468, -112.7843)

Roadcut at milepost 280 of diktytaxitic texture in basalt with Table Legs Butte, a shield volcano in the distance. (43.4338, -112.7638)

The northwesternmost farm along the highway is reached at milepost 284, southeast of which the highway descends toward the Snake River as it moves off the crest of the axial volcanic zone. In the distance to the east and south, you can see the linear ranges of the Basin and Range consisting of mainly Paleozoic and Mesozoic rocks of the Idaho thrust belt.

The southernmost toe of the 2,000-year-old Hells Half Acre lava field invades the agricultural area between milepost 300 and 301. The first canal carrying Snake River water is crossed at the base of the lava-cored hill near milepost 301. South of this canal, US 26 crosses alluvium of the Snake River near Blackfoot, home to the Idaho Potato Museum. This small city, developed along the Union Pacific Railroad on the banks of the Snake River, remains an important potato processing center.

US 30
Bliss—Twin Falls
44 miles

Prior to the construction of I-84, US 30 was the principal east-west highway along the Snake River Plain, linking towns large and small. Passing through a picturesque region and generally paralleling the Snake River, this section of US 30, part of the Thousand Springs Scenic Byway, offers views of nearby Hagerman Fossil Beds National Monument and units of the Thousand Springs State Park system. East of Buhl, the highway heads eastward through irrigated fields and several small towns before arriving in Twin Falls.

Leaving I-84 at exit 137, US 30 heads east for a few miles before entering the small town of Bliss, situated about 600 feet above the Snake River. At milepost 173, US 30 veers southward away from I-84. Just south of milepost 175, the highway begins its descent into the Snake River Canyon, passing through dark basalt that erupted from McKinney Butte, about 10 miles to the northeast. A large pullout on the west side of the highway near milepost 176 provides a fine view of the Hagerman Valley and the mighty Snake River about 400 feet below. The abrupt black cliffs of the canyon's east side contrast markedly with the subdued, grassy slopes to the west. Westward across the canyon lie pale bluffs of soft sedimentary rock of the Pliocene-age Glenns Ferry Formation in Hagerman Fossil Beds National Monument. Local slides, indicated by the lumpy and irregular topography, occur due to the saturation of weak, clay-rich layers within the monument.

Between mileposts 177 and 178, the highway crosses the Malad River, the name for the conjoined Big Wood and Little Wood Rivers where they merge west of Gooding, about 12 miles upstream. A large roadcut on the east side of the highway and just north of the bridge over the Malad River exposes rounded pillows of black, glassy basalt, part of the lava dam that formed where McKinney Butte lava met the cool waters of the Snake River. Close inspection of these rocks reveals white, needlelike crystals of plagioclase feldspar with lesser amounts of green olivine crystals. See the road guide for I-84: Mountain Home—Twin Falls for a discussion of Malad Gorge, located just upstream.

US 30 crosses Billingsley Creek between mileposts 179 and 180. This short stream is fed by springs of the Snake River Plain aquifer, emerging from basalt cliffs a few miles east. The temperature and clarity of the springs are ideally suited to support a thriving local fish-farming industry, and the nearby town of Buhl hails itself as the

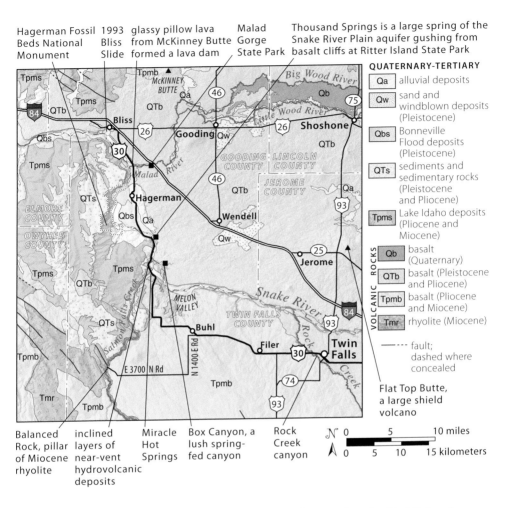

"trout capital of the world." These springs also provide critical wetland habitat such as that south of Hagerman near milepost 184.

Between mileposts 184 and 185, large, rounded boulders, or Melon Gravels, from the Bonneville Flood lie scattered on either side of the highway, a testament to the power of this cataclysmic flood. US 30 crosses the Snake River near milepost 185. The river looks more like a lake here due to the impoundment of the river at Upper Salmon Falls Dam, a hydroelectric power dam about 2 miles downstream.

Basalt along the Snake River Canyon's east rim near Thousand Springs came from Flat Top Butte, a large shield volcano about 22 miles to the east that erupted about 330,000 years ago. Numerous holes in the cliffs near milepost 188 are small lava tubes. West of the highway are older (Pliocene or Miocene) basalt flows and brown, bedded deposits of hyaloclastite, which formed where water interacted with lava, producing locally explosive eruptions. An impressive outcrop west of the highway near milepost 189 displays an inclined sequence of layered hyaloclastite that formed where fragments of rock accumulated around the vent of a small volcano. The fragments exploded upward from the subaqueous volcano, became airborne, then settled back into the water.

Rounded boulders, or Melon Gravels, deposited by the Bonneville Flood near milepost 185. (42.7668, -114.8839)

The inclined hyaloclastite layers near milepost 189 erupted from a subaqueous volcano. (42.7192, -114.8555)

BLISS SLIDE

Aerial view to the north of the Bliss Slide and the Snake River. The slide diverted the river about 100 feet, pushing it to the south and creating the narrow constriction and whitewater rapids.

Just across from the Bliss post office and near milepost 172, Old Bliss Grade/River Road heads southward, descending into the Snake River Canyon. The road skirts the eastern edge of a large landslide that occurred on July 24, 1993. The 100-acre slide moved sediment and debris into the Snake River, temporarily blocking the river and displacing it several hundred feet south of its former course.

The existing geology of the area was a major contributing factor of this mass wasting event. About 52,000 years ago, lava erupted from McKinney Butte, a local shield volcano, and traveled downhill to the rim of the Snake River Canyon near Bliss. The lava spilled over the edge and into the river, forming a conflicting cacophony of lava and water. The cold river water chilled the lava quickly, forming pillow basalt, and also flashed much of the water into steam, creating minor explosive conditions that fragmented and shattered the lava. Eventually, the continuous supply of lava into the river formed an immense dam of rubbly basalt nearly to the canyon rim, and river water ponded behind the dam to form a lake. Sediment carried in suspension by the Snake River slowed as it entered the calm water above the dam, allowing clay, known formerly as the Yahoo Clay, to drop and settle to the bottom of the lake. Geologists believe the dam was somewhat leaky, allowing the lake to maintain a semi-stable level over several years. Eventually the dam either failed or was breached by the river, but remnants of the lava dam and the lake-bottom clay deposits still cling in places to the canyon walls along the river's corridor despite extensive erosion from the Bonneville Flood 17,500 years ago. South of Bliss, the Yahoo Clay forms an extensive layer draped over the existing canyon walls.

The 1993 slide event mobilized a large portion of the Yahoo Clay into an earthflow, a slow-moving fluidized mixture of soil, sediment, and water that formed a lumpy, irregular landscape. The earthflow slid all the way into the Snake River, temporarily blocking the river for several hours before a new path was cut through the material. The narrow new path is now a formidable section of rapids just upstream of the Shoestring Road bridge.

In addition to the underlying geology, several other factors likely contributed to the 1993 event. Steady undercutting by the Snake River at a prominent westward bend at the base of the slope undermined the weak Yahoo Clay. Higher than average precipitation during the 1992–93 winter and heavy rain the day before the slide saturated the clay, making it prone to failure. Today, the slope is considered unstable and future slide events are expected.

Roadcut exposing Yahoo Clay along River Road south of Bliss. Frisbee for scale. (42.9176, -114.9509)

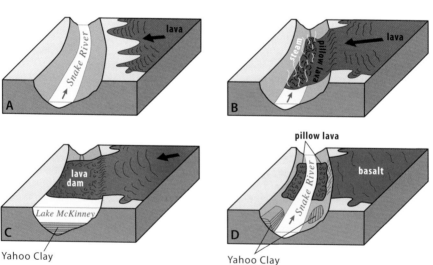

Formation of Lake McKinney near Bliss. (A) Lava from McKinney Butte approached the Snake River Canyon about 52,000 years ago. (B) The lava poured over the rim and into the Snake River, forming pillow lava and steam. (C) The hardened lava formed a lava dam across the river, backing water upstream to form Lake McKinney. Clay was deposited along the lake bottom. (D) The river eventually breached the lava dam, leaving outcrops of clay and pillow lava along sides of canyon. —From Willsey, 2017

HAGERMAN FOSSIL BEDS NATIONAL MONUMENT

Hagerman Fossil Beds National Monument, an area of bluffs and steep slopes on the west side of the Snake River, was set aside in 1988 to preserve one of the world's premier localities for Pliocene fossils. The fossils lie within the Glenns Ferry Formation, sediment deposited in lakes, streams, deltas, and marshes associated with Lake Idaho 4 to 3 million years ago. The monument's most famous fossil and the Idaho state fossil is the Hagerman horse (*Equus simplicidens*), a shorter version of the modern horse. Other fossils include mastodon, camel, beaver, snake, vole, otter, peccary, and fish. Most of the monument's acreage is off-limits to visitors to protect the fossils, but interpretive exhibits and fossils can be observed at the park's visitor center in downtown Hagerman.

Fossil reconstruction of the Pliocene-aged horse, Equus simplicidens, *Idaho's state fossil, in Hagerman Fossil Beds National Monument visitor center.*

RITTER ISLAND AND BOX CANYON SPRINGS NATURE PRESERVE

Units of Thousand Springs State Park

Thousand Springs, crystal clear water gushing from the lush eastern wall of the canyon around milepost 187, is a collection of springs emanating from the eastern Snake River Plain aquifer. Much of the groundwater originates as snowfall high in the mountains of central and eastern Idaho, and even as far as Yellowstone National Park. The melted snowpack feeds rivers and streams that carry the water into the Snake River Plain, where much of the water percolates downward into the permeable basalt. The groundwater then follows the pressure gradient westward through cracks and other voids in the rock. It is estimated that it takes a few hundred years for groundwater to travel through the basalt until it finally emerges at these springs.

The westward flow of groundwater within the aquifer largely parallels the path of the Snake River. An abrupt northward turn of the Snake River between Buhl and Hagerman allows the groundwater to intersect the canyon's eastern wall, forming an impressive array of springs including Thousand Springs. In addition to providing scenic and recreational opportunities, the multitude of springs are also used to provide hydropower and water for fish farms.

To get a close look at Thousand Springs and Ritter Island, head east on Wendell Road (2900 South), located between mileposts 183 and 184 south of Hagerman. Drive about 3 miles and turn south on 1300 East for another 3 miles to a westward turn into the canyon. Look for springs emerging from the canyon wall next to the road as you descend.

Springs emerging from Pleistocene basalt near Thousand Springs State Park. (42.7460, -114.8511)

If you are exploring the east side of the river and Ritter Island, consider also taking some time to check out Box Canyon Springs Nature Preserve, located about 3 miles southeast. From the top of the grade down to Ritter Island, head south on 1300 East as it curves east and becomes 3300 South. After a mile heading east, turn south on 1500 East for 1 mile and turn west onto a small gravel road (there are signs here for Box Canyon). Proceed about 0.6 mile to a large parking area near the head of the canyon. A small overlook offering impressive views of the head of the canyon is nearby. A trail into the canyon and a small, scenic waterfall along the creek are accessed about 0.5 mile to the west.

Box Canyon is a surprisingly lush oasis surrounded by a vast, sagebrush desert. It is a nearly 2-mile-long, 200-feet-deep canyon with sheer basalt walls, connecting to the Snake River on its western edge and terminating in a large amphitheater of stone on its eastern side. Box Canyon is fed by a large spring at the head of the canyon but lacks an extensive drainage basin and a flowing stream above. This unique landscape perplexed geologists for decades. Only recently has the canyon's geologic story been mostly resolved. If you walk around to the head of the canyon, notice the basalt is polished and scoured, evidence of sandblasting by water in a massive flooding event. Isotopic dating of these surfaces yields an age of about 45,000 years, too old for the Bonneville Flood. More likely sources of this megaflood are either the Big Wood and Little Wood Rivers to the north or the Big Lost River to the northeast. Although still under debate, some geologic evidence suggests that both rivers were prone to massive flooding events during the Pleistocene Epoch, likely caused by collapse of glacial dams across drainages.

SNAKE RIVER PLAIN 65

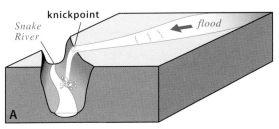

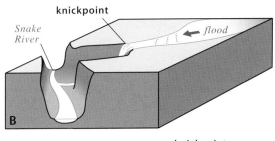

Box Canyon may have formed when a large flood poured into the Snake River, creating a knickpoint where erosion was concentrated. Over time, the knickpoint (a waterfall) migrated upstream to form the deep, narrow Box Canyon. —From Willsey, 2017

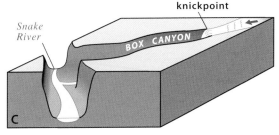

The impressive Box Canyon and its lush, spring-fed creek joining the Snake River in the distance.

Near milepost 190, US 30 enters a paleochannel carved by the Bonneville Flood. The Snake River lies about 1.5 miles to the east of this now-dry valley. The 200-foot-tall ridge between the river and the highway split much of the flood's flow, directing water around the ridge on either side.

Between milepost 190 and 191, the highway crosses Salmon Falls Creek, a lengthy tributary sourced from the highlands of northern Nevada. A few hot springs in this area tap heated groundwater in Pliocene and Miocene volcanic rocks, a separate and deeper aquifer than the one at Thousand Springs. Southward, the highway climbs out of the Snake River Canyon, affording fine views to the east of Melon Valley, a wide section of the canyon where large, rounded boulders were deposited by the Bonneville Flood.

In the agricultural fields above the Snake River Canyon, the highway turns east as it approaches Buhl. Near Twin Falls, the highway crosses Rock Creek and its lush canyon at milepost 216. Rock Creek begins about 30 miles south of Twin Falls in the Cassia Mountains (known locally as the South Hills). The creek gradually carved out a small canyon entrenched in 2- to 3-million-year-old basalt.

US 93
Jackpot, Nevada—Shoshone—Arco
156 miles

A major north-south highway, US 93 transects the breadth of the Snake River Plain, making its southern entrance into Idaho just north of Jackpot, a small Nevada border town with a handful of casinos. The route exposes a variety of volcanic rocks spanning two major, yet different eruptive phases of volcanism. Between the state line and milepost 9, roadcuts and nearby outcrops consist of rhyolite, ash, and other rocks associated with explosive eruptions of the Yellowstone hot spot between 10 and 7 million years ago. Recent research of these rocks delineates two possible sources, the Bruneau-Jarbidge and Twin Falls volcanic fields, a reasonable conclusion as this area lies nearly equidistant from each. Farther north, the highway cuts directly across the axis of the Snake River Plain, where it encounters younger shield volcanoes and their basaltic lavas.

Just south of milepost 2, the road crests at 5,636 feet, beginning a nearly steady decline of almost 2,000 feet toward the town of Twin Falls and the Snake River. From this vantage point, you can observe the effects of Basin and Range extension. Two prominent uplifts, one to the east and one to the west are readily apparent. The prominent north-trending ridge visible along the skyline to the west (called Browns Bench) is composed of an immense mass of stacked rhyolite lava and tuff bounded by a large normal fault on its east side. To the east, normal faults uplift the Cassia Mountains (locally called the South Hills), where similar layers of rhyolite are exposed along with Permian and Pennsylvanian sedimentary rocks. The highway is located in the Rogerson graben, a down-dropped block lying between the two faults. West of the highway, Salmon Falls Creek lies tucked below the horizon, out of view, within a narrow canyon in the vast lowlands. This stream drains northward from Nevada, ultimately converging with the Snake River west of Buhl.

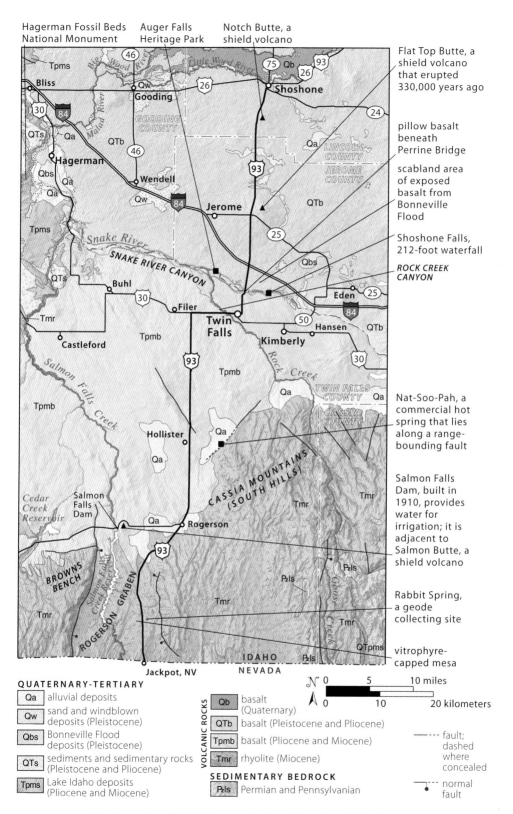

Geology along US 93 between Jackpot, Nevada, and Shoshone.

A distinct black-topped mesa lies just west of the highway between milepost 4 and 5. From a distance, the black rock looks much like basalt, a common rock in southern Idaho. However, this small hill is capped by vitrophyre, a glassy volcanic rock with conspicuous crystals. Vitrophyre is a hybrid of obsidian and rhyolite: it contains crystals like rhyolite, but the crystals are embedded in a glassy matrix, similar to obsidian. It forms where silica-rich, felsic magma initially cools slowly while in magma chambers at depth, creating the large crystals in the magma, but is then subsequently erupted onto the surface where it is quickly quenched, forming the glassy part of the rock.

Close-up of black vitrophyre near milepost 5. Penny for scale. (42.0641, -114.6754)

In this and other parts of the Snake River Plain, much of the vitrophyre and rhyolite formed where pyroclastic flows, hot (1,800°F), fast-moving (up to 400 miles per hour) clouds of ash, rock, and gas explosively erupted from silica-rich volcanoes. Where these eruptions were sufficiently hot, the ash particles welded together and flowed as a thick, pasty mass. Superficially, the ensuing rock resembles an ordinary rhyolite lava flow, but geologists have used subtle clues to unravel the true explosive history of these complex deposits.

Just east of the highway is Rabbit Spring, a favorite collecting area for local rockhounds in search of geodes. Most of the geodes here are scattered on the ground or require some digging to locate. The geodes form spherical masses where whole and cusp-like shapes where broken. The interior of a geode displays an attractive rind of white, creamy chalcedony (microcrystalline quartz) surrounding a central void. These geodes formed where silica-rich groundwater passed through the rhyolite and precipitated chalcedony within voids in the surrounding volcanic rock. Later the resistant geodes weathered out of the relatively softer rock.

A fantastic roadcut on the east side of the road between mileposts 5 and 6 displays a typical volcanic sequence for this area. The pinkish material along the base of the roadcut is an ancient soil, or paleosol. Directly above this is a white, well-layered deposit of ash that settled out of the air at the onset of an explosive volcanic eruption. The white ash is overlain by a dark-gray to black vitrophyre, which caps the roadcut. The black vitrophyre is the pyroclastic flow that swept across the air-fall ash and was hot enough for the ash particles to weld together before finally solidifying and cooling. Close analysis of these volcanic deposits suggest they were erupted from the Twin Falls volcanic field about 8 to 7 million years ago.

Near milepost 9, the first products of basaltic volcanism come into view on the dark ridge to the west. This basalt flow emanated from a nearby shield volcano, Salmon Butte, about 6 miles to the northwest, sometime during the Pleistocene. The basalt ridge blocks a clear view of the volcano until milepost 12, where its low-profile shape is visible, capped with a few communication towers. The section of the highway a few miles south of Rogerson also marks the contact between the rhyolitic rocks to

The white material lining the interior of these geodes is chalcedony, a type of microcrystalline quartz.

The roadcut between mileposts 5 and 6 exposes a pink paleosol overlain by white ash that in turn is overlain by black vitrophyre. Rock hammer on ash layer for scale. (42.0771, -114.6776)

the south and the basaltic rocks that form much of the vast Snake River Plain to the north. At Rogerson, a road heads west, providing access to Salmon Butte and the 217-foot-tall Salmon Falls Dam and reservoir. The concrete dam was built in 1910 to impound water for irrigation, but the porous volcanic rocks allow some of the water to escape around the dam.

Near milepost 20, the highway crosses a small, shallow canyon entrenched in basalt. Nat-Soo-Pah, a commercial hot spring east of Hollister, indicates the existence of subsurface heat from either past volcanic episodes or a deeper source. A fault near the base of the highlands provides a conduit for the hot water to rise to the surface. As the highway continues north, the sagebrush plains give way to large fields of crops aided by irrigation.

Just north of milepost 43, US 93 swings east, becoming Pole Line Road as it approaches the city of Twin Falls. The highway crosses Rock Creek Canyon at milepost 46, a narrow, 130-foot-deep canyon cut into Pleistocene basalt. Rock Creek originates in the Cassia Mountains (locally known as the South Hills) south of Twin Falls and joins the Snake River about 3 miles north of this point.

After a few miles skirting through the northwestern part of Twin Falls, US 93 heads north again, crossing the majestic Snake River Canyon via the iconic Perrine Bridge, 486 feet above the river. The bridge and its visitor center, nearby overlooks, and a rim trail are good places to observe the canyon landscape, the exposed volcanic rocks, and the effects of the catastrophic Bonneville Flood.

SNAKE RIVER CANYON

The Perrine Bridge area near Twin Falls is an excellent place to observe evidence from three major geologic events. The lower half of the Snake River Canyon contains 10- to 8-million-year-old rhyolite lava erupted within the Twin Falls volcanic field as southern Idaho drifted across the underlying hot spot. Rhyolite is exposed downstream of the bridge for another mile and upstream as far as the iconic Shoshone Falls (about 3 miles upstream), where it forms the resistant rimrock of the falls. The limited exposure of rhyolite along this canyon stretch is due to the rhyolite lava piling upon itself and forming a

The outcrop immediately below the north side of Perrine Bridge near Twin Falls exposes 95,000-year-old pillow lava. The pillow above the rock hammer was fed by the cylindrical tube above the pillow. (42.6024, -114.4552)

higher topographic surface. Farther away from this section, the rhyolite is located more deeply below the basalt and below the level of the canyon floor.

The upper half of the canyon walls exposes the second major geologic episode, a series of stacked basalt lava flows that erupted from local shield volcanoes over the past 2 million years. Some of these lavas poured into the smaller, ancestral canyon and interacted with river water to form rounded blobs of basalt known as pillow lava. A good place to see these is immediately beneath the bridge on the north side of the canyon, easily accessed by a pullout and a short set of stairs.

The final geologic episode to affect the Snake River Canyon was the Bonneville Flood, which occurred about 17,500 years ago. Spectacular erosional features are present throughout the canyon and stand as testament to the sheer power and scale of this colossal flood. This section of the flood's path is unique because it marks the convergence of two separate flood channels. The Eden Channel, which runs parallel to I-84, poured back into the main canyon, merging with the flood channel along the Snake River (see the map on page 39 in I-84 road log). The confluence generated massive and complex hydraulics of eddies, vortices, cascades, and rapids that ripped large blocks of rock from the canyon's walls, sculpting massive amphitheater-shaped alcoves, such as that observed just west of the highway on the north side of the canyon. Farther downstream where the canyon widens, these boulders were deposited on the canyon floor and lie strewn across the landscape such as at Auger Falls Heritage Park, about 3 miles downstream on the south side of the river.

The north wall of the Snake River Canyon at Perrine Bridge. The lower half of the canyon is 10- to 8-million-year-old rhyolite. The upper half is stacked Pliocene to Pleistocene basalt. The prominent, light-colored layer in between is windblown silt filling a low area atop the rhyolite. (42.5988, -114.4552)

North of Perrine Bridge, near milepost 50, the roadside scenery abruptly transitions from an urban environment to a stark landscape of sage and undulating basalt outcrops. The landscape over the next 3 miles to the intersection with I-84 is known as scabland, a term used to describe a region stripped of topsoil, exposing its underlying bedrock. The scabland here is the handiwork of a secondary path of the Bonneville Flood, the Eden Channel, flowing back into the Snake River. During the flood, the power and speed of the water was sufficient to denude the land of loose sediment and scour the uppermost rock surface. The flood's erosive power exposed the irregular surface of a lava flow that erupted about 95,000 years ago from Rocky Butte, a shield volcano more than 15 miles to the northeast.

Passing farms and dairies, US 93 crests the western shoulder of Flat Top Butte at milepost 59. This broad shield volcano, whose summit is easily recognized by the array of communication towers, erupted about 330,000 years ago, sending runny lava as far as 20 miles westward. Around milepost 65, another shield volcano, again adorned with towers, looms ahead to the north. Notch Butte erupted sometime between 95,000 and 52,000 years ago. Irregular mounds surrounding the volcano, called tumuli (or tumulus if singular), range from 3 to 20 feet tall and commonly have a large fracture splitting the axis of the mound. They form where the cooled, brittle crust of a lava flow buckles as the underlying lava pushes upward. A particularly good example exists near milepost 69, just east of the highway.

North of the railroad tracks in Shoshone, US 93 merges with US 20 and US 26, and milepost numbers jump forward. The combined highways head east initially, then northeast, closely paralleling the nearby Little Wood River. Fresh-looking basalt visible a few miles north of the river and highway emanated from Black Butte Crater about 11,700 years ago. The lava flow disrupted the Little Wood River, forcing it from its existing channel. The river reestablished a portion of its course along the margin of this young lava field. Black Butte Crater is about 18 miles north of Shoshone along ID 75. Views of these youthful lavas can be seen to the north across the river near mileposts 177 and 178. Older Pleistocene basalt lines the highway from mileposts 186 to 196, offering occasional outcrops of tumuli and eroded lava surfaces.

Just north of Carey, near milepost 205, the highway crosses a small northwest-trending normal fault that has uplifted a 4.2-million-year-old basalt flow on its east side. The fault's offset abruptly diminishes to the south. Carey Lake, a spring-fed lake nestled in a low area between lava fields, lies another mile eastward on the south side of the highway. At this point, US 93 has transected the vast breadth of the Snake River Plain and now heads northeastward along the base of the foothills leading into the rugged Pioneer Mountains. The hills immediately north of the highway contain Miocene rhyolite erupted from volcanic fields associated with the Yellowstone hot spot. The rhyolites dip southward into the Snake River Plain, a consequence of the plain sinking after being dragged off the hot spot. Good views up Fish Creek valley and into the Pioneer Mountains are found near milepost 213.

Between mileposts 214 and 215, the highway climbs onto aa lava erupted from Sunset Cone, one of about twenty-five cinder cones within Craters of the Moon National Monument and Preserve. The stark lava appears fresh due to the arid climate of the area but is actually about 12,000 years old.

Outcrops of Mississippian sedimentary rocks of the Copper Basin Group rise above the lavas at several places between mileposts 216 and 223. Where outcrops

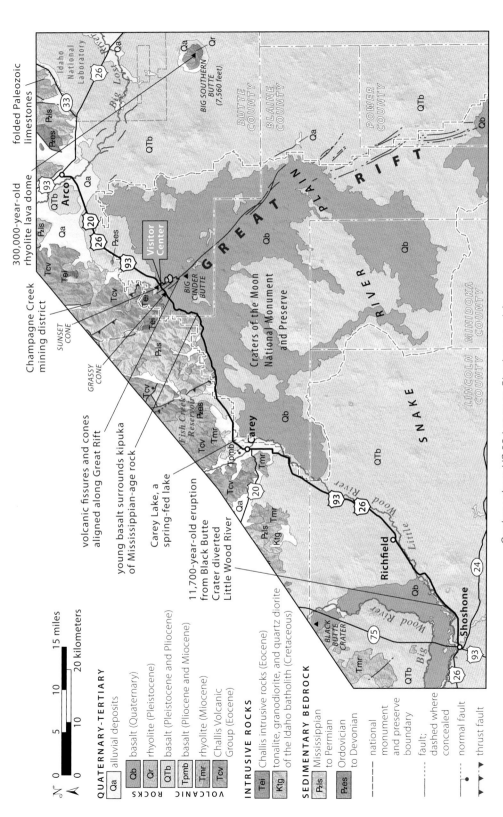

Geology along US 93 between Shoshone and Arco.

CRATERS OF THE MOON NATIONAL MONUMENT AND PRESERVE

Hosting the largest young basaltic lava field in the contiguous United States, the Craters of the Moon region is a volcanic wonderland of cones, vents, and flows. Southern Idaho's dry climate combined with the relatively recent age of the eruptions converge to create a landscape that looks nearly as fresh as it did after the lava erupted. The monument was originally set aside in 1924 by President Calvin Coolidge to "preserve the unusual and weird volcanic landforms." It was expanded to its current size in 2000 by President Bill Clinton.

Craters of the Moon National Monument and Preserve encompasses about 1,200 square miles and includes three discrete Holocene lava fields: the large Craters of the Moon lava field in the north and the Wapi and Kings Bowl lava fields near the monument's southern boundary. The most accessible part of the monument along US 93 is also the most scenic, diverse, and visited section. An informative visitors center hosts interactive displays about the geology and ecology of the area, and a nearby campground provides a unique overnight experience among the lavascape. The monument's scenic loop drive winds around a cluster of cinder cones, allowing access to a variety of exceptional volcanic features via overlooks, paved walkways, and hiking trails.

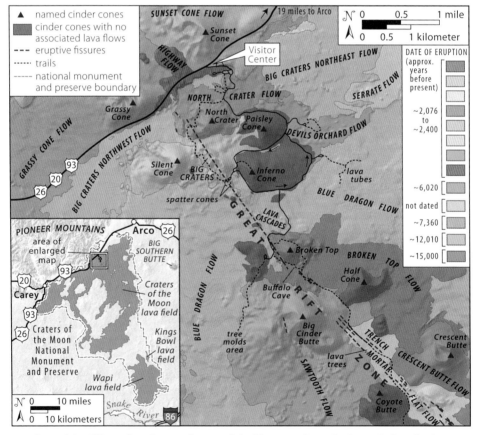

Ages of the different lava flows at Craters of the Moon are shown in the gradation of colors.

By using field relationships such as overlapping lava flows, along with carbon-14 radiometric dates derived from charcoal remains of burned trees found within the lava flows, geologists have pieced together the eruptive history. The landforms at Craters of the Moon formed during eight eruptive periods over the past 15,000 years, with the most recent eruption occurring a mere 2,000 years ago. Each eruptive period was separated by relatively tranquil periods that lasted a few hundred to as much as 3,000 years. The frequency of eruptions over the past 15,000 years and the age of its most recent eruption indicate that future outpourings of lava in this area are likely.

Eruptions in the Craters of the Moon lava field consist of two main varieties: (1) effusive eruptions of hot, runny lava from elongated fissures; and (2) locally explosive eruptions where small, gas-rich clots of lava (called cinders) are ejected from the vent, where they pile up to form nearly symmetrical hills called cinder cones. Some cinder cones also erupted lava from their bases, with the lava often tearing away and carrying a rafted block of the cone. Some of the lava flows in the Craters of the Moon lava field flowed as much as 30 miles from their vent.

Many of the cones and fissures in the monument are aligned in a northwesterly direction, forming a nearly 60-mile-long chain of vents across the monument known as the Great Rift. The trend of these volcanic vents is similar to the orientation of normal faults in the Basin and Range Province, suggesting that Basin and Range extension may have played a role in forming the Great Rift.

In addition to the cones, fissures, and lava flows, Craters of the Moon also contains tree molds, lava trees, lava tubes, and spatter cones. During eruptions, lava laps up against and around tree trunks, cooling quickly against the cool, moist wood to form a solidified ring of rock around the base of the tree. Eventually, the tree is incinerated and all that remains is a hollow cylinder, a tree mold, that often preserves the texture of the tree's bark. If the lava subsides or drains elsewhere, the hardened ring around the tree is left standing above the ground to form lava trees. Lava tubes form where the surface of a lava flow solidifies, remaining molten below. Eventually the lava supply wanes or drains elsewhere, emptying the tube. Visitors can explore several lava tubes such as Buffalo Cave or the cluster of lava tubes east of the loop road. Spatter cones are small, steep mounds that form where blobs of sticky lava (called spatter) are ejected tens of feet into the air and fall around the vent where the spatter collects to form the cone.

This tree mold preserves bark texture where the lava cooled quickly against the bark.

Conglomeratic turbidites on the north side of the highway at milepost 222. Park a few hundred yards west of the roadcut. (43.3906, -113.6503)

of these older rocks are completely surrounded by lava, they are called kipukas, a Hawaiian term. Sedimentary rocks of the Copper Basin Group were deposited in a foreland basin east of the now-eroded mountains formed by the Antler orogeny during the Mississippian Period. You can get a close look at these rocks at the milepost 222 roadcut (park at west end) where an eastward-tilted section of Copper Basin Group is exposed. The roadcut is a series of turbidite deposits, alternating sequences of sandstone (now quartzite) and shale laid down as an underwater sediment slurry that moved down a slope on the seafloor. As the turbidites came to rest, the largest and heaviest particles settled first and the lighter, smaller particles settled later. Look for pebbles along the base of many of the quartzite layers.

The highway crosses basalt from Grassy Cone, a cinder cone that erupted about 7,400 years ago, near milepost 221. Fantastic views of several other cinder cones at Craters of the Moon occur near milepost 225. The tallest and southernmost cone visible from this point is Big Cinder Butte, rising over 700 feet above the basalt at its base. Scattered across the lava field near the highway are large blocks of rock. These pieces of cinder cone walls were detached and rafted away by lava pouring from the base of the cone.

Grassy Cone lies north of the highway at milepost 228. A nice roadcut here allows direct observation of the contents of a cinder cone. The cinders are generally smaller than a baseball and incredibly lightweight due to the abundance of gas bubbles (known as vesicles). Because Grassy Cone is a young volcano in an arid climate, many of the cinders are quite pristine, displaying an iridescent play of colors. As the iron in the cinders absorbs more oxygen, the cinders become more rust colored.

Just south of the highway and Grassy Cone is Silent Cone. Compare the tree-covered slopes of Silent Cone to the grass and sage on Grassy Cone. South-facing slopes are drier and hotter, making it harder for more water-dependent plants to grow and thrive.

In contrast, the north-facing slopes, facing away from the sun, hold soil moisture longer, allowing limber pines and Douglas firs to grow. Notice that many of the trees lean eastward due to strong westerly winds. The turnoff to Craters of the Moon National Monument and Preserve lies between mileposts 229 and 230, south of Sunset Cone.

Big Southern Butte, a large rhyolite dome rising over 2,400 feet above the Snake River Plain, is visible to the southeast at milepost 230. A scenic pullout is located on the south side of the road. Unlike its surrounding basaltic neighbors, Big Southern Butte is made of rhyolite and formed about 300,000 years ago when sticky, silica-rich magma forced its way upward to create a steep-sided dome.

North of the highway at milepost 235 lies a kipuka of Ordovician quartzite and dolostone surrounded by Pleistocene basalt. Champagne Creek, where gold and other metals were mined intermittently since the early 1900s, exits the mountains

Close-up of iridescent cinders from the Grassy Cone cinder cone. (43.4504, -113.5862)

Pleistocene basalt flowed around and completely surrounded this hill of Ordovician quartzite. (43.3666, -113.7614)

behind the kipuka. The geology of this mining district consists of Paleozoic sedimentary rocks overlain by Eocene Challis lava rocks. Around milepost 238, the south end of the Lost River Range looms to the northeast. These fault-bounded mountains are the product of Basin and Range extension and include many of Idaho's tallest peaks. At milepost 240, the highway leaves the younger basalt and enters farm fields at the mouth of the Lost River Valley near Arco. Mississippian limestones of the Scott Peak and Surrett Canyon Formations form the cliffs east of Arco, upon which are painted high school graduation years.

US 95
Oregon Border—Weiser
82 miles

Formerly known as the North and South Highway, US 95 runs along the extreme western edge of Idaho and stretches nearly the full length of the state. This section focuses on the southernmost part of the highway, beginning at the Oregon border then cutting across the Owyhee Mountains and into the agricultural region of the Treasure Valley to Weiser. The highway also intersects several major westward-flowing rivers of southwestern Idaho: the Snake, Boise, Payette, and Weiser.

From the Oregon border, US 95 heads northeast, ascending the western base of the Owyhee Mountains. Roadcuts in the first few miles expose colorful layers of Miocene ash and lake sediments that accumulated in valleys during Basin and Range extension. A good example is the large roadcut between mileposts 3 and 4. Much of the volcanic ash has been altered to clay. The lack of vegetation in the layers is due to the clay's swelling when wet and shrinking when dry, making it a difficult material for plant roots to become established. Small erosional gullies, caused by runoff and called rills, are spaced every few feet or so on the roadcut and add another attractive textural element. As US 95 climbs higher into the Owyhee Mountains, dark craggy outcrops of rhyolite lava appear, sometimes overlying beds of ash and lake deposits, such as in the hill to the north from milepost 8.

Perhaps the most striking and complicated roadcut along this section lies near milepost 14 at the crest of the highway near French John Hill. Roadcuts on both sides of the highway expose a jumbled, chaotic mess of rock and sediment. The dark rock is rhyolite and vitrophyre, part of the Jump Creek Rhyolite, which erupted about 11.2 million years ago from nearby vents. A close look at these rocks reveals one of its distinguishing characteristics, large crystals of white plagioclase feldspar. Much of the rhyolite is brecciated, or broken, into angular chunks, a common trait of the outer surface of rhyolite lavas. Deformed lenses and layers of ash, stream gravel, and other sediments in and among the rhyolite make this roadcut a real head scratcher. The messy situation was produced when the 700-foot-thick Jump Creek Rhyolite flowed over wet sediments and ash of the Succor Creek Formation. The sheer weight and load of such a massive body of lava exerted extreme downward pressure, causing the sediments to liquefy, deform, and move in a complex manner.

Near milepost 19, northbound travelers enter the nearly flat Snake River Plain and its vast tracts of agricultural fields. US 95 crosses the Snake River near milepost 35 at Homedale. The highway imperceptibly crosses a low divide near Wilder and then

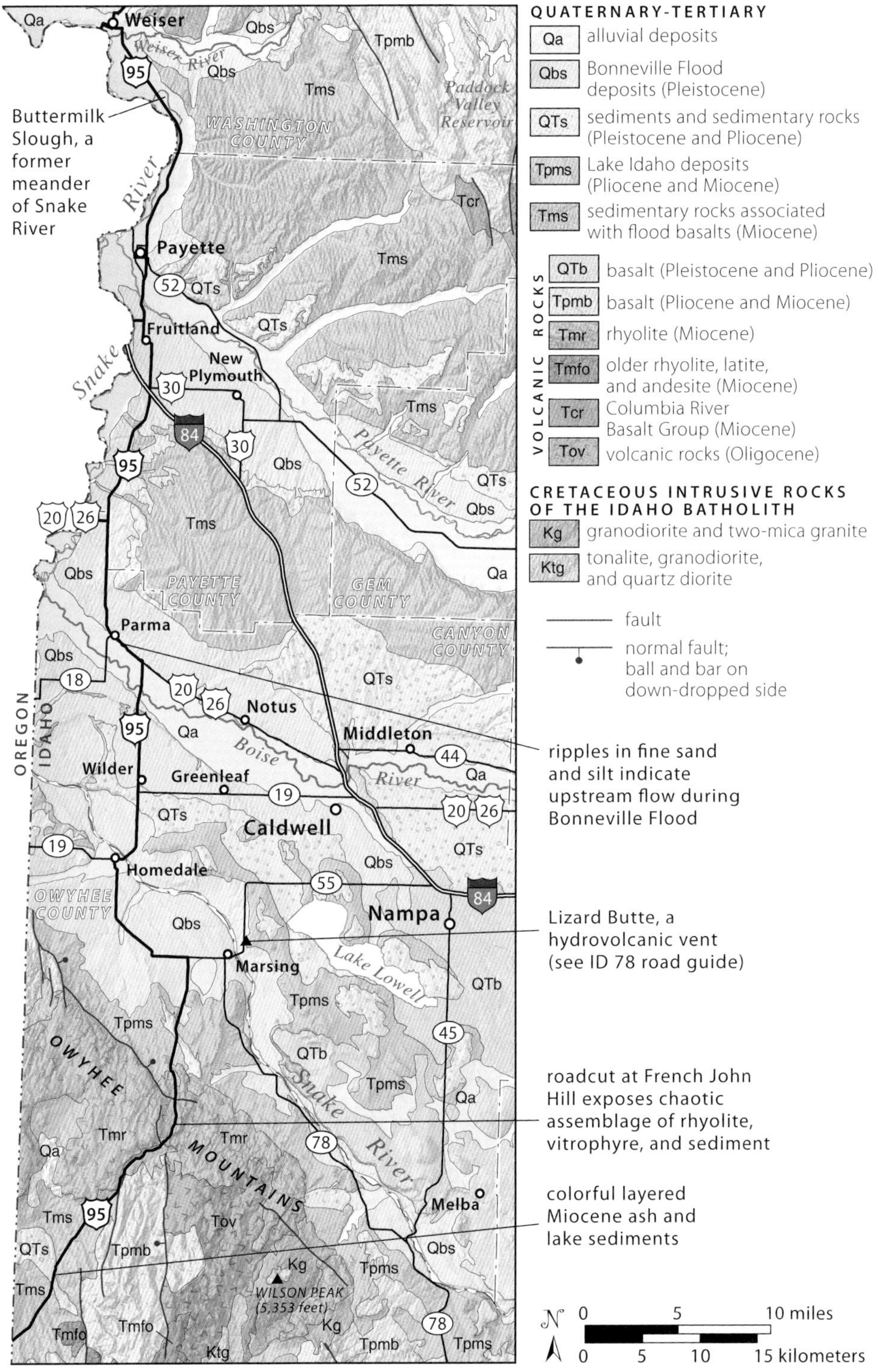
QUATERNARY-TERTIARY
Qa alluvial deposits
Qbs Bonneville Flood deposits (Pleistocene)
QTs sediments and sedimentary rocks (Pleistocene and Pliocene)
Tpms Lake Idaho deposits (Pliocene and Miocene)
Tms sedimentary rocks associated with flood basalts (Miocene)
VOLCANIC ROCKS
QTb basalt (Pleistocene and Pliocene)
Tpmb basalt (Pliocene and Miocene)
Tmr rhyolite (Miocene)
Tmfo older rhyolite, latite, and andesite (Miocene)
Tcr Columbia River Basalt Group (Miocene)
Tov volcanic rocks (Oligocene)
CRETACEOUS INTRUSIVE ROCKS OF THE IDAHO BATHOLITH
Kg granodiorite and two-mica granite
Ktg tonalite, granodiorite, and quartz diorite
fault
normal fault; ball and bar on down-dropped side
Buttermilk Slough, a former meander of Snake River
ripples in fine sand and silt indicate upstream flow during Bonneville Flood
Lizard Butte, a hydrovolcanic vent (see ID 78 road guide)
roadcut at French John Hill exposes chaotic assemblage of rhyolite, vitrophyre, and sediment
colorful layered Miocene ash and lake sediments
Weiser
Weiser River
Paddock Valley Reservoir
WASHINGTON COUNTY
Snake River
Payette
Fruitland
New Plymouth
Payette River
PAYETTE COUNTY
GEM COUNTY
CANYON COUNTY
Parma
Notus
Middleton
Boise River
Wilder
Greenleaf
Caldwell
Homedale
OWYHEE COUNTY
Nampa
Marsing
Lake Lowell
OREGON
IDAHO
OWYHEE MOUNTAINS
Melba
WILSON PEAK (5,353 feet)
0 5 10 miles
0 5 10 15 kilometers
N

Geology along US 95 between the Oregon border and Weiser.

descends to intersect the Boise River at milepost 44. The Boise River's confluence with the Snake River lies about 8 miles to the west.

US 95 merges with US 20/US 26 between mileposts 46 and 47. Small roadcuts on the north side of the intersection expose thin beds of beige silt and fine sand with noticeable ripples formed by flowing water. The shape of the ripples indicates an eastward-moving current, opposite that of the westward-flowing Boise River. Researchers

Colorful Miocene lake and ash beds in a roadcut near the Oregon border. (42.9580, -117.0081)

The chaotic outcrop near milepost 14 exposes dark rhyolite and vitrophyre with light-colored sediments of the Succor Creek Formation. (43.4051, -116.8705)

These beige fine-grained sand and silt deposits contain ripples (wavy layers above the yellow notebook) from the Bonneville Flood that backed up into the Boise River drainage. The ripple-forming currents flowed from left to right. This roadcut is between mileposts 46 and 47. (43.7689, -116.9143)

determined that the rising waters of the Bonneville Flood 17,500 years ago backed up into the Boise River drainage, causing a reversal of flow in this stretch of the river.

About 1 mile south of its intersection with I-84, the highway crests another slight topographic divide separating the Boise River and Payette River drainage basins. US 95 crosses the Payette River near milepost 66. North of the town of Payette and the confluence of the Snake and Payette Rivers, the highway passes along the eastern edge of the Snake River. Near milepost 78, the road crosses Buttermilk Slough, a former meander of the Snake River that was cut off from the river's path due to erosion. Today, the stagnant water in the slough is fed by groundwater and agricultural runoff. Just south of Weiser around milepost 82, ID 95 crosses the Weiser River, a 100-mile-long river that begins in the mountains west of McCall.

ID 32: Teton Scenic Byway
Tetonia–Ashton
28 miles

This lovely highway, paralleled by an Idaho State Parks rail-to-trail bicycle path, traverses rolling dryland wheat fields on soils developed from rhyolite ash deposits of the Yellowstone volcanic field. Much of the surface here is covered by the 2.1-million-year-old Huckleberry Ridge Tuff, the first and largest of three major eruptions of the Yellowstone volcanic field. Its eruption blasted out around 600 cubic miles of ash and produced the Island Park caldera.

The high Teton Range, uplifted along its eastern side by a steep normal fault, is the primary focus along ID 32 and is always in view to the east. A fine, commanding view of the Tetons rising above the Teton River lies at an overlook on ID 33, three miles west of its junction with the southern end of ID 32. The overlook lies on a high point made of Huckleberry Ridge Tuff. From north to south, the three prominent peaks dominating the skyline are the Grand Teton, Middle Teton, and South Teton. Much of the Tetons, composed of Archean basement rocks, were extensively glaciated during the Pleistocene to form jagged, pyramidal peaks. West of these high peaks lie west-dipping Paleozoic sedimentary rocks that form the smaller peaks in front of the Tetons. These rocks were tilted westward as the steep Teton fault, on the east side of the range in Wyoming, uplifted the Tetons. Glaciers descending from the high peaks deposited glacial till that forms rolling hills along the east side of the valley south of Tetonia.

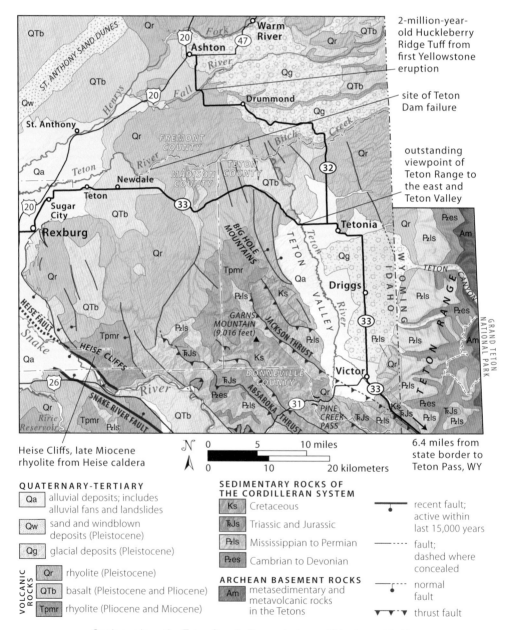

Geology along the Teton Scenic Byway between Tetonia and Ashton.

Teton Range from potato fields along ID 32 about 7 miles north of Tetonia. The snow-covered mountains below the towering Archean gneiss of the Tetons are west-dipping Paleozoic sedimentary rocks.

Northward from Tetonia, ID 32 passes through rolling farmlands for several miles. At milepost 8, the highway crosses Bitch Creek, which drains the west side of the Tetons. A roadcut on the north side of the drainage exposes Huckleberry Ridge Tuff. Around milepost 11, the highway affords fine views of the Yellowstone Plateau to the northeast as the highway begins to head westward. More hilly farmland ensues with the highway crossing Fall River, which drains the southwest corner of Yellowstone National Park, around milepost 24. Outcrops on the west side of the creek are Huckleberry Ridge Tuff. The highway swings north one last time, offering distant views on clear days of the Beaverhead Mountains and Lemhi Range across the Snake River Plain to the northwest.

ID 33
US 26 Junction—Howe—Rexburg
79 miles

East of the junction with US 20/US 26, ID 33 travels northeast, following the northern edge of the eastern Snake River Plain and passing the south ends of the big Basin and Range mountains of east-central Idaho. East of the highway junction, the mountains to the north are the Lost River Range, home to several peaks over 12,000 feet. The Lost River Range is mainly composed of marine Paleozoic sedimentary rocks, in places rich in invertebrate fossils. These rocks were intensely folded and cut by thrust faults during the Sevier orogeny in the Cretaceous Period. Between mileposts 8 and 10 are spectacular views of folds in late Paleozoic carbonates. Resistant beds of limestone and dolostone form steeply inclined and vertical fins of rock, part of a train of tight anticlines and synclines whose axes run northwest-southeast.

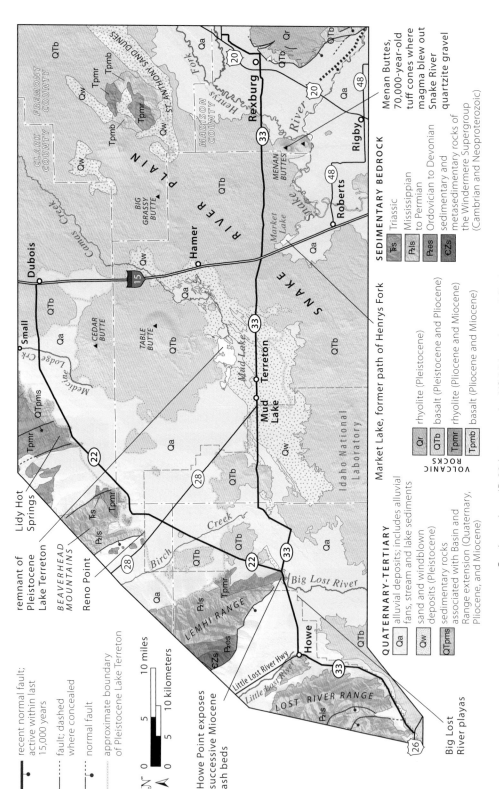

Geology along ID 33 between the US 26 junction and Rexburg.

View to the north of a large train of anticlines and synclines in Mississippian limestone at the south end of the Lost River Range. (43.6812, -113.0393)

At Howe, ID 33 turns east and travels across the flat and irrigated southern end of the Little Lost River Valley. The Little Lost River and the Big Lost River both empty into the Big Lost River playas, one of which lies south of the highway and out of view. Any water that reaches the playa infiltrates into the subsurface, leaving a dry lakebed. This water contributes to the eastern Snake River Plain aquifer, a vast underground water resource that supplies irrigation and drinking water to many farms and communities in the Snake River Plain. The groundwater flows west, eventually resurfacing as dozens of large springs on the north wall of the Snake River Canyon west of Twin Falls.

East of Howe, the Lemhi Range looms north and east of the highway. The western face of the range contains one of the most complete successions of Paleozoic rocks in the United States, from the Cambrian to the Permian. A few miles to the north of ID 33, these strata rest unconformably on Mesoproterozoic rocks of the Lemhi arch, a broad northwest-trending promontory or uplift that formed in late Neoproterozoic and Cambrian time.

The highway climbs over the end of Howe Point, the southern tip of the Lemhi Range at milepost 24 and descends to the junction with ID 22. Near the highway, Howe Point exposes four Miocene tuffs, the two lower tuff units have been dated as 10.2 and 8.8 millon years old and originated from the Picabo volcanic field. The upper two tuffs are 6.6 and 6.2 million years old and were erupted from the Heise volcanic field.

From the ID 22 intersection east of Howe, ID 33 heads east across the Snake River Plain to the farming communities of Mud Lake and Terreton. To the south of the highway just west of Mud Lake are vegetated sand dunes. The agriculture takes advantage of the fertile lake sediments of ancient Lake Terreton, a large freshwater lake fed by the Big Lost and Little Lost Rivers, Birch Creek, Medicine Lodge Creek, and other local streams during the cooler, glacial conditions of the Pleistocene Epoch. Ancient peoples, believed to be ancestors of today's Shoshone-Bannock tribe, inhabited the lakeshore, hunting wooly mammoths and large bison herds. Today, shallow Mud Lake, about 2 miles north of Terreton, is all that remains of ancient Lake Terreton.

ID 22 TO DUBOIS

The mostly flat ID 22 traverses the floor of Pleistocene Lake Terreton between the Birch Creek valley and Medicine Lodge Creek west of Dubois. East from the ID 28 intersection (east of milepost 38), ID 22 traverses along the southern end of the Beaverhead Mountains, a northwest-trending range in the Basin and Range region. Its rocks are the same late Paleozoic carbonates as present to the west in the Lemhi Range. Near milepost 35, the prominent pyramidal peak in the Lemhi Range to the northwest is Diamond Peak, rising to an elevation of 12,201 feet. The sharp ridge lines extending outward from the peak formed during the ice ages of the Pleistocene when alpine glaciers scoured and sculpted these airy mountains.

In the foothills south of the Beaverhead Mountains are more exposures of Miocene rhyolitic tuff of the eastern Snake River Plain, which lap onto the tilted Paleozoic rocks. Rhyolite units present here from the Heise Group include, from oldest to youngest, the 6.6-million-year-old tuff of Blacktail Creek, 6.3-million-year-old Walcott Tuff, and 4.5-million-year-old Kilgore Tuff. These tuffs were erupted from the Heise volcanic field and overlie the Miocene-age Medicine Lodge Formation, which mainly consists of lake-deposited limestone with some interbedded tuff. Hydrothermal fluids have altered some of the limestone to travertine. The Medicine Lodge Formation is mined in the hills north of Lidy Hot Springs, near milepost 52.

The Miocene and Pliocene units dip southward because the Snake River Plain subsided as it was covered by numerous basaltic lavas and as the underlying crust cooled when it as moved off the hot spot. The high Beaverhead Mountains are also tilted southward, exposing Paleozoic limestones.

A small roadcut at milepost 53 exposes Kilgore Tuff. Around milepost 58, the summit of a well-defined shield volcano is visible about 6 miles south of the highway. The basalt-filled Medicine Lodge valley to the north contains a rough gravel road that climbs over the Continental Divide into Montana.

View to the west near milepost 35 of Paleozoic limestones forming the striking summit of Diamond Peak in the Lemhi Range, with low foothills of the Beaverhead Mountains in the foreground. (44.0073, -112.7176)

The hill north of milepost 51 is capped by Walcott Tuff, which erupted from the Heise volcanic field about 6.3 million years ago. (44.1186, -112.5652)

Between I-15 and the Henrys Fork of the Snake River west of Rexburg, ID 33 crosses a basalt lava flow thought to be between 10,000 and 20,000 years old. It blocked the westward course of the Henrys Fork. No farming occurs here due to the lack of sufficient soil above the young basalt. The hill where BYU–Idaho and the Rexburg Temple are located is uplifted along the Rexburg fault, a Quaternary normal fault.

MENAN BUTTES

A brief diversion from ID 33 will bring you to one of Idaho's most unique volcanoes. The Menan Buttes lie a few miles south of ID 33 and are easily visible from the highway. Turn south onto East Butte Road (N 3700 E) between mileposts 70 and 71. After 1 mile, turn west onto Twin Butte Road, contouring around the west side of the publicly accessible North Menan Butte to a parking area and trailhead. In a distance of less than 1 mile with an elevation gain of 650 feet, the trail climbs the flanks and summit of the volcano.

The twin Menan Buttes are spectacular examples of tuff cones, a type of volcano created where ascending magma mixes with groundwater, generating explosive conditions. Here, shallow groundwater and surface water from the nearby confluence of the Henrys Fork and South Fork of the Snake River provided ample water to chill the magma as it rose, flashing the water to steam and producing explosions of rock and ash, which collected around the vent. Both Menan Buttes have similar shapes with wide, shallow craters and broad, steep flanks. Both buttes are also slightly elongated in a northeast direction, suggesting winds were out of the southwest at the time of the eruption. Perhaps the most striking feature at Menan Buttes are the rounded quartzite gravels embedded in the volcanic tuff, evidence of the magma mixing with sediments in the nearby river during the eruption. The age of the Menan Buttes eruptions is bracketed as sometime between 140,000 and 10,000 years ago, quite a span of time. It is difficult to acquire dateable samples from lava altered by water.

Aerial view of Menan Buttes, a pair of tuff cones, from the south. —Photo courtesy of Google Earth

ID 46
WENDELL—FAIRFIELD
43 miles

This seldom used but scenic highway cuts across the northern portion of the central Snake River Plain. Heading north from I-84, it crosses basalt of the Snake River Plain before climbing through Miocene rhyolite in the eastern half of the Mt. Bennett Hills.

North of Wendell and its surrounding farmland, between mileposts 103 and 108, ID 46 travels over undulating basalt erupted from Notch Butte, a stately shield volcano barely visible on the horizon about 15 miles to the east. You get a much closer look at this volcano from US 93. Topped by communication towers, Notch Butte erupted sometime between 95,000 and 52,000 years ago. Its lava traveled westward about 8 miles beyond ID 46 and as far as the Snake River. Another shield volcano, Gooding Butte, lies about 2 miles west of the highway near milepost 110 and erupted about 373,000 years ago. Don't be fooled by its seemingly diminutive size. Much of its surface area was covered by younger flows such as those of Notch Butte, leaving only its summit area rising slightly above the landscape.

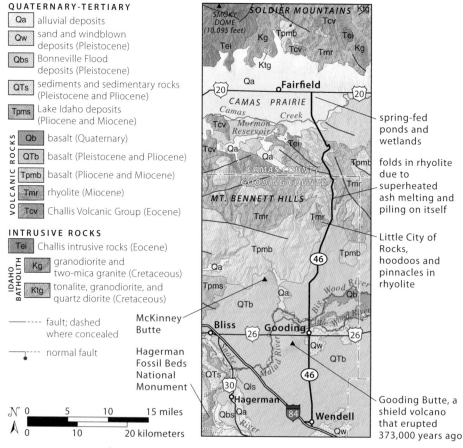

Geology along ID 46 between Wendell and Fairfield.

The highway begins its ascent of the Mt. Bennett Hills near milepost 122, following the surface of a basalt flow that erupted from a subtle vent about 7 miles north. On both sides of the highway, volcanic rock layers slope southward toward the Snake River Plain. Over the past 6 million years or so, the increasing weight of the large mass of injected basaltic magma into the crust, along with the pile of basaltic lava on the plain, caused the crust to subside, or sink, downwarping the rocks at both the southern and northern boundaries of the Snake River Plain.

Outcrops just east of the highway between mileposts 131 and 132 expose tightly folded gray rhyolite with an age of 10 to 9 million years. Erupted as hot pyroclastic flows, the superheated ash became a molten mass as it draped the landscape, rolling over itself to form intricate flow folds before crystallizing into rhyolite tuff.

Near milepost 135, the highway begins its descent into the Camas Prairie, an east-west valley bounded by normal faults along the edges of the Mt. Bennett Hills to the

LITTLE CITY OF ROCKS

A 1-mile-long, unimproved road heading west just north of milepost 124 leads to the Little City of Rocks and its fantastic fairyland of hoodoos. These strange towers, fins, and turrets of stone are sculpted from 8.2-million-year-old rhyolite tuff that initially erupted as hot clouds of ash from the Twin Falls volcanic field. The eruption was so hot that the ash particles welded together as they accumulated, reforming into rhyolitic lava, which flowed slowly before fully crystallizing. The rock fractured as it cooled and was broken even further when the Snake River Plain subsided. Freeze-thaw cycles of water in these fractures has further weathered the rock, helping to form the ethereal rock shapes.

Hoodoos of 8.2-million-year-old rhyolite tuff at Little City of Rocks. (43.1196, -114.6831)

Prominent flow fold in Miocene rhyolite (just right of author Shawn Willsey) is east of the highway between mileposts 131 and 132. (43.2602, -114.6680)

south and the Soldier Mountains to the north. The east-west orientation is perplexing because so many of southern Idaho's valleys trend in a nearly north-south fashion, generally perpendicular to east-west Basin and Range extension. It is thought that the Camas Prairie formed in response to the downwarping of volcanic rocks along the northern margin of the Snake River Plain. As the Snake River Plain dropped, the crust along the margins of the plain was stretched north-south, creating east-west faults and a down-dropped basin. Between mileposts 135 and 138, look for stairstep topography with numerous small benches, each bounded by faults that collectively lowered the Camas Prairie. With highlands on either side, a large amount of water is funneled into the Camas Prairie, creating numerous streams, spring-fed ponds, and wetlands such as those near milepost 142.

ID 51
Nevada Border—Mountain Home
94 miles

This lonely stretch of road links the highlands of north-central Nevada to the western Snake River Plain of Idaho. Beginning at the Nevada state line in Duck Valley, the highway cuts across the Owyhee River basin before snaking eastward into the Bruneau River drainage basin. Both desert rivers generally flow northward into the Snake River from highlands in northern Nevada.

A 1,400-foot-high escarpment capped by Pliocene-aged basalt lies east of the highway near the Nevada state line. This dramatic uplift parallels the highway northward for several miles and was formed by Basin and Range extension. A large,

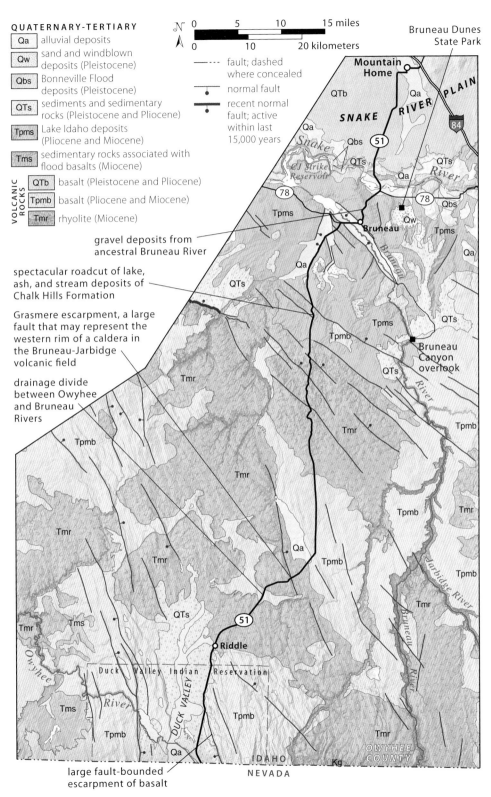

Geology along ID 51 between the Nevada border and Mountain Home.

north-trending normal fault lies near the base of the uplift, just east of the highway, and is responsible for the exceptional difference in elevation. A similarly oriented normal fault lies at the western edge of Duck Valley, where it has uplifted another, albeit smaller basalt bench. The two faults confine the down-dropped basin of Duck Valley, forming a graben. Occasional exposures of similar basalt dominate the next 20 miles of ID 51.

Between mileposts 25 and 26, roadcuts expose reddish-brown, highly fractured rhyolite that erupted from the immense Bruneau-Jarbidge volcanic field about 12 to 10 million years ago when this portion of Idaho sat above the Yellowstone hot spot. The Bruneau-Jarbidge volcanic field produced at least nine major ash-forming eruptions early in its heyday and later spewed out several thick, pasty flows of rhyolitic lava. Some of the explosive eruptions defy comprehension. Colossal clouds of ash rose from this massive volcano, ascending tens of miles into the atmosphere and raining ash as far away as the Great Plains. Indeed, at Ashfall Fossil Beds State Park in Nebraska, beautifully preserved fossils of rhinos, camels, horses, and other large mammals lie entombed in thick ash deposits erupted from the Bruneau-Jarbidge volcanic field.

A large normal fault lies along the base of the prominent basalt bench east of ID 51. (42.1242, -116.1435)

The highway drops over the Grasmere escarpment near milepost 26, a descent of approximately 500 feet over a few miles. This significant topographic slope has been interpreted as the western rim of a caldera within the Bruneau-Jarbidge volcanic field. At the base of the slope, the highway heads north through low-lying terrain with occasional outcrops of rhyolite. The elevation continues to drop northward, and rhyolite gives way to lighter-colored sedimentary layers deposited in the Snake River Plain during the reign of Lake Idaho in the Pliocene Epoch.

Around milepost 58, a prominent roadcut on the east side of highway exposes alternating layers of cream-colored silt and light-gray sand of the Chalk Hills Formation. The silts were deposited in Lake Idaho, when the lake level was high. During lower lake levels, streams flowed into the lake, depositing sand. Small lenses of gravels within some of the sand layers indicate periods of stronger stream flow, such as floods. A small normal fault cuts through the middle of the roadcut and drops the south side relative to the north side by a bit less than 1 foot. The upper portion of the outcrop consists mainly of white to light-gray layers of ash, possibly from the Bruneau-Jarbidge or the Twin Falls volcanic fields.

Agricultural fields irrigated by the Snake River and groundwater appear near milepost 63. The highway merges with ID 78 near milepost 70, then heads east across the lower reach of the Bruneau River and its namesake town near milepost 72. A fantastic overlook into the 800-foot-deep and narrow confines of Bruneau River canyon lies about 15 miles south of the town on Hot Springs Road. Follow signs for the canyon overlook. At the overlook, Late Miocene and Pliocene basalt is exposed in the canyon walls. Collectively, these stacked lavas emanated from numerous shield volcanoes that covered the earlier rhyolite eruptions within the the Bruneau-Jarbidge volcanic field.

North of Bruneau, ID 51 turns north, cresting a small rise separating the Bruneau River basin from the Snake River basin. Roadcuts near milepost 73 contain gravels deposited by the ancestral Bruneau River. ID 51 crosses the Snake River at milepost 77. During the Bonneville Flood, this broad and wide river valley allowed the floodwater to slow, depositing fine sand and silt. North of the Snake River, the highway climbs about 600 feet to the basalt rimrock that continues to Mountain Home and I-84.

Gravels from the ancestral Bruneau River near milepost 73. (42.9000, -115.7921)

A vertical normal fault (just to the right of the rock hammer) with about 1 foot of offset cuts through the middle of this large roadcut through alternating beds of cream-colored silt and light-gray sand of the Chalk Hills Formation near milepost 58. (42.7316, -115.8983)

ID 75
Shoshone—Timmerman Junction
28 miles

The southern end of ID 75 begins in the sleepy railroad town of Shoshone at an intersection of highways. Just north of Shoshone, the highway crosses the Shoshone lava field, an outpouring of basaltic lava from about 11,700 years ago and one of the youngest eruptions in the central Snake River Plain. The source of the lava is Black Butte Crater, a shield volcano not visible here, but good views are found farther north along ID 75. Near Shoshone, the lava flowed west, terminating near Gooding. When it erupted, the lava poured into existing streambeds like the nearby Big Wood and Little Wood Rivers, filling them with lava and forcing their paths to the outside margin of the Shoshone lava field. Today, the rivers merge near Gooding, much farther downstream

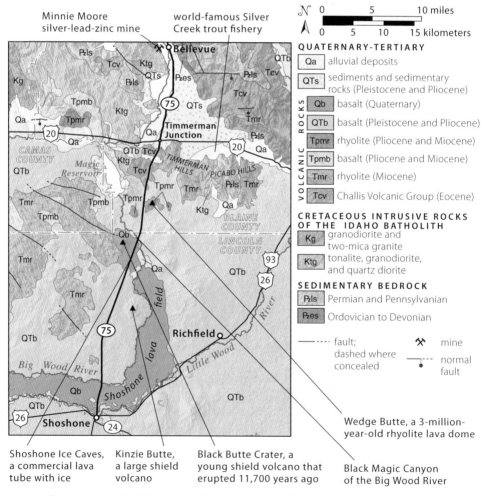

Geology along ID 75 between Shoshone and the Timmerman Junction at US 20.

than they originally did. The highway crosses the Big Wood River at milepost 77, but the riverbed here is seasonally dry due to upstream diversions for irrigation.

Kinzie Butte looms ahead, just east of the highway beginning around milepost 81. Another classic shield volcano, the exact age of Kinzie Butte has not been determined, but it is certainly older than the much younger Black Butte Crater to the north. A turnoff to the west for Mammoth Cave, a commercially operated lava tube, lies between mileposts 81 and 82. Basalt outcrops near the highway belong to the same lava that formed the cave and emanated from an unnamed vent a few miles northwest.

As the highway crests the western flank of Kinzie Butte at milepost 85, you can see the jagged peaks of the Pioneer Mountains to the north. Reaching elevations over 12,000 feet, the upper portions of these mountains were extensively glaciated and sculpted during the Pleistocene ice age. To the northwest, the dark, brooding summit of Black Butte Crater, the source of the lava just north of Shoshone, finally comes into view. Around milepost 89, the highway again crosses the youthful Shoshone lava field where lavas flowed southeast from Black Butte Crater. A collapsed pit on the east side of the road near milepost 90 indicates the presence of a lava tube. A turnoff for the Shoshone Ice Caves, another commercially operated lava tube, lies just ahead on the west. Rain and snowmelt seep downward through the permeable basalt and freeze in the caves during the winter. Rock is an effective insulator and allows the ice to persist through the summer. North of the ice caves turnoff, ID 75 bisects a large aa lava field. Hot basaltic lava flowing from the summit of Black Butte Crater first formed smooth pahoehoe lava, but as these lavas traveled downslope, they progressively cooled, becoming more viscous and slower. The change in viscosity caused the lava to run over itself, becoming stickier and eventually morphing into the jagged, blocky aa texture seen near the highway.

The surface of the Shoshone lava flow collapsed into a subterranean lava tube near milepost 90. (43.1618, -114.3372)

BLACK MAGIC CANYON

A surreal landscape of fluted, scoured, and polished basalt lies entrenched in the narrow slot canyon of the Big Wood River near milepost 92. Informally called Black Magic Canyon, intermittent sections of this fantastic wonderland of stone lie along the Big Wood River from just downstream of Magic Reservoir to near the town of Shoshone. The closest access point to ID 75 is a small parking area reached by turning west onto West Magic Road just south of where the highway crosses the river.

The geologic story of Black Magic Canyon starts with the eruption of nearby Black Butte Crater about 11,700 years ago. Lava filled the original channel of the Big Wood River and diverted the river to its current path along the outside edge of the lava field. Since then, floods carried gravel and cobble-sized chunks of chert, quartzite, and quartz-rich volcanic rocks along the streambed. These hard rocks have eroded, gouged, and deepened the basalt bedrock channel, creating a bizarre array of sculpted flutes, arches, windows, fins, and potholes along the canyon. Note that the canyon is often filled with water from late spring to early fall as it is used to deliver water for irrigation. When dry, it is advised that interested visitors call the Big Wood Canal Company at 208-886-2331 to ensure that no release of water from the upstream dam is planned.

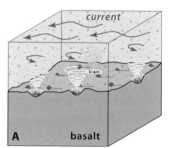

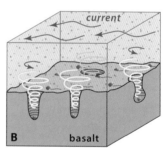

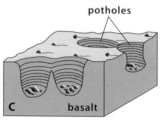

The surreal labyrinth of Black Magic Canyon, a narrow slot cut by floods on the Big Wood River over the past 11,700 years. (43.0778, -114.3011)

Swirling currents in the Big Wood River during flood events dragged streambed gravels in circular paths, eventually boring down into bedrock to create potholes in the bedrock floor. Adjacent potholes may merge, creating sculpted fins of rock between them. —From Willsey, 2017

Wedge Butte, a steep-sided dome, erupted rhyolite lava about 3 million years ago. (43.2296, -114.3127)

The large, steep hill just east of the highway near milepost 94 is Wedge Butte. Unlike the broad, low shield volcanoes that dominate the Snake River Plain, this volcano presents a markedly different profile. Its steep slopes and sheer height (over 700 feet tall) contrast sharply with Black Butte Crater and other nearby basaltic volcanoes. Wedge Butte formed where silica-rich rhyolitic lava oozed upward, piled upon itself, and constructed a steep dome about 3 million years ago. Similar rocks exist west of the highway atop Dinosaur Ridge, an apt name for the spiny crest of this high point.

Near milepost 97, ID 75 exits the Snake River Plain, slicing through the Timmerman Hills. The uplands east of the highway are covered by the Picabo Tuff, rhyolite ash deposited 9 to 8 million years ago by explosive eruptions from the now buried Picabo volcanic field. The tuff also caps some of the high points west of the highway, where it lies atop older rocks of the Challis Volcanic Group and Cretaceous granodiorite of the Idaho batholith. Around milepost 101, the highway descends northward into the Wood River valley, home of the Big Wood River.

ID 78
Marsing—Hammett
99 miles

Also known as the Owyhee Highway, ID 78 follows the southern edge of the Snake River and passes beneath the eastern foothills of the uplifted Owyhee Mountains. Geologically, the highway mainly intersects light-colored sediments from various phases of Lake Idaho. Sections of highway along the river also intersect Bonneville Flood deposits. A smattering of volcanic buttes, vents, and flows adds to the diversity.

ID 78 begins just west of the Snake River in Marsing, a productive agriculture and viticulture region. Looking eastward across the river between mileposts 4 and 6 provides views of Liberty Butte, one of several hydrovolcanic vents exposed near the

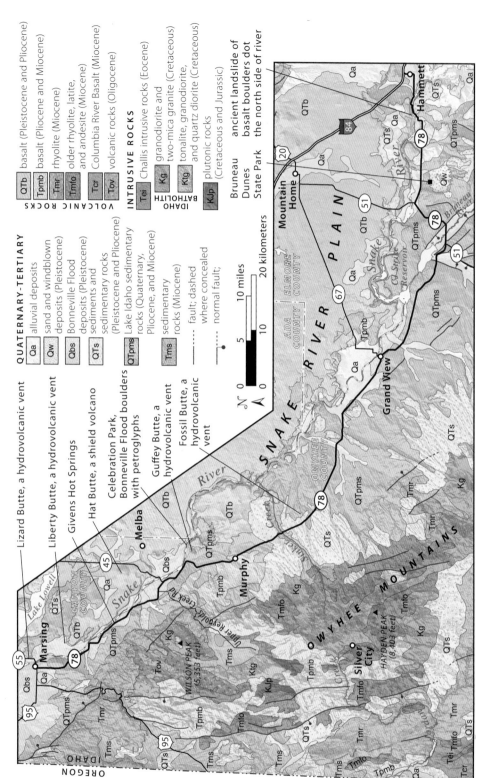

Geology along ID 78 between Marsing and Hammett.

highway. These volcanic features formed where ascending basaltic magma interacted with either shallow groundwater or the waters of Lake Idaho, flashing the water to steam and shattering the lava and surrounding rock into a variety of particle sizes. These ejected deposits accumulated around the vent and were cemented together to form yellow-brown layers of tuff. Harder than the surrounding lake sediments, the tuff forms many of the stately buttes, pinnacles, and other landscape features near ID 78. This area of the western Snake River Plain has been lauded as one of the best places in the world to study the wide variety of features produced when lava interacts with water.

LIZARD BUTTE

A short (1 mile) side trip from Marsing across the Snake River on ID 55 will bring you close to the base of Lizard Butte, which resembles a lizard basking in the sun. This interesting butte is the eroded remnant of a hydrovolcanic vent that erupted through wet Lake Idaho sediments of the Pliocene Glenns Ferry Formation. The light-colored lake sediments are obvious below the overlying dark volcanic rock, which consists of layers of ash and lapilli that were blasted from the vent as the basaltic lava interacted with water and wet sediment, flashing the water to steam and forming explosive conditions that ripped lava into pieces. The uppermost caprock, which resembles typical basalt, is actually welded spatter formed from clots of lava thrown from the vent and then stuck together upon impact. The volcano was heavily eroded by the Bonneville Flood 17,500 years ago, leaving only Lizard Butte as stately evidence of its existence.

Lizard Butte, a hydrovolcanic vent just east of Marsing. Light beds of Glenns Ferry Formation lie beneath basaltic ash and lapilli that is capped by basaltic spatter. (43.5572, -116.7922)
—Photo by Danielle Marquette

Around milepost 11, the highway passes Givens Hot Springs, an area used by ancient Native Americans for thousands of years and later developed by settlers. Today, it is a commercial hot spring and campground open to the public. Across the river to the north and west, stacked basalt lavas from the nearby Hat Butte shield volcano cap the prominent mesa and lie atop Lake Idaho sediments of the Glenns Ferry Formation deposited from about 4 to 3 million years ago. A small roadcut at milepost 15 exposes silt and sand layers of the Glenns Ferry Formation. Across the Snake River in

UPPER REYNOLDS CREEK ROAD

Near milepost 20, Upper Reynolds Creek Road heads west into the Owyhee Mountains. The short side journey described here ends at the cattle guard about 10 miles from ID 78, but the road continues southward for those who wish to see more of this area. No mile markers appear along the road so reset your odometer at the intersection with ID 78.

Beige sedimentary deposits from Lake Idaho dominate the first 3 miles along the road. About 3.5 miles into the drive, the first outcrops of red-brown rhyolite appear. The rhyolite is about 11.5 million years old and formed as a thick, pasty lava flow erupted from a vent about 4 miles to the southwest. About 5 miles from ID 78, light-colored, banded sedimentary deposits of the 10- to 6-million-year-old Chalk Hills Formation appear in the large hillside to the west. A mixture of silt, sand, gravel, and some ash, this unit was deposited in and around Lake Idaho.

Near the Kane Springs parking area about 5.7 miles from ID 78, the first outcrops of light-gray granite appear. Underlying much of the Owyhee Mountains, the Cretaceous granite is part of the Idaho batholith and formed when subduction along the west coast, which was located near the Idaho-Oregon border at the time, generated vast magma chambers beneath Idaho. The Owyhee granites are separated from their central Idaho siblings by extensional faulting that formed the western Snake River Plain. About 9.5 miles into the drive, the hillside west of the road beautifully exposes the granite with red rhyolite directly above.

Vertically fractured, red Miocene rhyolite lies atop light-gray Cretaceous granite from the Idaho batholith along Upper Reynolds Creek Road in the Owyhee Mountains. (43.2629, -116.7077)

CELEBRATION PARK

Celebration Park, Idaho's only archaeological park, showcases exceptional petroglyphs inscribed on massive basalt boulders deposited by the powerful Bonneville Flood. To get there, turn north onto ID 45 between mileposts 19 and 20, cross the Snake River, and then turn right (east) onto Ferry Road. After 2 miles, turn right (south) onto Hill Road and follow it for 2.2 miles as it winds and becomes Warren Spur Road. Turn left (south) onto Sinker Road and continue 2.8 miles to a final left (east) turn to the parking area. The park includes a visitor center, information kiosks, and short trails.

Celebration Park lies on Walters Bar, an immense gravel bar deposited by the Bonneville Flood about 17,500 years ago. Walters Bar lies just downstream of an 11-mile-long constriction in the canyon. During the flood, the raging water plucked and eroded large chunks of the basalt cliff faces along the narrow canyon's corridor. These rocks were tumbled along the streambed and smashed into other boulders, rounding them as they moved. As they entered the wide valley around Celebration Park, the flood's velocity slowed, and the large boulders dropped to form Walters Bar. Some of the boulders are over 15 feet in diameter, testifying to the extraordinary power, size, and scale of the Bonneville Flood.

Large Melon Gravel boulders from the Bonneville Flood at Celebration Park. (43.3043, -116.5316)

Petroglyphs inscribed on a dark weathered surface of desert varnish of a boulder at Celebration Park. (43.3002, -116.5222)

BRUNEAU DUNES STATE PARK

The Bruneau Dunes lie in a large semicircular basin formed by an ancient meander of the Snake River. Between mileposts 84 and 85 on ID 78, take Bruneau Sand Dunes Road southward to access this state park, which also features a visitor center, several campgrounds, and an observatory. The dunes here are touted as the tallest single structured dune (a dune not braced against other dunes or other structures) in North America. It

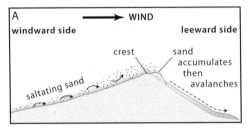
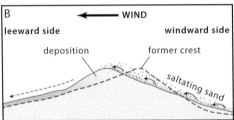
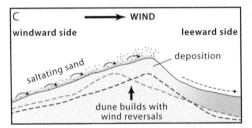

Reversing wind directions piles sand higher, forming the tall dunes at Bruneau Dunes State Park.
—From Willsey, 2017

Salt-and-pepper sand of Bruneau Dunes State Park. Dark grains are basalt and light grains are mainly quartz. Beetle is about 0.25 inch long.

The main dune at Bruneau Dunes State Park rises about 470 feet above the valley floor. (42.8969, -115.6959)

rises about 470 feet above the valley floor. Unlike other dunes that are formed by one dominant wind direction, the Bruneau Dunes are a reversing dune, formed from two dominant wind directions (northwest and southeast) that prevail both before and after strong cold fronts roll through the area. The reversing wind pattern deposits sand atop itself, piling up to form the lofty dune.

Another striking feature of the Bruneau Dunes is the composition of its sand. Most sandy environments are dominated by quartz, a hard and common mineral found in many rock types. At Bruneau, the sand is much darker, and close inspection shows that it is a salt and pepper mixture of dark basalt grains and lighter quartz grains. The basalt comes from the numerous recent eruptions in the Snake River Plain. The quartz is derived from some or all of the following sources: (1) older eruptions of rhyolite as the area passed over the Yellowstone hot spot, (2) weathered Cretaceous granitic rocks from the central Idaho mountains or Owyhee Mountains, and (3) quartz-rich metasedimentary Proterozoic rocks in eastern Idaho.

the distance, a dark basalt lava flow from the Hat Butte shield volcano caps the mesa and protects the underlying, light-colored Lake Idaho deposits from erosion.

Another hydrovolcanic vent, Guffey Butte, rises to the east about 500 feet above ID 78 near milepost 23 as the highway swings south, away from the river. The vent erupted about 1 million years ago when rising basaltic magma mixed with groundwater and generated a series of violent steam-driven explosions that blasted blocks of rock up to 10 feet across.

South of Murphy, near milepost 36, the highway descends to cross Sinker Creek. Roadcuts here expose brown to gray basalt stacked in thin sheets, indicative of hot pahoehoe lava at the time of eruption. Closer inspection reveals numerous gas bubbles or vesicles in some layers. Many of the vesicles have been partially or wholly filled with other minerals, forming amygdules. The coloring of the rock and the amygdules suggests hydrothermal alteration, where fluids moved through the basalt sometime after it formed, depositing other minerals into the vesicles and changing the overall color of the rock.

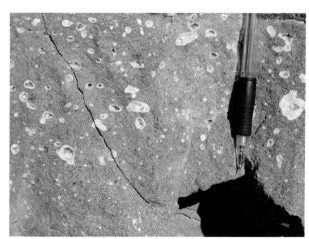

White zeolite minerals fill vesicles to form amygdules in the basalt of the roadcut near milepost 36. (43.1377, -116.4880)

Fossil Butte, yet another hydrovolcanic vent, appears north of the highway around milepost 40. Gravel carried by older stream systems draining the Owyhee Mountains caps the small ridges along the highway. ID 78 reenters the agricultural fields along the Snake River around milepost 50.

ID 78 merges with ID 51 at milepost 76 and mile markers jump back by 6 miles. The highway cuts across the floodplain of the Bruneau River, a large river system stemming from mountains in northern Nevada. The river and its tributaries flow through spectacularly deep, narrow canyons. East of Bruneau, the highway climbs through rolling hills and bluffs of beige Glenns Ferry Formation before heading east just south of the Snake River.

Between mileposts 93 and 94, massive blocks of basalt scattered on slopes across the Snake River to the north are the product of a large landslide. Hard basalt caps the mesa to the north but is underlain by relatively soft sediment of the Glenns Ferry Formation, creating an unstable situation where slopes often fail. Here, undercutting of these softer materials by the Snake River likely caused the slide. The highway crosses the Snake River between mileposts 94 and 95. A roadcut near milepost 96 exposes tan silts and slightly darker cross-bedded sands of the Glenns Ferry Formation.

Landslide north of Snake River near milepost 93. Undercutting by the Snake River into soft Pliocene-aged Glenns Ferry Formation caused the overlying basalt to break, topple, and slide. (42.9327, -115.5611)

The Lost River Range in the Basin and Range of east-central Idaho forms a linear wall of high peaks north of Mackay. The Lost River fault, a large active normal fault, lies along the western base of the range and has incrementally pushed the range up while dropping the valley down through thousands of earthquakes over at least the past 10 million years. (43.93, -113.63)

BASIN AND RANGE

The Basin and Range Province, a massive region of the Intermountain West, measures about 1,000 miles north to south and as much as 500 miles west to east. Best known for its unique topography, this seemingly endless series of generally north-trending mountains and intervening valleys stretches from the Sierra Nevada of California to the Wasatch Mountains of Utah and from southeastern Arizona to southwestern Montana. The mountains are typically 8 to 12 miles wide and separated by valleys of comparable width. The uplifted mountains soar to lofty heights and include Borah Peak, Idaho's high point at 12,662 feet.

East-west extension that began about 17 million years ago created the exceptional landscapes of the Basin and Range Province. The exact cause of the extension is controversial, but possible ideas involve (1) a wholesale change in the plate tectonic setting as the San Andreas fault system developed to the west, or (2) the subduction of a divergent plate boundary, the northern continuation of the East Pacific Rise, beneath western North America. The East Pacific Rise is the spreading ridge that forms the Gulf of California. No matter the mechanism, the extended region has substantially thinner crust and higher heat flow in the subsurface than other portions of North America.

During the east-west extension, which continues today, the brittle upper crust breaks along normal faults in which the block of rocks below the fault (the footwall) rises, while the block above the fault (the hanging wall) drops. As mountains form along the footwall side of the fault, basins develop on the other side. These basins fill with sediment eroded from adjacent mountains, so the vertical offset on any one fault is more than twice the topographic expression of the mountains. Each fault movement produces an earthquake and creates greater topographic relief, a process that has played out thousands of times over millions of years.

The geologically youthful Snake River Plain bisects the northern edge of the Basin and Range Province in Idaho.

The faults dip at moderate to high angles beneath the basins, so fault movement also causes horizontal separation, meaning that rocks on either side of the fault move farther away from each other as the fault accommodates the stretching. As extension along these faults occurs, the fault angle dictates how much vertical and horizontal offset is produced. High-angle normal faults dip between 30 and 60 degrees and commonly define steep mountain fronts, like the west face of the Malad and Lost River Ranges or the east side of the Sawtooth Range. The steep orientation of these normal faults produces nearly equal amounts of vertical and horizontal offset.

In the long period between earthquakes, the uplifted mountains are eroded by water and wind. The low basins gradually collect sediment shed off the mountains or deposited by through-going rivers that snake around the mountain ranges. In geologic lingo, the valleys produced by normal faulting are called grabens, and the intervening uplifted blocks are called horsts.

Extension in the Basin and Range is manifested today with each earthquake. GPS measurements of fixed landmarks move farther apart with time, both through earthquakes and through aseismic creep. Large historic earthquakes such as the 1983 Borah Peak quake are uncommon, but smaller earthquakes occur each year, detected by seismic networks run by the University of Utah and the US Geological Survey.

BASIN AND RANGE 107

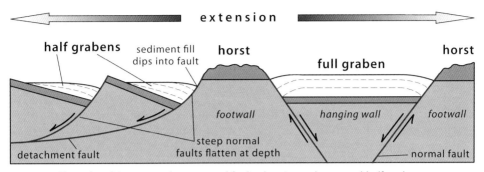

Extensional forces produce normal faults, horsts, grabens, and half grabens.

View to the south of the steep, vegetation-free fault scarp from the 1983 Borah Peak earthquake in the Lost River Valley. The fault scarp is where the fault broke through to the Earth's surface. You can see the scarp extend across the base of the range to the south just above author Willsey's head. Borah Peak lies just to the left of this photo. The photo was taken about 1 mile southeast of the Borah Peak earthquake interpretive site north of Mackay. (44.1485, -113.8472)

The distinct topography of the Basin and Range in Idaho exerts a strong influence on human habitation. Roadways largely follow valleys, working their way around, rather than over, the intervening mountain ranges. Many of the valleys harbor small towns and communities where ranching and farming dominate. Within these valleys, surface water in streams and groundwater in aquifers support these industries. The mountains offer opportunities for recreation, exploration, and geologic investigation.

In southern Idaho, geologically young (Pliocene to Holocene) volcanism of the Snake River Plain overprints some of the Basin and Range Province, bisecting it into two regions. No sedimentary rocks older than Miocene age are observed within the Snake River Plain. This suggests that the intense volcanism associated with the passage of the region over the Yellowstone hot spot both surrounded and totally disrupted the older rocks. Pervasive east-west Basin and Range extension in the eastern Snake River

Plain appears to have controlled the orientation of some Holocene volcanic vents and fissures, such as the Great Rift that begins west of American Falls and extends north through Craters of the Moon National Monument.

Because Basin and Range faulting began in the middle Miocene about 17 million years ago, the extension occurred throughout the period of Miocene to Pleistocene volcanism. Explosive rhyolite eruptions from the Snake River Plain produced volcanic ash that settled in the basins. Ash beds interlayered in basin sediments can be radiometrically dated and thus provide constraints on the time of deposition of those sediments, and thereby on the age of faulting that produced the basins.

NEOPROTEROZOIC TO MESOZOIC SEDIMENTARY ROCKS

Idaho's Basin and Range normal faults have uplifted huge blocks of the crust, exposing a variety of mainly sedimentary rocks deposited from the Proterozoic Eon to the Mesozoic Era. By carefully analyzing these rocks, including their grain size, distinctive features, and the occasional fossil, geologists have pieced together an extraordinary sequence of events in the Basin and Range of Idaho.

The Neoproterozoic sedimentary rocks record the rifting of the Proterozoic supercontinent Rodinia between 700 and 550 million years ago. The opening rift eventually created a growing ocean basin that would become the paleo-Pacific Ocean. The eastern side of the rift became the passive continental margin of ancient North America. The coast is called passive because the continental crust transitions to oceanic crust without an active plate boundary. A modern example is the eastern seaboard of North America.

The sedimentary rocks deposited in the growing rift in Neoproterozoic to early Cambrian time include formations and groups collectively mapped as the Windermere Supergroup. This distinctive succession from rift-related volcanic rocks to river deposits to marine sediments is recognized from the Yukon, Canada, south to Sonora, Mexico, and is interpreted to represent the rifted and subsiding passive margin on the western side of ancient North America. Above the sandy rocks of the Windermere Supergroup are mainly carbonate rocks of Cambrian age (540 to 500 million years old). These are overlain by Ordovician, Silurian, and early Devonian limestones and dolostones of a marine carbonate bank.

In the older part of the rift sediments, exposed at Portneuf Gap east of Pocatello and in Oxford Mountain northeast of Preston, are poorly sorted pebbly to cobble-bearing mudstones. These rocks, called diamictites, were deposited during glacial episodes when melting icebergs released stones that sank to the seafloor. The prevalence of late Neoproterozoic diamictites throughout the world suggests that the Earth was entirely or nearly entirely encased in ice that extended south to equatorial latitudes. This episode of extreme glacial conditions is sometimes referred to as Snowball Earth and is officially called the Cryogenian Period. Ages on volcanic tuffs from the Pocatello Formation on Oxford Mountain and at Portneuf Gap suggest that two stratigraphic horizons of diamictite were deposited between about 685 and 665 million years ago.

In the Lemhi Range and Beaverhead Mountains, the distinctive rift-to-passive-margin succession is missing. The Ordovician Kinnikinic Quartzite lies directly on the Mesoproterozoic Belt Supergroup. The lack of Neoproterozoic to Cambrian marine rocks suggests much of this area was above sea level from 700 to 500 million

	PERIOD	UNIT	SEDIMENTARY UNITS IN SOUTHEAST IDAHO	
CENOZOIC	TERTIARY — PLIOCENE and MIOCENE	QTpms	Salt Lake/Starlight Formations	
	TERTIARY — EOCENE	Tes	Wasatch Formation	
MESOZOIC	CRETACEOUS	Ks	Frontier Formation	
			Sage Junction Formation	
			Wayan Formation	
			Smiths Formation	
			Gannett Group	
	JURASSIC	Js	Stump Formation	
			Preuss Sandstone	
			Twin Creek Formation	
			Nugget Sandstone	
	TRIASSIC	₸Rs	Ankareh Formation	
			Thaynes Formation	
			Dinwoody Formation	
PALEOZOIC	PERMIAN	Pzls	Phosphoria Formation	
	PENNSYLVANIAN		Wells Formation/Oquirrh Group	
	MISSISSIPPIAN		Aspen Range Formation	
			Mission Canyon Limestone	Madison Group
			Lodgepole Limestone	
	DEVONIAN	Pzes	Bierdneu Formation	
			Hyrum Dolomite	
	SILURIAN		Laketown Dolomite	
			Fish Haven Dolomite	
	ORDOVICIAN		Swan Peak Formation	
			Garden City Limestone	
	CAMBRIAN		St. Charles Formation	
			Nounan Formation	
			Bloomington Formation	
			Elkhead/Blacksmith Limestones	
NEOPROTEROZOIC		∈Zs (Windermere Supergroup — Brigham Group)	Gibson Jack/Lead Bell Formations	
			Camelback Mountain Quartzite	
			Mutual Formation	
			Inkom Formation	
			Caddy Canyon Quartzite	
			Papoose Creek Formation	
			Blackrock Canyon Limestone	
		Zp (Pocatello Formation)	upper member	
			Scout Mountain Member	
			Bannock Volcanic Member	

Symbols: unconformity; conglomerate; sandstone or quartzite; limestone; dolomite; diamictite; chert; phosphatic rock; argillite or shale; siltstone; nodular limestone; fossiliferous clastic limestone; mafic volcanic rocks; tuff.

Scale: 1 mile / 2 kilometers.

Generalized stratigraphic column for southeast Idaho. See the stratigraphic column on page 151 for the Lost River and Lemhi Ranges.

years ago. Called the Lemhi arch, this area of elevated land corresponds generally with two suites of granitic rocks, the 650-million-year-old Big Creek suite in the Central Mountains east of Edwardsburg, and the 510- to 490-million-year-old Beaverhead pluton in the Beaverhead Mountains southeast of Leadore. The intrusion of these very low-density granitic rocks likely made this part of the North American continental crust more buoyant than the surrounding crust. The thinning of sedimentary

PHOSPHATE MINING

The Phosphoria Formation was deposited in Permian time in a sea basin that was restricted from open ocean currents but received nutrients from the upwelling of deep marine water. The basin became chemically stratified at times, and organic-rich sludge accumulated under chemically reducing conditions. Organisms thrived in the surface water, but when they died, their bodies were not destroyed because the mud at the bottom of the basin lacked oxygen. The mud accumulated carbon and organic phosphate from the dead organic matter, plus some heavy metals. When the mud was deeply buried, elevated pressures and temperatures generated hydrocarbons that migrated in the subsurface into Wyoming about 80 million years ago. Thus, Wyoming strata ended up as the reservoir for oil made in Idaho. While oil had a low enough viscosity to migrate, phosphate formed a waxy substance that stayed in Idaho. It forms a huge ore reserve, the source of mining's largest annual contribution to the Idaho economy.

The Phosphoria Formation is mined in open pits from folded strata of the Meade thrust plate that was thrust eastward during the Sevier orogeny. The Meade thrust plate runs from east of Montpelier northwest through the mountains east of Soda Springs to east of Fort Hall, so most of the mines lie east of those towns. The ore is crushed, and either elemental phosphorous or phosphoric acid is separated and produced for herbicide, fertilizer, and a host of chemical uses.

Among paleontologists and fossil enthusiasts, the Phosphoria Formation is well known for hosting a unique fossil of the genus *Helicoprion*, an extinct sharklike fish with a fantastic spiral array of serrated teeth set in its jaw like a circular saw. This fish-slaying beast, commonly known as the buzz saw shark, grew up to 35 feet long and used its peculiar arrangement of teeth to chop through its prey. Fossils of *Helicoprion* are occasionally uncovered in phosphate mines in southeastern Idaho.

An artist's rendition of Helicoprion, *with its main food, squid, in the Permian sea. The largest predator of its time, the buzzsaw shark grew up to 35 feet long and used a peculiar spiral arrangement of teeth in the lower jaw to chop through its prey.* —Art by Ray Troll

layers over the Lemhi arch in the Beaverhead Mountains is recognized in rocks as young as Devonian.

Throughout much of the Paleozoic, eastern Idaho's position along the passive margin of North America resulted in thick accumulations of sediment. Coastal areas or beaches were dominated by quartz-rich sand which later became sandstone (or quartzite if highly compacted by the pressure of overlying rocks). Shallow offshore regions typically deposited mud to form shale. In clear-water settings, the calcite exoskeletons of tiny marine organisms accumulated on the seabed to form limestone and dolostone. As sea level fluctuated throughout the Paleozoic, these environments shifted landward or seaward, resulting in stacked sequences of rock. Observing these changes in rock types and therefore environments over time allows geologists to reconstruct the geologic history along passive margins.

The Mississippian Period is noteworthy as it marks a time of high sea level globally. Much of North America was covered by warm, shallow ocean water where organisms thrived, forming a carbonate platform or bank. Mississippian limestone in the Lost River and Lemhi Ranges and in the Beaverhead Mountains was part of this large carbonate bank, as was the widespread Madison Group, exposed in southeastern Idaho. Later, in Pennsylvanian time, the thick Oquirrh and Sun Valley Groups were deposited in a large basin that extended from northern Utah to central Idaho.

As the Paleozoic Era drew to a close and the Mesozoic Era began, subduction developed along the west coast of North America. In response, much of western North America was uplifted, forming highlands. Sediments shed off these mountains collected in basins. The Triassic, Jurassic, and Cretaceous sedimentary rocks are now mostly found in mountains of southeast Idaho and western Wyoming.

Beginning in the Jurassic Period, rising sea levels encroached on the low topography east of Idaho in what is now the Great Plains and Rocky Mountain region. By the Cretaceous Period, seawater from the Gulf of Mexico and the Arctic Ocean merged to completely flood the continent's interior, forming the Cretaceous Interior Seaway. Rivers flowed east from the western and central Idaho highlands toward the vast inland seas. The streams carried sand and gravel in their channels and deposited mud in their floodplains. Small lakes, marshes, and swamps formed in the lowlands, creating organic-rich shales, mudstone, and coal. Shoreline deposits of sand and mud formed at beaches and tidal flats along the seaway's margin at the Idaho-Wyoming border.

SEVIER FOLD-AND-THRUST BELT

More than 100 million years prior to the onset of Basin and Range extension, much of western North America experienced east-west compression as part of the Sevier orogeny, a series of mountain building events that lasted from 140 to 60 million years ago. These periods of uplift occurred due to the collision between the continental plate of North America and the Farallon Plate, the once larger predecessor of today's Juan de Fuca Plate along the Pacific Northwest coast. Throughout much of Idaho, the Sevier orogeny left dramatic evidence on Idaho's rocks and landscapes. With an immense stack of Paleozoic and Mesozoic sedimentary rocks already in place throughout much of western North America, the east-west vice of the Sevier orogeny squeezed these layered rocks into folds and offset them by thrust faults. For example, the rocks in the Lost River and Lemhi Ranges belong to the Hawley Creek thrust plate and contain spectacular cliff-scale folds.

Where weak rock layers like shale existed, the rocks broke, forming thrust faults parallel or at a low angle to bedding that shoved huge stacks of rocks eastward, sometimes by tens of miles. These thrust faults put older rocks over younger rocks. The end result of this prolonged period of deformation is the Sevier fold-and-thrust belt, a zone of folded and thrust faulted rocks east of the Cretaceous batholiths, that extends from southeastern California to Canada. In Idaho, the belt runs through

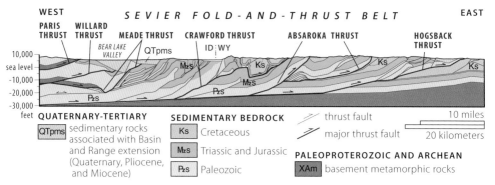

Compression during mountain building can push sheets of rock over other rock along large thrust faults. —Modified from Coogan, 1992

The Sevier orogeny produced spectacular folds in the Mississippian-age Scott Peak Formation (part of the Mississippian carbonate bank) on Leatherman Peak in the Lost River Range. (44.0749, -113.7251) —Derek Percoski photo

eastern Idaho from its border with Wyoming to just west of Twin Falls and Ketchum. Within the Basin and Range Province, telltale signs of the Sevier orogeny are abundant. Steeply tilted beds within folded sedimentary rock layers later became fins of rock exposed in the mountains.

Many of the Sevier thrust faults run northwest, similar to the alignment of most Basin and Range normal faults. In many places, the normal faults reactivated the older thrust faults, reversing the direction of movement. Today's mountain ranges, however, are mainly products of uplift on Basin and Range normal faults. Were it not for the extensive overprinting of the younger Basin and Range extension, there would be no mountains today, and the Sevier fold-and-thrust belt would qualify as its own unique geologic province.

The origin of the thick, thrust-faulted stack of sedimentary rock between the exotic terranes to the west and the continental craton to the east has been the subject of much debate. A proposal by Robert Hildebrand in 2009 holds that some of these sedimentary rocks, which extend from Alaska south through Canada to Mexico, were part of a Cordilleran ribbon continent that was accreted to ancient North America and is now preserved in the western part of the thrust belt. In southeast Idaho, this hypothesis is disproven by the presence of detrital zircon grains that were eroded from the Cambrian Beaverhead pluton of North America and deposited southward and eastward in latest Cambrian sedimentary rocks of the Idaho fold-and-thrust belt.

METAMORPHIC CORE COMPLEXES

Low-angle normal faults, also known as detachment faults, have also impacted the Basin and Range. With dip angles less than 30 degrees, these faults accomplish much more horizontal displacement, often as much as tens of miles, than do steeper normal faults. Despite their shallow dip, detachment faults cut deep into the Earth's crust, dividing cooler brittle rocks above the fault from warmer, more pliable metamorphic rocks below. As the crust is thinned due to continued movement along the detachment fault, older metamorphic rocks below the fault rise and are brought to the surface. Pulling away some of the upper crust removes weight from the crust below,

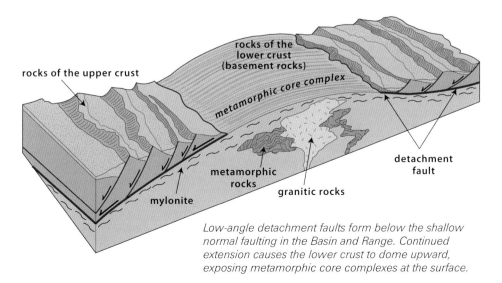

Low-angle detachment faults form below the shallow normal faulting in the Basin and Range. Continued extension causes the lower crust to dome upward, exposing metamorphic core complexes at the surface.

causing the lower crust to buoyantly dome upward through the remnant detachment fault. This blister of lower crustal rocks at the surface is called a metamorphic core complex. Only a few dozen metamorphic core complexes exist in the Basin and Range of North America, and Idaho is home to several. Late Eocene and Oligocene sedimentary basins formed within and between the exhuming core complexes, such as in the Donkey Hills east of Borah Peak.

Metamorphic core complexes in Idaho get younger to the south. Those in Northern Idaho began forming about 50 million years ago in Eocene time. North of the Snake River Plain, the Pioneer Mountains east of Ketchum expose an Eocene-age metamorphic core complex. A fantastic one in the Albion Mountains near Burley experienced three pulses of rapid extension and uplift, one about 40 million years ago and two during the Miocene Epoch at about 20 and 10 million years ago. The combined faulting exposed Archean metamorphic rocks and Proterozoic metasedimentary rocks, along with Oligocene granite. In southeast Idaho, the low-angle Bannock detachment fault underlies Cache Valley and the Malad Range. It moved about 10 million years ago. This area of extended rocks may overlie a deeper core complex not yet exposed.

LAKE BONNEVILLE

The cool climate of the Pleistocene allowed a large freshwater lake, Lake Bonneville, to form in the land-locked basin where today's Great Salt Lake of northern Utah resides (see map of the lake on page 25). As the climate fluctuated throughout the Pleistocene, this lake rose and fell accordingly. During cool, wet periods the lake rose, inundating low valleys in southeastern Idaho such as the areas around Malad City and Preston with as much as 500 feet of water. When the lake was stable for hundreds of years or so, shorelines developed along its margins where the wave action of the lake eroded the land to form nearly level benches. These shorelines can still be seen today. The two most prominent ones are the Bonneville shoreline (about 5,200 feet in elevation), which marks the highest stand of the lake before it overtopped one of its natural divides and caused the huge Bonneville Flood about 17,500 years ago (discussed in Snake River Plain section), and the Provo shoreline (about 4,800 feet in elevation), which marks the elevation at which the lake stabilized following the flood.

I-15
Utah border—Pocatello
68 miles

Travelers headed north on I-15 enter southeast Idaho in the Malad Valley, which was dropped down along the active Wasatch fault to the east. This normal fault runs along the front of the Wasatch Range from Nephi, Utah, northward into Idaho to Malad City. The fault dips westward at about 60 degrees and is divided into six or so discrete segments. The fault is classified as active, with major earthquakes (as large as magnitude 7.5) occurring on one of the segments once every 350 years on average. The Wasatch fault presents a serious geologic hazard to residents of northern Utah and southeastern Idaho. The 5.7-magnitude earthquake of March 2020 occurred on the western subsurface trace of the Wasatch fault on the southern edge of the Great Salt Lake.

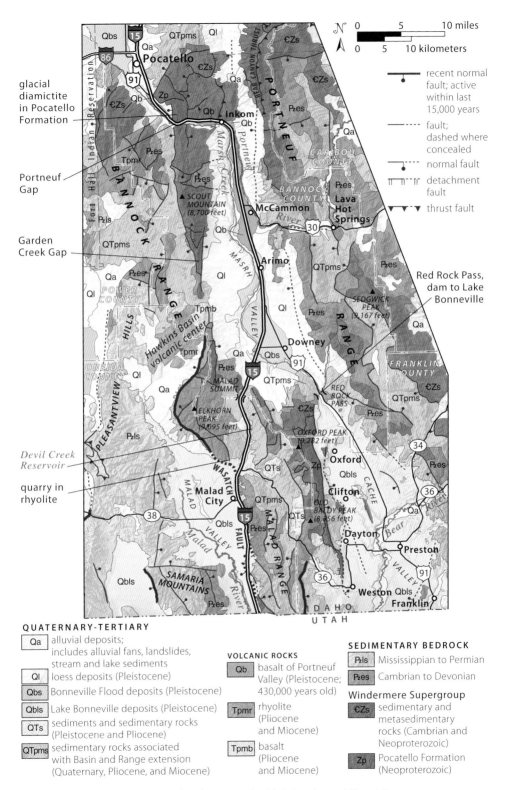

Geology along I-15 between the Utah border and Pocatello.

At the Utah-Idaho border, travelers would have been 500 feet beneath the surface of Lake Bonneville, which covered the entire Salt Lake City region with up to 1,000 feet of water. The Malad Valley was most recently inundated by a northern arm of the lake about 20,000 years ago. Two prominent shorelines are evident on hillsides above the Malad Valley: the Bonneville level, which was the highest level of the lake at about 5,200 feet, and the Provo level at about 4,800 feet, the level to which the lake fell after the Bonneville Flood 17,500 years ago.

I-15 and the Malad Valley follow the Wasatch line between Ogden and Pocatello. This line represents the western edge of the ancient North American craton (Laurentia) from Neoproterozoic through Paleozoic time. Areas on the continent to the east of the line did not receive much sedimentation. Areas to the west were on the continental shelf where a thick sequence of marine sediments accumulated. North of Tremonton, Utah, the highway crossed the Late Cretaceous-age Willard thrust fault, which carried the western Proterozoic and Paleozoic continental margin strata east over the top of the craton's metamorphic basement and thin overlying Cambrian sandstone. I-15 parallels the north-south surface trace of a large syncline above the Willard thrust, formed where the thrust abuts rocks of the underlying craton.

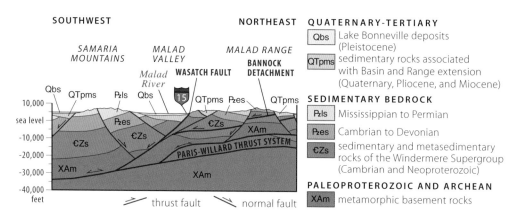

Section across the Malad Valley, showing the Wasatch normal fault, along with deeper thrust faults of the Willard thrust fault system. The Wasatch fault is concave upward and flattens into this basal fault system.

The Precambrian-Cambrian boundary, at 541 million years ago, marks the beginning of complex life forms. Rocks deposited just after this time crop out east of the highway north of milepost 8, south of Malad City. This same important stratigraphic horizon is mapped to the northwest on the top of Elkhorn Peak.

For several miles north of milepost 13, north of Malad City, northbound travelers can see a red quarry low on the mountain northwest of the highway. The quarry is in late Miocene-age volcanic rocks of the Malad volcanic center. Rhyolite is quarried and produced for a variety of uses by Hess Pumice, whose plant is at the south end of Malad City.

I-15 climbs uphill north of Malad City, and travelers poke their heads out of Lake Bonneville at Devil Creek Reservoir, elevation 5,180 feet, east of the highway at milepost 19. The outlet to the Pleistocene lake was at Red Rock Pass, at the north end of Cache Valley, one range east of I-15 (see US 91 road guide).

Oxford Peak, on the skyline east of Devil Creek Reservoir, contains diamictites of the Pocatello Formation that record glaciation about 685 million years ago. Diamictites are marine mudstones that contain large cobbles, some of which dropped from icebergs floating in the sea. Red quartzites of the Brigham Group in the high part of Oxford Peak are latest Neoproterozoic, deposited just before the beginning of the Cambrian Period at 541 million years. Cambrian strata are mainly limestones, exposed east of the highway at Malad Summit.

North of Malad Summit (north of milepost 24), the highway descends into broad Marsh Valley, another basin in the Basin and Range. In Marsh Valley, as with the Malad Valley, the biggest normal fault is on the east side. In southern Marsh Valley, east-dipping normal faults are present on the west side of the valley, making this part of the valley a true graben. Expansive views in Marsh Valley show the Portneuf Range to the northeast and the Bannock Range to the northwest, both containing Neoproterozoic and early Paleozoic rocks, generally dipping east. These rocks and the overlying late Miocene Salt Lake Formation are offset down to the west along the normal fault bounding the east side of Marsh Valley.

Oxford Peak, when viewed looking south from Marsh Valley, resembles a volcano thanks to a low-angle normal fault that cuts the Neoproterozoic quartzite, forming a bench near the summit. In reality, Oxford Peak is the northern end of a north-trending ridge.

Marsh Valley hosted the colossal Bonneville Flood about 17,500 years ago. Marsh Creek follows the flood path, and the highway crosses the path just north of milepost 30, about 5 miles north of Malad Summit, and again just north of milepost 55 at the

Oxford Peak viewed from the north. The mountain looks like a volcano but is in fact a north-trending ridge of sedimentary and metamorphic rock. (42.601, -112.193)

north end of Marsh Valley, just south of Inkom. From there west through Portneuf Gap and Pocatello, boulders up to 10 feet in diameter, mainly of basalt eroded from the Inkom area, were moved by the floodwaters. Much of the highway route in central Marsh Valley is just east of the flood path.

Garden Creek Gap, a low dip in the ridge to the west of Arimo (exit 40), was cut by a superposed stream, a waterway that existed prior to the uplift of the Bannock

Garden Creek Gap, viewed looking west from McCammon with a telephoto lens. The east-flowing Garden Creek cuts through a ridge of west-dipping Scout Mountain Member of the Pocatello Formation. (42.617, -112.199)

This hill of Cambrian limestone immediately southeast of Inkom is an island surrounded by the basalt of Portneuf Valley (dark horizontal rock at center right). This hill was also an island in the Bonneville Flood. The Portneuf Range, composed of east-dipping Neoproterozoic and Cambrian quartzite, forms the background. Marsh Creek is in the foreground. (42.7793, -112.2299)

Range. Ancestral Garden Creek flowed east on volcanic fill from the Hawkins Basin volcanic center. The stream maintained its course as the volcanic fill was removed, even when it cut down into hard quartz sandstone of the steeply east-dipping Scout Mountain Member of the Pocatello Formation of Neoproterozoic age. On the west side of Garden Creek Gap is a west-dipping normal fault that bounds the northeastern side of the Hawkins Basin volcanic center. You can see sandstones of the Scout Mountain Member up close if you drive west on West Arimo Road.

The Portneuf River enters the valley from the east at McCammon (exit 44), but it is confined to the low area to the east of I-15. Likewise, Marsh Creek is confined to the west side of the valley. In the middle is the basalt of Portneuf Valley, which consists of two flows of basalt that flowed north and west about 430,000 years ago. The basalt flows filled what used to be the valley of the Portneuf River and Marsh Creek, so the streams eroded new channels on each side of the basalt. The former low area is now a high area, an example of a process called inverted topography. The streams join at Inkom, where the Bonneville Flood scoured out much of the basalt lava. I-15 cuts past a pointed hill of Cambrian limestone that pokes up through the basalt east of the highway just south of the Portneuf River (between mileposts 55 and 56).

The source of the basalt of Portneuf Valley is in Gem Valley, east of the Portneuf Range. The lava followed the course of the ancestral Portneuf River and flowed as far northwest as Ross Park in Pocatello. The west-flowing ancestral Portneuf River (the modern Bear River) was dammed in northern Gem Valley by a basalt flow about 55,000 years ago, diverting it south into the Bonneville basin. As I-15 drops down off the top of the basalt, it cuts through roadcuts in the flat-lying dark rock on both sides of the highway.

Chaotic columnar jointing from irregular cooling in the basalt of Portneuf Valley at milepost 57 on the south side of Rapid Creek in Inkom. Clasts of this basalt make up most of the boulders found beneath Pocatello and the Michaud Flats to the north, carried there by the Bonneville Flood. (42.7966, -112.2488)

The Neoproterozoic sedimentary rocks in Marsh Valley slope southward and are 10,000 feet below the surface near Downey at the south end of the valley. In the canyon of the Portneuf River from Inkom (exit 57) west through Portneuf Gap, however, the thick stack of Neoproterozoic to Cambrian sedimentary rocks is exposed at the surface. These rocks, first named and defined here, include the Pocatello Formation, Blackrock Canyon Limestone, and Brigham Group of mainly sandstones. The rocks are folded into a shape resembling the letter *V* on its side with the hinge of the V pointed east. Most of the exposed rocks dip to the east and are right-side-up, but local west-dipping overturned sections exist, especially north of Portneuf Gap along the ridgeline, within Blackrock Canyon, and directly north of the Inkom post office. The east-tilted Bear Canyon thrust fault cuts the rocks north of Inkom.

In the Portneuf Gap, the rock layers north of the highway are structurally overturned. They dip gently west, while the rocks south of the highway are right-side up and dip east. These rocks are on a limb of an eastward-overturned Late Cretaceous fold above a thrust fault. The Scout Mountain Member of the Pocatello Formation is exposed both north and south of the highway at Portneuf Gap. Interested people can get access to these rocks from the Blackrock Canyon road (take the Portneuf exit and head east to the first road that goes north under the freeway. Then take the first left to access the canyon just north of the highway). North of the highway, near the top of the ridge, a 666-million-year-old rhyolitic tuff provides a minimum age for the black

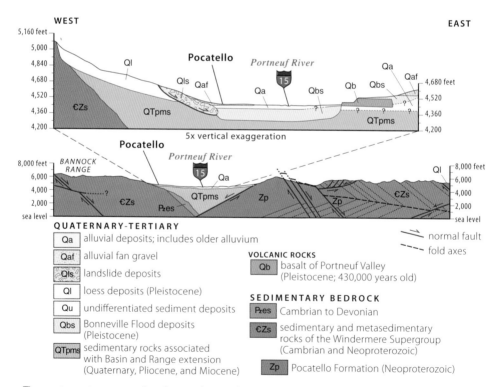

The east-west cross section (bottom) runs through the hills north of Inkom west to Pocatello, passing north of Portneuf Gap. Note the overturned fold in the hills north of Portneuf Gap. In the Pocatello Valley closeup (top), the basalt of Portneuf Valley occupied a valley likely cut by the Bear River (the ancestral Portneuf).

cliffs of diamictite that crop out on the west face of the mountain. South of Portneuf Gap, in the type area of the Scout Mountain Member, two horizons of glacially influenced diamictite are present above a basal basaltic volcanic unit that has produced dates of about 710 million years. These ages place the diamictites of the Scout Mountain Member (about 685 to 665 million years old) within the first (Sturtian) phase of the Cryogenian glaciation. The Sturtian was an early phase of the Neoproterozoic glaciations lumped as Snowball Earth.

On the west side of Portneuf Gap, the Pocatello urban area opens to the north and west near milepost 66. In the distance are volcanic rocks of the Snake River Plain. Though it poured through and partially deepened the gap, the Bonneville Flood did not cut the canyon through the Bannock Range. The 430,000-year-old basalt of Portneuf Valley at Ross Park in southern Pocatello flowed through the gap, too, hundreds of thousands of years before it was scoured by the floodwaters. A large northwest-flowing river, the ancestral Portneuf (Bear) River, cut the gap. The West Bench, prominent from the highway looking west near milepost 70, was incised by this river, perhaps 1 million years ago. Today, the Bear River flows southward into the Great Salt Lake basin, having been diverted near Grace, to the east of Pocatello along US 30, by lava flows about 55,000 years ago.

Light-colored tuff composed of ash from eruptions on the Snake River Plain and red conglomerate and shale beds of the late Miocene Starlight Formation crop out along the east side of the highway as it passes uphill from Idaho State University.

View to east from Sterling Justice mountain bike trail in hills south of Pocatello. Portneuf Gap is in the left center, with the Portneuf Range beyond and Pocatello out of view to the left of photo. Mountain maple color is at its peak on this early October afternoon. (42.759, -112.394) —Photograph by Howard Burnett

I-84
I-86 Junction—Utah border
54 miles

I-84 abandons the Snake River Plain near the I-86 junction, making a southeasterly beeline toward Utah and into classic Basin and Range country. The flat topography and young volcanoes of the Snake River Plain give way to mountains mainly composed of late Paleozoic sedimentary rocks and valleys covered in alluvium shed from nearby highlands.

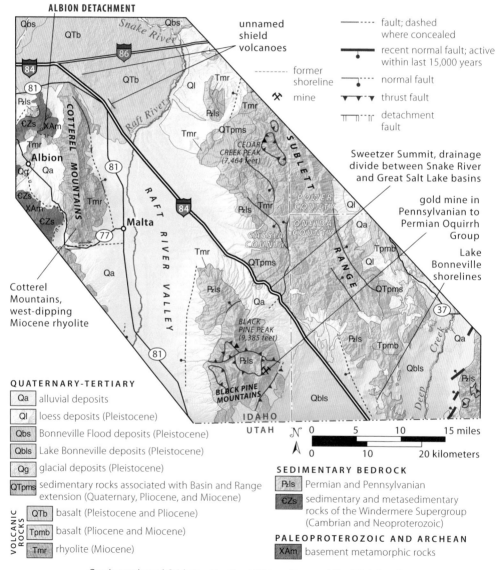

Geology along I-84 between the I-86 junction and the Utah border.

From the interstate junction, I-84 heads southwest across the Raft River Valley, a classic north-trending valley formed by east-west extension. A few shield volcanoes that erupted along the southern margin of the Snake River Plain during the Pleistocene Epoch lie just east of the freeway near mileposts 225 and 229.

Around milepost 232, look west to see the prominent ridgeline of the Cotterel Mountains, a north-south range of uplifted Miocene rhyolite lava and ignimbrite. There rocks were produced by the Yellowstone hot spot as this part of Idaho passed over it. West of the Cotterel Mountains are the airy peaks of the Albion Mountains, where the highest summits exceed 10,000 feet and alpine glaciers resided during the last ice age. The entire sequence of rhyolite in the Cotterel Mountains is tilted westward along an east-dipping detachment fault west of the range. A steeper east-dipping normal fault lies along the east side of the Cotterel Mountains, forming the pronounced escarpment that faces the freeway. During the past 8 million years, extension along the detachment fault transported these rocks eastward and down from off the Albion Mountains, while the older Precambrian rocks beneath the fault rose.

The interstate gradually climbs as it heads southeast, slowly exiting the Raft River Valley. The highlands to the east are the Sublett Range with the Black Pine Mountains to the south. Both ranges are mainly composed of the Oquirrh Group, a thick sedimentary sequence that filled a large basin (the Oquirrh–Wood River basin) stretching from central Idaho to central Utah from Pennsylvanian to Permian time. Several outcrops of gray limestone from these periods occur along the east side of the freeway between mileposts 254 and 256. These older sedimentary rocks are covered by white to light-gray Miocene tuff from Snake River Plain volcanoes. The tuff is exposed at milepost 257.

Fragmented Paleozoic limestone of the Oquirrh Group in an outcrop east of I-84 near milepost 255. (42.2213, -113.0741)

East of milepost 257 lies a small, interesting roadcut on Juniper Road, the frontage road accessed from either exit 254 or 263. The roadcut reveals an orange paleosol, or ancient soil horizon, overlain by light-gray layers of ash and capped by black vitrophyre, a glassy welded tuff. These volcanic deposits were erupted from the Twin Falls volcanic field between 10 and 6 million years ago. The initial eruption of ash fell from the sky and blanketed the surface to form the light-gray layer, which was then subsequently covered by a hot pyroclastic flow, which cooled quickly to form the black vitrophyre.

Near milepost 257, the orange paleosol (bottom left) is covered by light-gray, layered ash and capped by black vitrophyre from an explosive eruption in the nearby Snake River Plain. Rock hammer for scale. (42.2120, -113.0468)

Between mileposts 257 and 258, the freeway crests Sweetzer Summit (elevation 5,530 feet), the divide between the Snake River drainage basin and the Great Salt Lake basin. North of this point, surface water heads to the Snake River, then the Columbia, before ultimately spilling into the Pacific Ocean. South of the summit, water drains into the Great Salt Lake in Utah, a large, enclosed basin with no outlet. The dry climate of the region evaporates much of the lake's water, precipitating salt and creating a water body ten to twenty times saltier than the ocean. Southbound travelers have expansive views of the Great Salt Lake basin near milepost 263.

Prominent terraces and pits at the south end of Black Pine Mountains to the west mark the location of a gold mine within the sandstone and limestone of the Pennsylvanian to Permian Oquirrh Group. Similar to other modern gold mines found throughout northern Nevada, gold at Black Pine is not visible to the naked eye. It exists as microscopic particles disseminated within the sedimentary rock, and its presence can only be confirmed using chemical analyses. Geologists think the gold was carried and deposited by hot groundwater circulating through the rock some 40 million years ago in the Eocene Epoch.

Near milepost 264, the freeway crosses 5,200 feet, a noteworthy elevation marking the highest shoreline of Lake Bonneville. Nearly as big as Lake Michigan, Lake Bonneville existed periodically during the cooler climate cycles of the Pleistocene ice ages. About 17,500 years ago, the lake was at its highest level, lapping onto hillsides in southern Idaho. Wind-driven waves smashed onto the slopes, eroding them and creating shorelines or benches still visible today along the lake's ancient margin in southernmost Idaho and northern Utah. Look for more of these prominent wave-formed benches from Lake Bonneville if your journey continues southward into Utah.

US 26
Idaho Falls—Swan Valley—Wyoming Border
69 miles

From Idaho Falls, US 26 runs east from the Snake River Plain and follows the Snake River, also called the South Fork of the Snake River here, through the southeast-trending Swan Valley to Alpine, Wyoming. The Swan Valley is a true graben because there are inward-dipping normal faults on both the southwest and northeast sides. The Grand Valley fault on the northeast side is the larger one, and it flattens downward and merges with the Absaroka thrust fault. This Cretaceous-age thrust of the Sevier fold-and-thrust belt crops out in the Snake River Range. South of Alpine, the Grand Valley fault is called the Star Valley fault. Rocks in mountains on both sides of the Swan Valley are thrust-faulted Paleozoic and Mesozoic sedimentary rocks. The Absaroka thrust places older rocks (Paleozoic limestones) over younger rocks (red-colored Mesozoic mudstones and sandstones).

Just east of milepost 352 and near the intersection with the road that heads north to Kelly Canyon Ski Resort and Heise Hot Springs, look for the cliffs north of the highway on the north side of the Snake River. They are made of late Miocene rhyolite lavas and ignimbrites erupted within the Heise volcanic field, associated with the Yellowstone hot spot, between 6.5 and 4.5 million years ago. The rocks in the Heise cliffs were erupted within a caldera, a collapsed area where the rhyolite tuffs and lavas are thicker than those that had enough eruptive energy to escape the caldera rim. Side roads to Ririe Reservoir south of US 26 (east of milepost 338 and past 342) allow access to excellent outcrops of these volcanic rocks along Meadow Creek Road east of the reservoir.

From the Clark Hill (Antelope Valley) rest area on the north side of US 26 just east of milepost 357 is a beautiful view into the Snake River canyon. The Pleistocene basalt volcanic rocks that line the canyon erupted into the already formed Swan Valley graben. During the past 2 million years, lava and water have feuded over the floor of the Swan Valley. Repeated eruptions of lava from vents within the graben have rerouted the river and its tributaries at times. A large, striking roadcut near milepost 369 records one of the most recent episodes. Dark-gray basalt lies atop orange hyaloclastite, a layered accumulation of broken and altered basalt fragments that formed where lava and water mixed. The hyaloclastite is part of a lava dam that formed about 900,000 years ago. A local volcanic vent produced basaltic lava that poured into the river, quickly cooling and quenching the lava and shattering it into mostly small, angular shards. As the lava dammed the river, it formed a reservoir upstream. Continued input of lava into the reservoir formed the thick sequence of hyaloclastite here and farther east where the highway drops into the Snake River floodplain.

The highway crosses to the north side of the Snake River east of milepost 373. Roadcuts on the north side of the road from here to milepost 375 expose basalt with some pillow lava. Look for rounded, bulbous shapes with glassy rinds, formed when

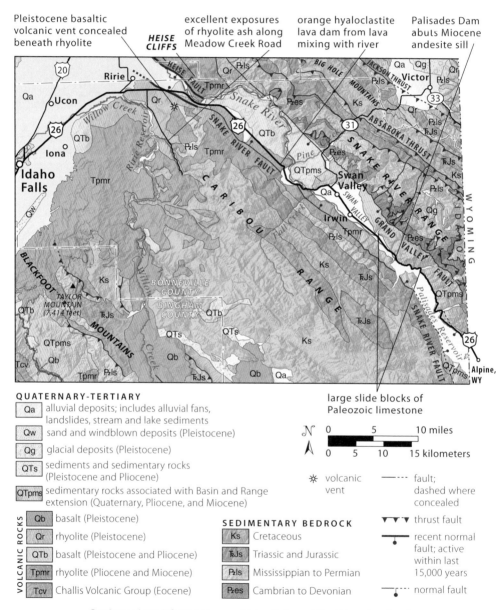

Geology along US 26 between Idaho Falls and the Wyoming border.

View northwest from the Clark Hill rest area down to the Snake River. Pleistocene basalt lava forms benches that lap against rounded buttresses of reddish late Miocene rhyolite. The Heiss cliffs rise at the upper right and the Beaverhead Mountains lie in the distance in the upper left. (43.5882, -111.6239)

Roadcut at milepost 369 of orange hyaloclastite, a shattered accumulation of broken volcanic material formed where lava interacted with the Snake River. The overlying gray rock is typical basalt that erupted on land. (43.4886, -111.4490)

basalt entered a standing body of water and cooled quickly as the lava was quenched by the water. Near the junction with ID 31, a prominent ledge of Huckleberry Ridge Tuff that erupted from the Island Park caldera is exposed near the top of the hill north of the town of Swan Valley. Miocene conglomerate below the tuff is exposed at the west end of the hill.

South of Irwin the valley narrows. The prominent bench with a gray cliff on the north side of the valley south of Irwin is Huckleberry Ridge Tuff. Craggy cliffs of Cambrian limestone near milepost 384 and the mouth of Palisades Creek are part of a slide block, a large mass of rock that slid off the Snake River Range to the northeast, likely in one large catastrophic event. Near milepost 387, US 26 begins its ascent to the top of Palisades dam. The reservoir is one of the largest on the Snake River. The prominent conical hill, Calamity Point, at the west side of the dam is part of a Miocene andesite sill that is also exposed along the large pullout on the east side of the dam near milepost 388. The resistant rocks on both side of the canyon made solid footings for the dam.

Resistant ledge of the 2.1-million-year-old Huckleberry Ridge Tuff lines the hillside north of Swan Valley at the intersection of US 26 and ID 31. (43.4539, -111.3363)

Palisades Reservoir looking southeast from between mileposts 388 and 389. The rocks along the north side of the road are part of a Miocene andesite sill. (43.3260, -111.1847)

US 26 follows the north side of the Palisades Reservoir. Late Paleozoic sedimentary rocks form the Snake River Range to the north, part of the Idaho fold-and-thrust belt. Roadcuts and outcrops of gray Mississippian limestone near milepost 391, where the highway crosses an arm of the reservoir, are part of another large slide block. Another slide block is exposed near milepost 398 in a big chaotic roadcut containing white ash, alluvial gravels of the Salt Lake Formation, and Paleozoic sedimentary rocks.

Hidden beneath the full reservoir, but exposed at low water, are beds of the late Miocene Teewinot Formation, which contain eruptive products from the Heise volcanic field. Some rhyolite ignimbrites have been traced to vents within the volcanic field. The Teewinot Formation also contains vertebrate fossils of some of the large mammals that lived here in late Miocene time.

US 30
McCammon—Soda Springs—Wyoming Border
82 miles

US 30 traverses Idaho's fold-and-thrust belt as it crosses five Basin and Range valleys in its southeastward route from McCammon east to Wyoming. It follows the Portneuf River to Lava Hot Springs, then follows the Bear River from Soda Springs to the Wyoming border. Pioneers on the Oregon Trail and its cutoffs also followed portions of these rivers through southeast Idaho.

East of exit 47 from I-15 and north of McCammon, US 30 crosses the Portneuf River, flowing north here toward its confluence with the Snake River north of Pocatello. The first toll bridge in eastern Idaho, built in 1870, was located just north of where US 30 crosses the river. At McCammon, US 30 and the river skirt along the east side of basalt lava that erupted about 430,000 years ago. Two lava flows, known as the basalt of

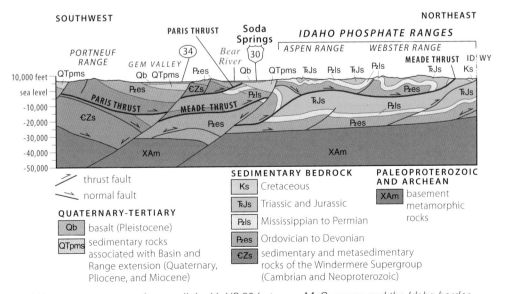

West-to-east cross section parallel with US 30 between McCammon and the Idaho border.

130 BASIN AND RANGE

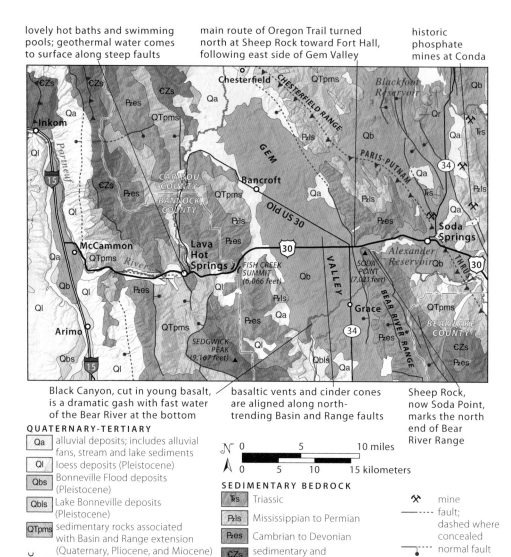

Geology along US 30 between McCammon and Soda Springs.

the Portneuf Valley, were erupted from vents in northern Gem Valley between Bancroft and Chesterfield. The lava flowed westward down the Portneuf River to Pocatello. It can be seen next to the river between McCammon and Lava Hot Springs.

Between McCammon and Lava Hot Springs, the highway passes Cambrian and Ordovician sedimentary rocks of the Paris thrust plate, the body of rock that moved east along and above the Paris thrust fault.

Lava Hot Springs is located at the junction of two Neogene-age faults, one trending north and one trending west. The almost mineral-free water comes up along the

Quartzite of the Ordovician Swan Peak Formation is broken into blocks of breccia along a fault at the brow of the hill immediately north of the Lava Hot Springs pools, near milepost 371. If you stop here, park at the old historical marker on the south side of the road and be careful. (42.6202, -112.0075)

east-west fault at a scorching 110 degrees Fahrenheit. The extensive fault breccia of white quartzite fragments that allows the hot water to be transmitted to the surface along the fault can be seen near milepost 371. The hottest commercial pools are uncomfortable for soaks on hot summer days but are quite pleasant and soothing in the dead of winter.

At Fish Creek Summit east of Lava Hot Springs (milepost 376), US 30 passes through a big roadcut in whitish quartzite of the Ordovician Swan Peak Formation. The view east across the Gem Valley toward Sheep Rock (Soda Point) at the north end of the Bear River Range is wonderful. Gem Valley dropped down along a west-dipping normal fault on the west side of the Bear River Range. At Sheep Rock, the Bear River, which has flowed northwest from the Uinta Mountains on the Utah-Wyoming border, turns south and flows toward the Great Salt Lake. In 1849, Sheep Rock was the parting of the ways between pioneers who went north on the Oregon Trail and those who continued west across Gem Valley on the Hudspeth Cutoff.

Pleistocene basalt lava from the Blackfoot lava field north of Chesterfield cascaded south into Gem Valley some 55,000 years ago and dammed the Bear River's former westward path. Prior to the eruption, the river likely flowed to Pocatello and carved Portneuf Gap. Some of the basalt erupted from vents aligned along Basin and Range faults. Downstream from Grace, the river cut Black Canyon into the dark basalt. Despite its young age, the basalt is not the most recent geologic feature in the Gem Valley. Just east of the intersection with ID 34 that heads south to Grace, US 30 crosses an active west-dipping normal fault that offsets the basalt flow by at least 20 feet. A scarp of uplifted black basalt shows the trace of the fault across the agricultural fields.

As the highway passes Sheep Rock at the north end of the Bear River Range, Alexander Reservoir opens up on the south. The reservoir, which stores water for irrigation and power generation, inundated several pressurized and carbonated springs and hot pools that were noted by travelers on the Oregon Trail.

The Soda Springs Geyser is located in downtown Soda Springs, two blocks north of the highway. The geyser was capped in 1937 and is allowed to erupt every hour when the wind is calm so that nearby buildings are not coated with the mineral laden water. The geyser is surrounded by travertine deposits, calcium carbonate precipitated from the water. Dissolved carbon dioxide gas in the water of the confined aquifer provides the eruptive force for the geyser.

The Paris-Putnam thrust fault is crossed east of Soda Springs near milepost 407. From there eastward, the highway is in the Meade thrust plate, host to the black Permian phosphate rock mined today at Soda Springs and historically at Georgetown and Montpelier. The Paris thrust fault is called the Putnam thrust fault north of Soda Springs.

View east toward Sheep Rock from the northern end of ID 34, with a normal fault scarp of about 6 feet cutting the 55,000-year-old basalt in the foreground. (42.6450, -111.7301)

The geyser in Soda Springs erupts water rich in carbonate every hour on the hour, except when the wind is too strong, and the automatic switch is turned off. Repeated eruptions formed the red-brown travertine terraces in the foreground. (42.6570, -111.6049)

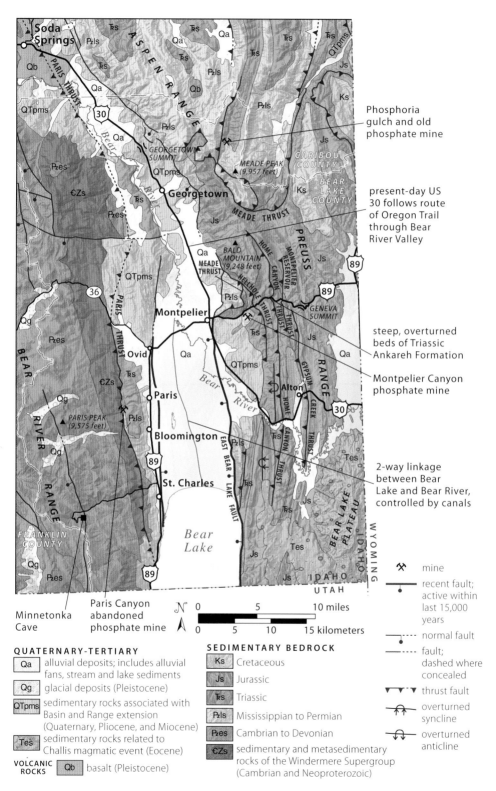

Geology along US 30 between Soda Springs and the Wyoming border.

Southeast of Soda Springs, US 30 parallels the active normal fault on the southwest margin of the Aspen Range. At Georgetown Summit (just south of milepost 419), the highway crosses east-tilted tuffaceous rocks of the Salt Lake Formation of late Miocene age. Much of the volcanic ash in these rocks was derived from rhyolite eruptions on the Snake River Plain. Look for large sloping roadcuts in this loosely consolidated, whitish rock.

In the Preuss Range east of Georgetown, folded rocks above the Meade thrust fault include the late Permian Phosphoria Formation, which contains two 20-foot intervals of minable phosphatic shale. An old railroad served a now-abandoned mine in the canyon north of Georgetown. Another abandoned mine is located east of Montpelier, south of US 89.

Between Montpelier and the Wyoming border, US 30 follows close to the same route that the Oregon Trail followed, as did the Overland Route of the Union Pacific Railroad. Montpelier hosts an Oregon Trail museum (near the junction with US 89) that is well worth a visit.

South of Montpelier, the highway drops onto the flat floodplain of the meandering Bear River at milepost 443. At milepost 445, the road swings east, exposing a roadcut of grayish-brown limestone and siltstone of the Triassic Thaynes Formation. From here to the Wyoming border, numerous folds and thrust faults repeat tilted Triassic and Jurassic sedimentary rocks, which were deposited in shallow seaways and adjacent mudflats.

The Home Canyon thrust cuts across the highway at milepost 446. Red and orange sandstone and siltstone of the Triassic Ankareh Formation is exposed in a large roadcut between mileposts 446 and 447. The salmon-colored Nugget Sandstone of Jurassic age is revealed in a roadcut at milepost 447, where US 30 heads northeast

The salmon-colored sandstone of the Jurassic Nugget Sandstone crops out just east of the Alton Road turnoff, halfway between mileposts 449 and 450. (42.2286, -111.1444)

and away from the Bear River, and also halfway between mileposts 449 and 450. The open pore spaces in this sandstone creates high permeability, making it a petroleum reservoir rock in several oil fields in Wyoming and Utah.

Complexly folded and faulted Jurassic Twin Creek Limestone borders the highway between mileposts 450 and 452. The Wyoming border is reached between mileposts 455 and 456. The Sublette Mountains are prominent to the east, with a normal fault on their west side. Uplifted Permian strata on the east side of the fault host an abandoned and reclaimed phosphate mine just east of the Wyoming line. Within the range, accessed from Raymond Canyon to the north, vertical Permian and overlying Mesozoic rocks are exposed.

US 89
Utah Border—Montpelier—Wyoming Border
44 miles
See map on page 133.

US 89 cuts through the southeast corner of Idaho, providing a link between Logan, Utah, and Wyoming's Salt River valley. The first leg of the highway's journey through Idaho follows the west edge of Bear Lake through a region tied to the history of Latter-Day Saints settlement in Idaho starting in the 1860s. The exquisite Paris Tabernacle was built between 1884 and 1889 using red Nugget Sandstone of Jurassic age. The quarry that produced the stone was 24 miles east of Paris in Indian Creek canyon, east of Bear Lake. The building stone was dragged across the frozen lake by draft horses.

Bear Lake occupies a full graben, with the major normal fault (the Bear Lake fault) on the east side of the lake. The fault has been active since middle Miocene time. A perennial lake has existed here for at least 10 million years, accumulating sediment that is carried to the lake by streams. The water in Bear Lake is only 15 feet deep on the western side, where people live, and 210 feet deep at its deepest, on the southeastern side. Uplifted on the east side of the fault is the Bear Lake Plateau, where sedimentary rocks of the Wasatch Formation are exposed. These were deposited in Eocene time after the Cretaceous folding and thrusting had ended and erosion filled basins with sediment.

View west from St. Charles across Bear Lake toward the Bear Lake Plateau and North Eden Canyon just across the border in Utah. (42.0064, -111.4092)

Bear Lake is known for its chalky, vivid-blue color, which is due to dissolved carbonate derived from the dissolution of limestone bedrock of the Bear River Range to the west. The Paris thrust fault, the western major thrust fault of Idaho's fold-and-thrust belt, is located on the eastern side of the Bear River Range, west of US 89. In late Cretaceous time, movement on the Paris thrust fault shoved Neoproterozoic and early Paleozoic rocks of the Paris thrust plate eastward over the top of late Paleozoic and Mesozoic rocks of the Meade thrust plate. Up St. Charles Canyon, Minnetonka Cave is located in Mississippian limestone of the Paris plate.

Oil exploration occurred in the foothills of the Bear River Range, searching for folds that trap oil in the late Paleozoic and Mesozoic rocks below the Paris thrust fault. Paris Canyon Road, on the south edge of Paris, heads west up Paris Canyon, where underground phosphate mines are located in the Phosphoria Formation in the Meade plate.

At Ovid, US 89 turns east, heads across the Bear Lake graben, and crosses two canals, one carrying water north and one carrying water south. The canals allow the regulation of the level of Bear Lake by controlling the inflow of water from the Bear River. The water rights are owned by power companies and used for electric generation. At the present time, the Bear River does not flow naturally into Bear Lake, though it surely did when Bear Lake was larger and extended farther north during the wetter climate of the Pleistocene ice ages.

The highway jogs north in Montpelier and then turns east again, just south of the National Oregon/California Trail Center. Eastbound travelers are greeted with a one-two fault punch. The highway crosses the East Bear Lake fault, an active Basin and Range normal fault responsible for the uplifted Preuss Range, near the intersection with North 2nd Street, between mileposts 26 and 27. About 0.15 mile farther east, the highway crosses the trace of the inactive Meade thrust, one of the major thrust faults that shoved huge sections of rock eastward during the Sevier orogeny. Craggy outcrops of Mississippian limestone lie on the hillsides immediately north of the canyon mouth.

MINNETONKA CAVE

Home to a plethora of diverse cave formations as well as five bat species, Minnetonka Cave is well worth a visit for those traveling through the Bear Lake area. Take St. Charles Canyon Road (Minnetonka Cave Road) from US 89 about 1 mile north of St. Charles. The cave parking area is reached about 10 miles up the canyon in the Bear River Range. A concessionaire runs summer tours of the cave. Despite the temperature outside, a jacket is advised in the cool cave.

Like most solution caverns, Minnetonka Cave formed when water percolating downward from the surface moved through fractures and along bedding planes in the limestone, gradually dissolving the rock and creating larger cavities. As the mountains rose higher and streams cut deeper, the water table lowered, leaving the cave above the water table where it was finally discovered in the early 1900s by a hunter searching for a fallen grouse. The cave lies in the Lodgepole Limestone, which formed in a warm tropical sea when southeastern Idaho straddled the equator in Mississippian time. Invertebrate fossils such as brachiopods, crinoids, and coral abound in the limestone and can be readily observed at many places within the cave. The path through the cave follows bedding planes that are tilted 20 to 40 degrees to the west, resulting in several sets of stairway sections.

Jagged cliffs of Mississippian limestone rise to the east of Montpelier in the upthrown block of the East Bear Lake fault at the mouth of Montpelier Canyon. (42.3230, -111.2892)

The abandoned Montpelier open-pit phosphate mine in Permian rocks lies in a fault slice of the Meade thrust on the south side of US 89 at milepost 29. Several folds and imbricate faults, parallel to and below the Meade thrust, crop out east of the abandoned phosphate mine. A roadcut of west-dipping beige sandstone of the Pennsylvanian-age Wells Formation lies across from the Caribou National Forest sign, between mileposts 29 and 30.

At Home Canyon, between mileposts 30 and 31, the highway crosses the Hellhole thrust fault, a secondary thrust fault within the Absaroka thrust sheet that lies beneath the Meade thrust. Big roadcuts just east of Home Canyon reveal gray limestone and siltstone of the Triassic Thaynes Formation. Another large thrust fault, the Home Canyon thrust, is crossed just east of milepost 32. Steep, overturned beds of Triassic Ankareh Formation are exposed in a dramatic roadcut about half a mile to the east. Still another thrust fault, the Montpelier Reservoir thrust, is crossed just east of milepost 33 at the turnoff to Montpelier Reservoir. A 100-foot-tall roadcut on the north side of the highway just east of this junction exposes salmon-colored Jurassic Nugget Sandstone in the lower plate of the thrust. From this impressive roadcut east to Geneva Summit and the tiny town of Geneva, the highway passes through folded and faulted sections of Jurassic Twin Creek Limestone and Jurassic Preuss Formation. Near the summit at milepost 35 is a cliff of tightly folded limestone overlain by a minor low-angle thrust fault with breccia immediately above it. The rocks are mapped as the Watton Canyon Member of the Jurassic Twin Creek Limestone.

East of Geneva, US 89 turns north. The range to the east is the Sublette Mountains, another fault-bounded range caused by Basin and Range extension. Like the Preuss Range, it also contains north-trending Paleozoic and Mesozoic rocks, folds, and thrust faults. The Idaho-Wyoming border lies at the mouth of Salt Canyon where ID 89 turns east.

Near-vertical beds of Triassic Ankareh Formation between mileposts 32 and 33 on US 89. (42.3403, -111.1870)

US 91
Utah Border—Preston—I-15 Junction
42 miles

Heading north through the Cache Valley from Logan, Utah, US 91 follows the Union Pacific Railroad and was once a major north-south artery. The highway's importance waned with the construction of I-15, but it is now making a comeback with urban growth. Cache Valley, which is mostly in Utah, extends across the Idaho border and is part of the Basin and Range Province, with a west-dipping normal fault on the east side of the valley. Some geologists interpret Cache Valley to lie above the mid-Miocene Bannock detachment fault, the basal extensional fault of this area that moved 12 to 10 million years ago.

In Idaho, US 91 passes through historic Mormon settlements of Franklin and Preston, located hundreds of feet below what was once the level of Lake Bonneville. Irrigation was provided by canals readily constructed within lake sediments below the 4,800-foot Lake Provo shoreline elevation. This shoreline, as well as the Bonneville shoreline, is easy to see on Little Mountain, northwest of Franklin (milepost 2). Little Mountain, the tip of a bedrock exposure of Cambrian limestone, is nearly buried by young sediments of Cache Valley. Across US 91 from Little Mountain (between mileposts 4 and 5) is another ridge with shorelines and three large gravel quarries in whitish Lake Bonneville shoreline sediment.

The lake level dropped 400 feet, from the Bonneville shoreline at 5,200 feet to the Provo shoreline at 4,800 feet, in a few months about 17,500 years ago when the Bonneville Flood occurred. After holding steady at the Provo shoreline for some time after the flood, the lake gradually fell as the climate warmed and became dryer.

BASIN AND RANGE 139

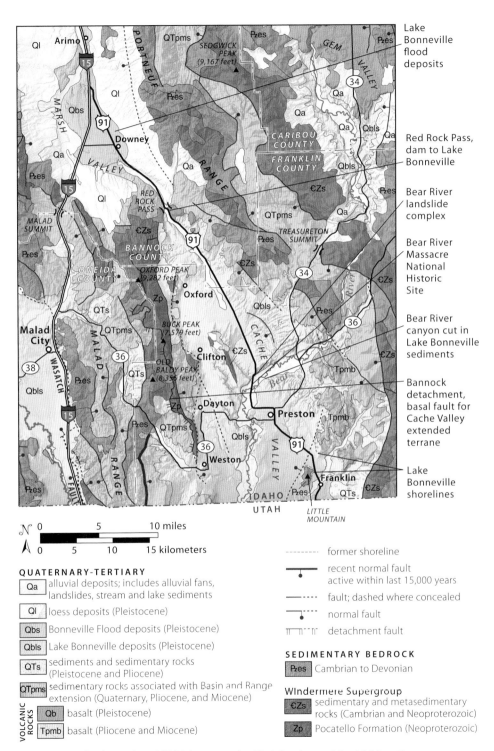

Geology along US 91 between the Utah border and the I-15 junction.

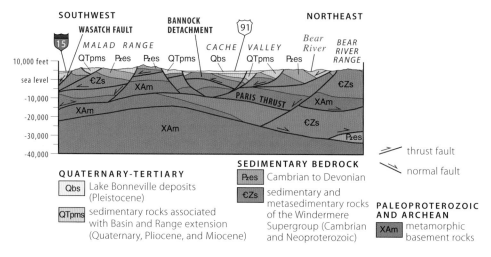

Cross section of the Malad Range and Cache Valley. Cache Valley lies above the Bannock detachment, a low-angle extensional fault that the steeper Basin and Range normal faults connect to in the subsurface.

Little Mountain viewed from the north a few miles south of Preston. The Lake Bonneville shoreline is near the top of the butte, above the "1860," which is the year of the founding of Franklin, Idaho's first city. The Lake Provo shoreline is 400 feet lower. (42.0224, -111.8058)

Preston, formerly called Worm Creek, was thankfully renamed after a Mormon bishop. Worm Creek itself flows about 4 miles south of town. For information east and west of Preston, see the road guide for ID 36.

North of Preston, the Bear River incised its 300-foot-deep and 1-mile-wide broad canyon into Lake Bonneville sediments in the last 17,500 years. These unconsolidated silts and sands are actively being eroded in the Bear River landslide complex northeast of the highway in the sandy cliff on the north side of the Bear River.

The Bear River Massacre National Historic Site is just east of the highway north of the Bear River crossing. There are three historical signs, each with a somewhat different interpretation of the "battle," which in later signs is called a "massacre" of the Indians. The actual Bear River Massacre site is west of the highway near milepost 13, at the mouth of Battle Creek. An extensive historic overlook of the Bear River Massacre site is located east of the highway at the top of the grade at milepost 14. The grade climbs through low cuts in Lake Bonneville's light-colored sediments.

An east-dipping normal fault is responsible for the steep east face of Oxford Peak on the west side of northern Cache Valley from Clifton north past Oxford. Uplifted rocks in the north-trending ridge are Cambrian and underlying Neoproterozoic sedimentary rocks, including glacial diamictites of the Pocatello Formation.

Red Rock Pass at milepost 30 is the site of a natural dam that held Lake Bonneville before it failed catastrophically 17,500 years ago. East of the parking area at the pass is a small Cambrian limestone butte with concrete steps to the top. This butte and the entire area of the pass was covered with Pleistocene alluvial fan gravel that formed the northern shore of Lake Bonneville. The rising lake put enormous pressure on the soft, unconsolidated alluvial fan. At some point, perhaps triggered by a lake wave produced by an earthquake on the Wasatch fault, the lake overflowed and then quickly eroded through the alluvium and exposed the limestone butte in the catastrophic Bonneville Flood. The level of the massive lake dropped about 400 feet in just a few months.

The time of the flood has been estimated with increasing precision with new geochronologic dating techniques in the past fifty years. Before 1980, the age of the flood was estimated at 30,000 years ago. Between then and about 2000, the timing of the breakout flood was placed at 14,500 years ago. The most recent estimate of 17,500 calendar years ago is based on cosmogenic nuclide ages that determine how long basalt boulders moved by the flood have been exposed to cosmic rays.

Marsh Valley, on the north side of Red Rock Pass, is a north-trending basin of the Basin and Range Province, its southern part filled with more than 10,000 feet of basin sediment. The Neoproterozoic and Cambrian rocks in the Portneuf Range on the east side of the valley are offset upward along a normal fault that is mostly covered by alluvium. Late Miocene and younger tuffaceous volcanic rocks and Pleistocene lakebeds are present in the basin fill. The Bonneville Flood swept northward down the valley

Red Rock Pass viewed looking north at milepost 30. The rock is Cambrian limestone. The highest part of the large butte in the background stood just above the Lake Bonneville shoreline. Where the top of the steps is today was 200 feet below the lake level. (42.3535, -112.0487)

and left flood features, including scoured basalt, along the course of Marsh Creek and north of Inkom. After the flood, a river from the remnant of Lake Bonneville flowed northward for a few thousand years (until about 14,000 years ago) while the lake remained at the Lake Provo level of 4,800 feet elevation.

When viewed to the south from Downey, Oxford Peak looks like a volcano, but it is actually a long north-trending ridge controlled by a normal fault on its east side.

US 93
Arco—Challis
80 miles

Between Arco and Challis, US 93 mostly follows the Lost River Valley, a big basin in Idaho's Basin and Range. The Lost River Range to the east has been uplifted along the Lost River fault, which lies at the base of the mountains. The central portion, north of Mackay, moved during the October 1983 Borah Peak earthquake. As you drive north through this wide-open basin, watch for signs to the many public fishing access points, picnic areas, and other sites that get you closer to the Big Lost River and the mountains. The rocks in the mountains are more closely related to those in the Central Mountains. See the introduction to that chapter for a discussion of the Challis magmatic event and the road guide for ID 75: Timmerman Junction—Stanley for a discussion of the Pioneer Mountains.

The cliffs above Arco to the east are Mississippian limestone on which high school graduating class years are painted. Horn corals can be collected from these rocks. Similar late Paleozoic folded rocks of the southern White Knob Mountains come close to the west side of the highway north of milepost 90. Antelope Creek Road, on the west

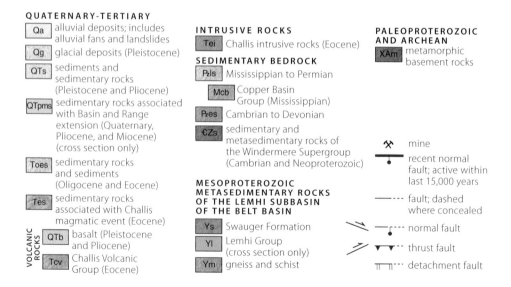

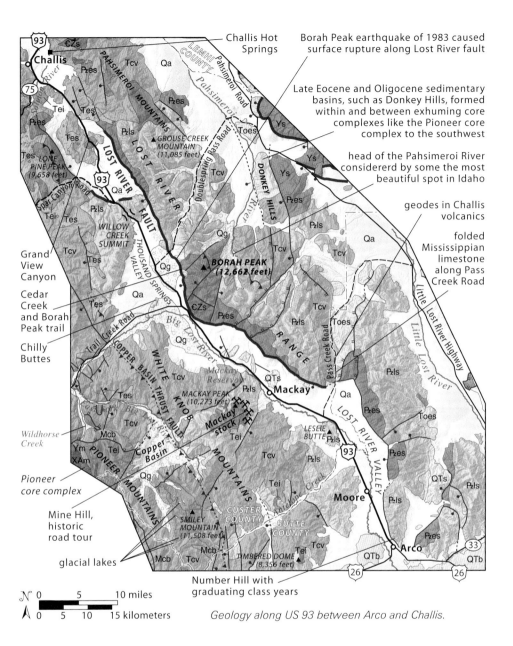

Geology along US 93 between Arco and Challis.

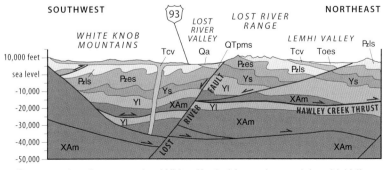

Cross section between the White Knob Mountains and Lemhi Valley, showing how the Lost River Range was uplifted along the Lost River fault.

Concentric anticline in Mississippian limestones on the north end of Leslie Butte, viewed from milepost 100. (43.8595, -113.4499)

side of the highway north of milepost 93, provides access to the high southern Pioneer Mountains, which lie west of the White Knobs. Between mileposts 99 and 100, folded Mississippian rock layers on Leslie Butte lie close to the road on the west. To the east, the southern Lost River Range contains Ordovician through Mississippian strata deposited on the western margin of Laurentia, the ancient North American continent.

Pass Creek Road, a gravel road on the northeast between mileposts 101 and 102, crosses through the Lost River Range and connects to the Little Lost River Valley. The low pass in the otherwise imposing range was eroded in Eocene Challis Volcanic Group rocks. A large landslide can be seen on the mountain slopes north of Pass Creek. Near the entrance to the canyon, the road crosses the Lost River fault and passes from the valley into folded Mississippian limestone on the uplifted side. About 1 mile farther up the road, you can look for geodes on the right in the Challis Volcanic Group. Get gas before heading up this road if you plan on circling around through the Little Lost River Valley.

Mackay was established in 1890 to support mining camps in the Mackay mining district. In the White Knob Mountains to the west, Mississippian limestone was intruded by Eocene plutons, including the Mackay stock. Hydrothermal fluids that accompanied the magma produced copper, silver, and gold mineralization along the stock's contact with the limestone. South of Mackay, look for the open pit visible on the range to the west. The pit was last actively mined in the 1960s, but exploration continues. A historic driving tour to the Mine Hill area can be reached by heading west on Main Street, the main intersection in Mackay, just south of milepost 109. Old buildings, mine equipment, and a tramway are among the ruins at this historic site.

North of Mackay near milepost 111, more knobs of Paleozoic rock poke up west of the highway, and at milepost 112, a large roadcut in Mississippian White Knob Limestone lies east of the highway. The mountains to the east of Mackay Reservoir are spectacular and steep, uplifted along the active Lost River normal fault at their base. Alluvial fans, deposited by streams flowing from the canyons, slope down toward the highway. The fans east of the road are notably steep because they formed along the rising Lost River Range. The junction with Trail Creek Road lies on the Elkhorn alluvial fan.

North of the Trail Creek Road junction, US 93 descends off the Elkhorn alluvial fan and skirts Thousand Springs Valley. The springs provide the water for Chilly Slough, a great bird watching area. Groundwater derived from the uplands to the west emerges as springs above the west-dipping Lost River fault. Chilly Buttes, rising above the flat expanse to the west, are made of folded Mississippian White Knob Limestone.

North of milepost 129, Birch Springs Road on the northeast provides access to the trail up Borah Peak, the highest point in Idaho at 12,662 feet. This steep climb can be completed in a day if you are in excellent shape and comfortable on steep, exposed terrain, including a section of "trail" called Chicken Out Ridge. During the Pleistocene ice ages, glaciers existed in this part of the Lost River Range, but today the glaciers are almost gone. Borah Glacier, on the north face of Borah Peak, may no longer flow, so it is technically not a glacier anymore.

North of milepost 131, Doublespring Pass Road (Forest Road 279) provides access to an interpretive site about the magnitude 6.9 Borah Peak earthquake of October 28, 1983. The quake produced strong shaking throughout central Idaho and was also felt throughout much of the western United States. The quake's rural location precluded any large-scale devastation, but the nearby towns of Mackay and Challis bore the brunt of the quake's wrath. Dozens of buildings and homes were damaged, and two children were killed in Challis as they walked to school that morning.

The quake also left its mark on the landscape of the Lost River Valley. At the interpretive site, a down-dropped graben up to 12 feet deep formed along the fault. The fault scarp can be seen as a sharp line running along the base of the mountains for several miles north and south of the site. Earthquakes are nothing new to this region because they occur every time there is movement along the Basin and Range faults.

TRAIL CREEK ROAD AND COPPER BASIN GROUP

Trail Creek Road to the west, which takes off north of milepost 124, follows the Big Lost River toward its headwaters in the Pioneer Mountains and then goes over the mountains to Sun Valley. (See the Trail Creek Road sidebar in the Central Mountains chapter for the west end of the road.) Heading west on this road from US 93 provides access to several wild places with complex geology. About 16 miles east of US 93, the East Fork of the Big Lost River enters from the east. Metamorphic rocks of the Pioneer Mountains core complex are exposed along Wildhorse Creek, which flows into the East Fork of the Big Lost River a few miles south of Trail Creek Road.

On the east side of the Pioneer Mountains, Eocene Challis volcanic rocks were erupted onto a surface eroded onto the Copper Basin Group of Mississippian age. This 10,000-foot-thick unit consists of turbidites, deepwater sediment gravity flows that were fed northward into a basin from the south. The coarser sediment, including boulders, was deposited rapidly, while the finer-grained sediment continued northward across the basin leaving deposits that become progressively finer to the north. Strike-slip and normal faulting formed the basin during the Antler orogeny. The rocks are named for an early copper mine in the Copper Basin to the south.

Borah Peak view south from milepost 134 south of Willow Creek summit. The Lost River fault scarp from 1983 cuts the lower slope of the mountain. The Borah Peak syncline, a Cretaceous-age fold of the Sevier orogeny, can be seen in brown sandstone beds within the early Paleozoic rock layers high on the range. (44.1752, -113.9244)

The cumulative vertical displacement on the Lost River fault is at least 1.6 miles, so if the 6 feet of slip during the Borah Peak earthquake is typical, more than 1,300 similar earthquakes have occurred during the past 10 million years.

US 93 heads over Willow Creek Summit between mileposts 138 and 139 and drops into Antelope Flats. The Twin Peaks of the Salmon River Mountains west of Challis are visible straight to the north. These peaks mark the Twin Peaks caldera, a major eruptive center of the Challis Volcanic Group. To the east are folded Mississippian limestones of the northern Lost River Range, called the Pahsimeroi Mountains.

At milepost 147, US 93 cuts through some of the Eocene-age volcanic rocks. About a half mile north of milepost 148, US 93 passes through Grand View Canyon cut in dolostone of the Grand View Member of the Devonian Jefferson Formation. The short, curvy path of this canyon was cut by a stream that existed prior to erosion of the Challis volcanic rocks that once covered the older rocks. The stream maintained its meandering path as it cut downward into the dolostone. The downcutting developed as central Idaho was uplifted over the past 10 million years.

SPAR CANYON ROAD TO WHITE CLOUD MOUNTAINS

North of milepost 144, Spar Canyon Road leads west, following the trail that ore wagons used in early days of mining in the 1870s. It crosses a moonlike landscape, devoid of vegetation, through Eocene Smiley Creek conglomerate, and then into red Challis Volcanic Group rocks. The gravel road descends into the East Fork of the Salmon River and if you turn right, you'll follow it north to the Salmon River along ID 75. If you turn left and follow the river upstream, you come to the Boulder and White Cloud Mountains. Trails lead up to the Railroad Ridge area of the White Cloud Mountains, which was the site of the proposed molybdenum mine that spurred action to establish the Sawtooth National Recreation Area in 1972. The proposed mine site is now protected within the recreation area.

Just south of milepost 159, Hot Springs Road leads east to Challis Hot Springs and also provides access to Leaton Gulch in the Pahsimeroi Mountains. The Leaton Gulch area is noteworthy because it reveals rocks associated with the largest confirmed meteor impact in the United States. Called the Beaverhead impact structure, this huge impact event took place sometime between 900 and 800 million years ago. The multitude and magnitude of subsequent geologic events such as the Sevier orogeny, Challis volcanism, and Basin and Range extension has obscured, destroyed, or buried much of the evidence—don't look for a gaping crater here. Despite this, Leaton Gulch exposes some impressive outcrops of quartzite breccia with blocks up to 6 feet in diameter. For more information and directions to Leaton Gulch and its outcrops, see *Geology Underfoot in Southern Idaho* by Shawn Willsey.

View looking south at Grand View Canyon south of Challis, where the now-dry creek bed was cut into east-dipping dolostone of the Grand View Member of the Jefferson Formation. (44.3715, -114.0794)

Twin Peaks, north of Challis, viewed from the south at milepost 156. These peaks are part of the caldera system from which Eocene Challis volcanic rocks, which form the mountains below the peaks, were erupted. Rocks in the foreground belong to the Challis Volcanic Group. (44.4453, -114.1407)

US 93 crosses the Salmon River at milepost 160. ID 75 follows the river upstream to its headwaters in the Sawtooth Range near Stanley. The Land of the Yankee Fork State Park, at the junction of US 93 and ID 75, features a historical mining museum about the Challis area. Extensive placer gold ore deposits are present in stream gravels. The gold, originating in hydrothermal systems driven by heat from Eocene plutons, was concentrated by streams during erosion. Silver-lead vein mineralization at the Bayhorse townsite occurs where Cambrian limestones were altered by hydrothermal fluids circulating through the rocks during Eocene time. See the road guide for US 93: Challis—Salmon—Montana in the Central Mountains chapter for more information about the Challis area.

ID 28
Mud Lake—Leadore—Salmon
122 miles

People and services are few along ID 28, which follows a big, dry valley between the Lemhi Range on the west and the Beaverhead Mountains on the east in the northern part of Idaho's Basin and Range. The big valley is a classic half graben, with the Beaverhead Mountains uplifted by a fault along its western base. The Lemhi Range also was uplifted along a fault along its western base. The rocks and thrust faults in the mountains are related to those in the Central Mountains. See the introduction to that chapter for a background on the Lemhi subbasin of the Belt Supergroup and the Lemhi arch.

North of Mud Lake, the highway heads straight northwest across the Snake River Plain. The extremely flat expanse was the floor of Lake Terreton in Pleistocene time. The lake sediments are productive agricultural soils. At milepost 21, the highway crosses a large canal, north of which there is no agriculture, not because the soil becomes less fertile but because the highway has entered the Idaho National Laboratory.

As northbound travelers get closer to the mountains, they can see they are approaching a large valley, the mouth of Birch Creek. At the south end of the valley is an archeological site where spear points and mammal bones have been found at hunting camps dated to about 10,000 years ago. Low buttes east of the road are Miocene volcanic rocks of the Heise volcanic field.

The southern Beaverhead Mountains to the east have been uplifted along the Beaverhead normal fault that lies at their base. Here, the mountains are composed of Mississippian limestones, spectacularly exposed at Skull Canyon, just south of milepost 46. East up Skull Canyon is an anticline cored by Mesoproterozoic Belt Supergroup overlain by the Kinnikinic Quartzite, a shallow marine sand deposited in the Paleozoic sea in Ordovician time. From Neoproterozoic to Ordovician time, this area was part of the Lemhi arch, an uplifted region of land above sea level.

North of the hamlet of Lone Pine and north of milepost 47, tilted Miocene basalt forms the canyon walls of Birch Creek. The basalt erupted sometime between 10 and 6.5 million years ago.

To the west is the Lemhi Range made of folded early Paleozoic strata. Diamond Peak is prominent to the west all along this route, but a particularly great view lies at milepost 49. North of Diamond Peak, about halfway between mileposts 55 and 56, is

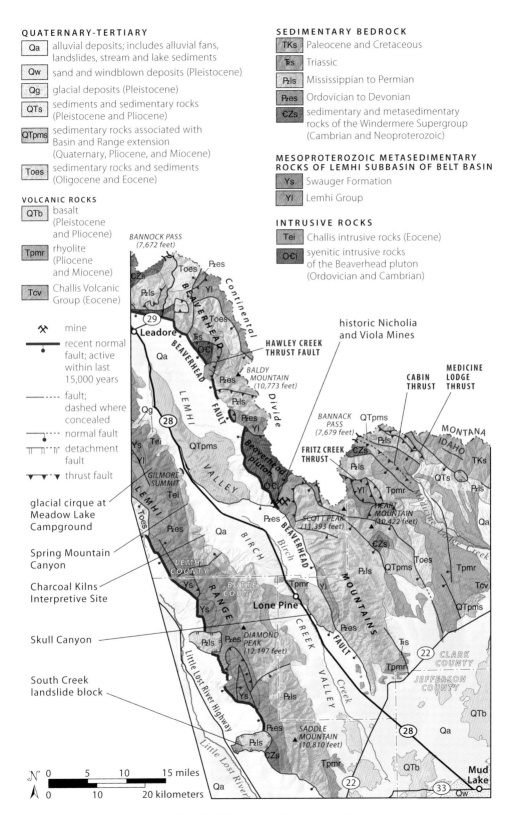

Geology along ID 28 between Mud Lake and Leadore.

West-dipping Mississippian limestone of the Scott Peak Formation forms a steep mountain front at the mouth of Skull Canyon. (44.1651, -112.9180)

Lone Pine basalt, erupted between 10 and 6.5 million years ago, rises above a wooden ranch house at milepost 48. View is to the east with Mississippian limestones of the Beaverhead Mountains in the background. (44.1809, -112.9329)

a fine view of distinctive Bell Mountain, a glacial horn carved by Pleistocene ice into the shape of a bell.

The Nicholia mining district, to the east on a gravel road just south of milepost 58, extracted silver and lead in the early twentieth century from Paleozoic limestones. The Beaverhead pluton, an isolated latest Cambrian to Ordovician intrusion, is exposed up Nicholia Canyon. The tectonic origin of this granitic rock is not clear, but its distinctive age, which is unique in this region, makes the rock incredibly useful for determining when highlands formed and were eroded. Detrital zircon grains eroded from these intrusions are found in uppermost Cambrian sedimentary rocks

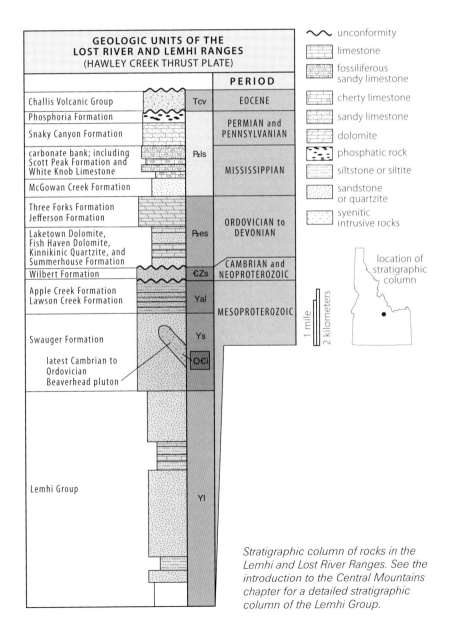

Stratigraphic column of rocks in the Lemhi and Lost River Ranges. See the introduction to the Central Mountains chapter for a detailed stratigraphic column of the Lemhi Group.

At milepost 61, a dirt road heads west 5 miles to the Charcoal Kilns Interpretive Site. These charcoal kilns burned logs to make charcoal to feed the silver-lead Viola smelter from 1886 to 1888. The kilns are at the foot of the Lemhi Range, where the trees grew, but the mine was across the valley to the east at Nicholia. (44.3115, -113.1789)

as far east as central Wyoming and as far west as Challis. The grains were eroded in Late Cambrian time, immediately after the granite was intruded, proving uplift of the Lemhi arch.

A gravel road to the west, halfway between mileposts 67 and 68, leads to the Hahn townsite and a steep ascent into the cirque of Spring Mountain Canyon. Ordovician Kinnikinic Quartzite and the overlying Devonian Jefferson Formation are exposed here, intruded by an Eocene stock.

The highway ascends toward Gilmore Summit (milepost 73), a low pass within the valley that divides the drainage basin of the south-flowing Birch Creek from that of the north-flowing Lemhi River. Exploring the Gilmore ghost town and the campground beyond in the glacial cirque of Meadow Lake Creek is a great side trip west of the highway. Gilmore, a silver-lead mining community in the early 1900s, was the terminus of the Gilmore & Pittsburg Railroad, the last rails built in Idaho. This railroad connected with the Union Pacific south of Dillon, Montana, and was completed in 1910, just before World War I, to service the Gilmore mines. The rails were torn up by 1940, but the old roadbed can be seen east of the highway north of Gilmore Summit. Along this railroad grade, 6.5-million-year-old tuff from the Heise volcanic field is exposed.

Leadore, a tiny settlement that services a huge area, was originally a mining community. Scars of the old mines can be seen on the Paleozoic limestones to the east of town in the Beaverhead Mountains. Silver, copper, and lead veins cooled from hydrothermal fluids that altered the limestone. Mississippi Valley–type lead and silver ore deposits were recognized by the first geologists who worked here in the early 1900s. These veins of galena and sphalerite follow sedimentary beds and formed when hot water migrated through the limestones.

The Hawley Creek thrust, putting early Paleozoic Beaverhead granite above late Paleozoic limestones, is exposed in the mountains east of Leadore. West of the thrust fault, in its upper plate, Mesoproterozoic Belt Supergroup is intruded by the latest Cambrian and early Ordovician Beaverhead granite. This region was part of the Lemhi arch, an area which stood high during Neoproterozoic rifting of ancient North America. The arch was eroded after the intrusion of the granite and was overlain in

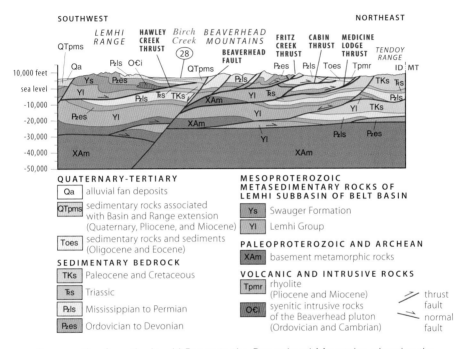

Cross section from the Lemhi Range to the Beaverhead Mountains, showing the major thrust faults of the Sevier fold-and-thrust belt in this part of eastern Idaho. This section crosses ID 28 just north of Lone Pine.

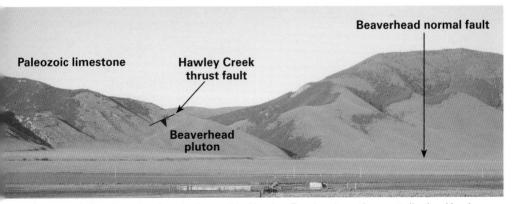

Annotated view from Leadore to the east to Hawley Creek, where the west-dipping Hawley Creek thrust places Cambrian plutonic rocks of the Beaverhead pluton over late Paleozoic rocks. An active strand of the Beaverhead normal fault forms the straight mountain front at Hawley Creek. (44.6575, -113.3345)

the middle Ordovician by seashore sands of the Kinnikinic Quartzite. Three other thrust faults are exposed farther east in the Beaverhead Mountains. One of these, the Cabin thrust, cuts Archean metamorphic rocks, indicating that the Sevier thrusting in Cretaceous time moved Paleozoic rocks of the continental shelf eastward over rocks of the ancient North American continent.

The northern half of the Lemhi Range is mostly Mesoproterozoic Belt Supergroup rocks. Glacial cirques, high in the mountains northwest of Leadore, were carved by

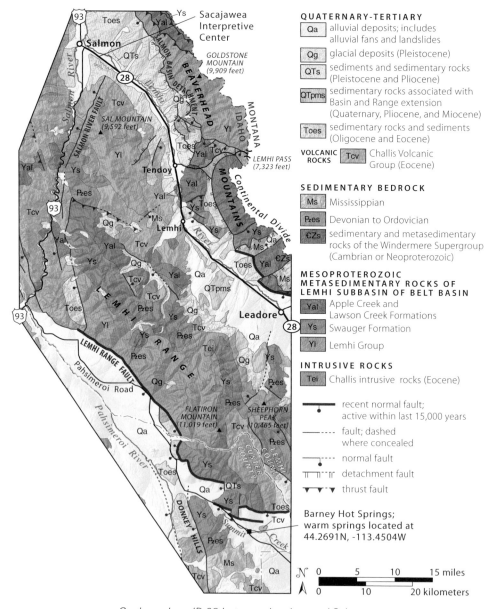

Geology along ID 28 between Leadore and Salmon.

Pleistocene glaciers. North of Leadore, the Lemhi River flows through a flat floodplain below low, flat-topped terraces, former floodplains of the Lemhi River. These terraces reflect a higher base level from which the river adjusted in Pliocene or Pleistocene time as it flowed north to the Salmon River.

The northern Beaverhead Mountains to the east contains a huge thickness, over 6 miles, of the upper part of the Lemhi subbasin of the Belt Supergroup. Detrital zircons in these rocks are mostly 1.7 billion years old, suggesting they were eroded from rocks of that age. The source of the zircons has been proposed to be the 1.7-billion-year-old Big White magmatic arc to the south in what is now Utah or Nevada.

The Apple Creek Formation and overlying Gunsight Formation of the Belt Supergroup form hills both east and west of the highway at the mouth of Hayden Creek (north of milepost 108).

Lemhi Pass, east of the highway in bare, grass-covered slopes of the Beaverhead Mountains, was eroded along the Lemhi Pass normal fault. In Eocene time, a major east-flowing river followed this fault, carrying Challis volcanic detritus southeastward all the way to the Green River basin in southwest Wyoming. Lewis and Clark came over the Lemhi Pass in 1804 in search of a path to the Pacific Ocean. They came down into the Lemhi Valley and were absolutely lucky to be able to trade with the Shoshone Indians for horses, which were essential to the expedition. The chief of the local tribe was Cameahwait, brother of Sacagawea.

A gravel route with historic signs heads up Lemhi Pass from the village of Tendoy. The rugged north end of the Beaverhead Mountains contains glaciated peaks composed of quartzites of the upper part of the Mesoproterozoic Belt Supergroup. The Lemhi Pass normal fault dips southwest north of the pass, such that the uplifted rocks to the north are older than Eocene Challis volcanic rocks to the south. The Lemhi Pass area contains anomalous concentrations of thorium and several trace metals.

Lemhi Pass over the Beaverhead Mountains, viewed to the east from Tendoy, was eroded along the Lemhi Pass normal fault. (44.9613, -113.6436)

As the highway approaches Salmon, light-colored Eocene lakebeds can be seen both east and west in the Lemhi Valley. These lakebeds contain leaf fossils demonstrating that 40 million years ago, the climate here was subtropical. Recent research established that an Eocene to Oligocene sedimentary basin formed here above the northwest-dipping Salmon Basin detachment fault. This fault forms the western front of the northern Beaverhead Mountains and was active from perhaps 46 to 30 million years ago. The town of Salmon, at the north end of ID 28, is its own world, in the big valley of the Salmon River, closer to Missoula, Montana, than to Idaho Falls, and with wilderness to the north and west.

ID 31
Swan Valley—Victor
21 miles

This short, 21-mile-long highway climbs over the Snake River Range via Pine Creek Pass, connecting the Swan Valley with the Teton Valley. Though windy and narrow, ID 31 is regularly used as a shortcut to a variety of outdoor recreation destinations. From the tiny town of Swan Valley, ID 31 ascends for a couple of miles to the top of Pine Creek Bench, a basalt bench capped by loess. This basalt lava flowed up Pine Creek from its source in the Swan Valley, and later the modern creek cut down through the basalt. North of milepost 5, ID 31 crosses Pine Creek on a high bridge over a basalt-rimmed canyon. Several pullouts in the first mile north of the bridge provide views into the narrow basalt-lined gorge.

As ID 31 slices northeast through the Snake River Range, it intersects several imbricated thrust faults and tight folds in the hanging wall, or upper plate, of the Absaroka thrust system, the youngest and most easterly of the Sevier fold-and-thrust belt in eastern Idaho. The rocks were squeezed in a southwest to northeast direction during the late Cretaceous Period, resulting in elongate slices of rock bounded by faults that run northwest-southeast. Many of the tributaries of Pine Creek follow this orientation, exploiting the faults and softer rocks within the thrust belt. Despite the thrust faults shoving older rocks from the southwest over younger rocks to the northeast and largely repeating the rock layers, the erosional surface exposes mostly older rocks of the thrust plates on the southwest side of the range and the younger rocks beneath them on the northeast side of the range. Thus, the rocks generally get younger as travelers head northeast on ID 31.

Craggy gray outcrops on the northwest side of the highway between mileposts 6 and 7 are the Mission Canyon Limestone, a unit of the Mississippian-age Madison Group. Thrust faults repeat Mississippian limestone for the next few miles, although roadcuts and outcrops are few and far between. Just west of the bridge over West Pine Creek, between mileposts 9 and 10, a small brown roadcut reveals near-vertical beds of the Triassic Woodside Shale on the downthrown side of a strand of the Absaroka thrust fault.

Between milepost 10 and North Fork Pine Creek, a roadcut on the northwest side of the road displays an overturned anticline within the Permian Phosphoria Formation. Black phosphate-rich shale layers are folded tightly near the center of the fold, forming steep southwest-dipping beds. Between this roadcut and North Fork Pine Creek, the Absaroka thrust fault cuts across the highway. Northeastward from North

Fork Pine Creek, rocks along ID 31 are in the footwall or lower plate of the Absaroka thrust fault and mainly consist of Cretaceous sandstone and shale.

A great roadcut and pullout at milepost 11 allows travelers to observe beautiful slickensides in the Cretaceous Frontier Formation. Slickensides form when fault movement grinds rocks on either side of the fault plane, forming a polished surface with distinctive lines indicating the direction of fault movement. Look for polished and striated planar surfaces in the resistant sandstone at the center of the roadcut.

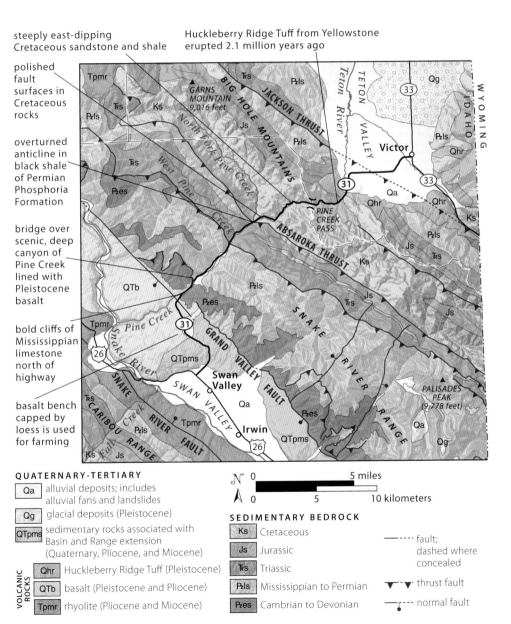

Geology along ID 31 between Swan Valley and Victor.

Tight overturned anticline in the Permian Phosphoria Formation. The distinct black, phosphate-rich layer is repeated on either side, or limb, of the fold. (43.5591, -111.2792)

Polished and striated slickensides from fault movement in a roadcut in the Frontier Formation at milepost 11 on ID 31. (43.5633, -111.2677)

Numerous and interesting roadcuts of steeply dipping Cretaceous sandstone and shale are found between mileposts 12 and 16. The highway crests Pine Creek Pass between mileposts 14 and 15 at an elevation of 6,764 feet and begins its descent into the Teton Valley, which lies on the west slope of the iconic Teton Range. Slabby outcrops of gently dipping Huckleberry Ridge Tuff, erupted from the Island Park caldera 2.1 million years ago, are encountered near milepost 16.

Huckleberry Ridge Tuff, erupted 2.1 million years ago from the Island Park caldera, along ID 31 near milepost 16. (43.5725, -111.1872)

ID 34
Preston—Soda Springs—Wyoming Border
56 miles

East of Preston ID 34/ID 36 heads north over hilly terrain in the late Miocene to Pliocene Salt Lake Formation. These river and lake deposits accumulated in a basin that formed as the region was stretched along the Bannock detachment fault, a regional low-angle extensional structure. See the road guide for ID 36 for more about the section of highway between Preston and the Bear River.

ID 34 diverges from ID 36 north of the Bear River crossing near milepost 13. It then heads uphill on the ridge west of Oneida Narrows, a deep canyon cut by the Bear River. A half mile north of the intersection, silts deposited in a delta of Lake Bonneville are exposed on the east as you begin the climb up to Treasureton Summit. Look for ripples in the silt. The highway passes through rolling country, all below the level of Lake Bonneville, before poking out of the lake about 7 miles north of the ID 36 intersection at milepost 20.

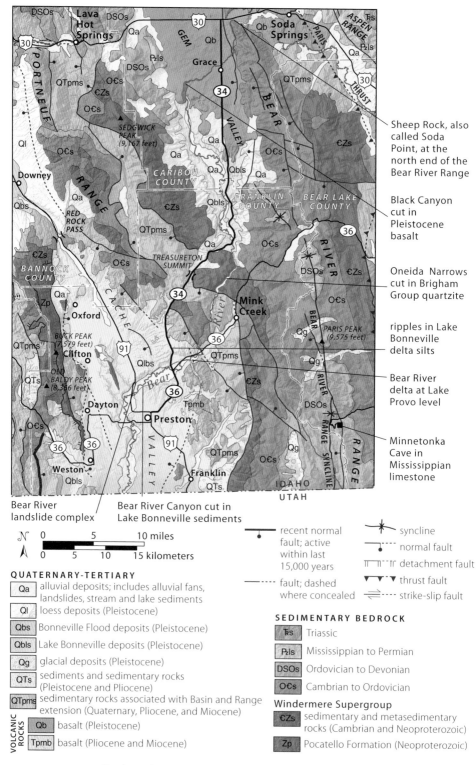

Geology along ID 34 between Preston and Soda Springs.

Ripples in silt were sculpted by currents moving to the left (west) near the Lake Provo shoreline. Outcrop is just east of the highway as it climbs out of the Bear River valley and above the West Cache Canal. Holes are swallow nests, 3 to 4 inches in diameter. (42.1741, -111.8368)

North of Treasureton Summit (and east of milepost 25), a steeply dipping exposure of purple pebbly sandstone of the Neoproterozoic Mutual Formation of the Brigham Group can be seen on the east side of the road. The highway heads down into southern Gem Valley, which is also called Mound Valley, so named for the hot spring mound visible south of the highway on the west side of the Bear River just south of milepost 29.

Lake Bonneville's highest stage at 5,200 feet occupied the southern end of the valley here. Before Lake Bonneville existed, another Pleistocene lake occupied southern Gem Valley. Known as Lake Thatcher, it existed 100,000 years ago and rose to 5,400 feet. At this time, the Bear River flowed northwest to the north end of Gem Valley and eventually west through Pocatello. Young basalt lava erupted 55,000 years ago to block the river's northward path, forcing the river south through the area of Lake Thatcher. The Bear River cut through a preexisting gap at Oneida Narrows and became a major tributary to Lake Bonneville, providing more than half of the lake's freshwater input.

The highway emerges from below Lake Thatcher at milepost 35, about 6 miles north of the bridge over the Bear River. From here north is a land of basalt. A north-trending line of basalt cinder cones runs from near Niter (south of Grace) northward across US 30. Red basalt cinders make up the volcanic cones. Dark basalt lava flows with ages as young as 55,000 years were erupted from these cones.

North of Grace (north of milepost 46), the highway crosses the Bear River upstream from a canyon it cut in the basalt. To the east of the highway is the north end of the Bear River Range, and Sheep Rock, a landmark among fur trappers in the 1830s. Between the highway and the Bear River Range is a normal fault scarp in the 55,000-year-old Gem Valley basalt, with fault movement down to the west. Thus, this young fault has moved since the basalt erupted. ID 34 joins US 30 east to Soda Springs. See US 30 for that section of road.

North of Soda Springs, ID 34 runs near phosphate processing plants. The Permian Phosphoria Formation is mined in open pits from folded strata of the Meade thrust

The Bear River flowing south at the north end of Oneida Narrows. Bedrock in the gorge is resistant quartzite of the Brigham Group, which spans the Precambrian to Cambrian time boundary. (42.3395, -111.7210)

plate in the mountains to the east. Elemental phosphorous and phosphoric acid is separated from the crushed ore and produced for herbicide, fertilizer, and a host of chemical uses. Old pits of the Conda Mine, which began as an underground mine in the 1930s, can be seen east of the highway at milepost 63.

Between Soda Springs and Grays Lake National Wildlife Refuge, ID 34 crosses into the Blackfoot lava field, which is in the headwaters of the Blackfoot River. These volcanic rocks are late Pleistocene and are thus younger than and not physically connected with nearby volcanic rocks produced by the Yellowstone hot spot. Rhyolite domes with eruptive ages of 57,000 years are prominent to the west and include China Hat and China Cap. Basalt lava flows form the surface along the Blackfoot River and the Blackfoot Reservoir. North of milepost 78, the highway cuts between outcrops of folded Mississippian limestone, with crinoid, brachiopod, and coral fossils.

Grays Lake National Wildlife Refuge occupies a basin dammed by Pleistocene basalt lava. Caribou Mountain looms large over the north shore of Grays Lake. A gold mining rush to placer deposits on Caribou Mountain established the Caribou mining district in the 1880s. A group of historical markers about 4 miles east of Wayan, east of milepost 93, mention early explorer John Gray, for whom the lake was named, and Dr. Ellis Kackley, who lived on Chippy Creek to the south of this marker. Silicified logs from Late Cretaceous tree ferns of the genus *Tempskya* are found in the Wayan Formation up Chippy Creek.

In the Tincup Creek drainage, where ID 34 is known as the Tincup Highway, the road passes several outcrops of folded Cretaceous sedimentary rocks of the Gannett Group and the Wayan Formation. Look for outcrops halfway between mileposts 106 and 107 and between mileposts 111 and 112. These units were deposited in river floodplains and lakes on the western side of the Western Interior Seaway. Fossils of dinosaurs that roamed its shores are found in this area.

34 BASIN AND RANGE   163

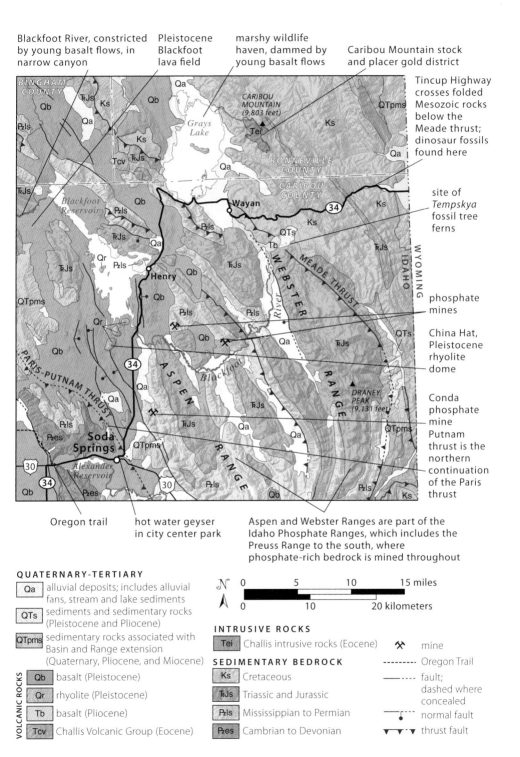

Geology along ID 34 between Soda Springs and the Wyoming border.

China Hat, part of a field of Pleistocene rhyolite domes, viewed southwest from Henry. (42.8098, -111.563)

Steeply east-dipping red shale and overlying white limestone of the middle Cretaceous Gannett Group at milepost 112 along ID 34. (42.9967, -111.0763)

ID 36
Malad City—Preston—Montpelier
79 miles

ID 36, a curvy secondary highway between Malad City and the Bear Lake Valley, crosses some of Idaho's most complex geology. Because the highway travels generally east-west, it crosses the low-angle extensional Bannock detachment fault and high-angle normal faults of the Basin and Range. This includes an active fault of the Bear Lake graben near Ovid. East of the summit of the Bear River Range, the highway crosses the late Cretaceous Paris thrust fault.

Pick up the west end of ID 36 at exit 17 of I-15, three miles north of Malad City in Malad Valley. For the first couple of miles east of the junction, you'll pass mostly vegetated light-colored sediments deposited in Lake Bonneville. An arm of the lake extended north into the Malad Valley during late Pleistocene time. The prominent bench to the south is the Bonneville shoreline.

At milepost 104, as the highway passes the lower Deep Creek Reservoir, ID 36 enters the Salt Lake Formation of late Miocene to Pliocene age. Although similar in color to the unconsolidated Bonneville sediments, the Salt Lake Formation is more consolidated. The Salt Lake sediments were deposited in a large basin that extended east to Cache Valley, and thus existed before the Bannock Range, which ID 36 crosses here, was uplifted. The 12- to 3-million-year-old Salt Lake Formation includes white rhyolitic tuffs derived from eruptions on the Snake River Plain, locally derived conglomerates, sandstones deposited by rivers, and lakebeds containing freshwater mollusk fossils. This entire area is on the upper plate, or hanging wall, of the regional Bannock detachment, a major low-angle fault that underlies the Salt Lake Formation basin between Malad City and Preston. Rocks above the fault moved west as the Basin and Range was stretched apart.

Near the Franklin county line, upstream from Weston Creek Reservoir and southeast of milepost 112, the highway enters Weston Canyon, where high cliffs of Cambrian-aged Blacksmith Limestone lie unconformably below the Salt Lake Formation. The canyon is narrow for about 3 miles while passing through this scenic limestone.

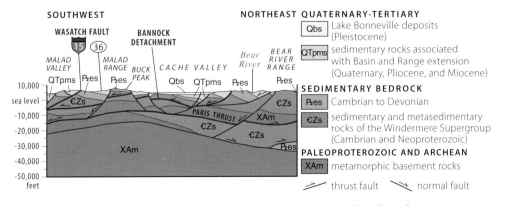

East-west cross section between the Malad Valley and the Bear River Range.

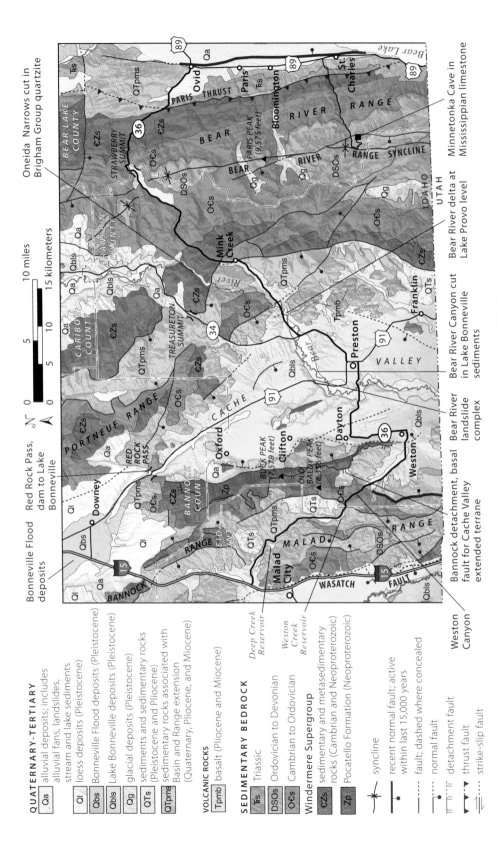

Geology along ID 36 between Malad City and Ovid.

Cambrian Blacksmith Limestone rises above the eastern mouth of Weston Canyon. (42.1132, -112.0964)

East of the bedrock in Weston Canyon, the highway crosses open country in the Salt Lake Formation. This valley was submerged by Lake Bonneville, and you can see the shoreline of the 5,200-foot water level (the Bonneville level) on the hillside to the west above the level of the highway as you descend toward Weston. As the highway turns east approaching Weston, the Bear River Range forms a wall of steep mountains to the east.

Weston, where the highway turns north, lies on the western edge of the Cache Valley, an agricultural area irrigated with water from the Bear River. Along Fivemile Road (2.5 miles west of the village of Dayton), you can see diamictites (42.1078, -112.0314) of the Pocatello Formation. These 680-million-year-old marine mudstones include reworked glacial till deposited during a Snowball Earth glaciation.

ID 36 crosses the Bear River east of Dayton and then climbs the grade to Preston. The entire central Cache Valley was under the waters of Lake Bonneville at times when the lake was large and deep. Preston, at 4,800 feet, lies at about the elevation of the Provo level to which Lake Bonneville fell after the Bonneville Flood drained the upper 400 feet of the lake 17,500 years ago. Upstream from Preston, the Bear River has cut down through sediments deposited in the lake. These unconsolidated clays are prone to slumping, and small landslides have formed along the river valley walls.

East of Preston, the highway climbs north past small reservoirs along the west margin of the Bear River Range. Near the top of this bench and at milepost 12 as you start heading back down to the Bear River, look northwest for views of the Bannock Range with the river's incised canyon in the foreground, cut though Lake Bonneville sediments.

View looking north from halfway between mileposts 11 and 12 at the Bear River valley. Above the floodplain, the flat area in the middle distance, is a pile of reddish, unconsolidated sediment deposited in Lake Bonneville. Junipers grow on steep slopes and form a line above a clay bed that restricts the downward flow of groundwater, making it available to the junipers. The flat farmed bench on the right side of the view was the bottom of Lake Provo. The Bear River cut down through the lake sediment in the last 14,000 years as the lake shrunk in size. Oxford Peak of the Bannock Range is in the distance. (42.1447, -111.8260)

ID 36 turns east at the ID 34 junction north of the Bear River crossing. The mileposts start at 0 here on ID 36 as it follows the Bear River upstream to where the highway crosses the river below the Oneida Narrows canyon. The river cut Oneida Narrows through hard Proterozoic and Cambrian Brigham Group quartzites. The final phase of cutting was in the last 55,000 years, after the Bear River was diverted south by a young basalt flow in Gem Valley to the north. After the flood, when the level of Lake Bonneville dropped dramatically, the gradient of the river increased. The river cut down even farther through the quartzite in Oneida Narrows as well as through the unconsolidated silts and sands deposited on the floor of the now-drained lake.

East of this Bear River crossing, ID 36 follows Mink Creek upstream toward the Bear River Range. Look for a Lake Bonneville shoreline above the fields on the north, halfway between mileposts 3 and 4. Platy tuff of the Salt Lake Formation is exposed in a small roadcut at milepost 5. The volcanic ash in this tuff was derived from rhyolite eruptions on the Snake River Plain in late Miocene time.

North of the community of Mink Creek, ID 36 begins the climb over the Bear River Range by heading up Strawberry Creek. The normal fault that drops the Salt Lake Formation down to the west and uplifts older rocks in the Bear River Range to the east is about halfway between mileposts 12 and 13. The exposures east of the fault are unique because they contain a continuous east-dipping, 7,500-foot-thick section of the earliest Paleozoic rocks in southeastern Idaho—early and middle Cambrian rocks that record the rapid diversification of life at the beginning of Paleozoic time. Fossils, however, are difficult to find here. The oldest rock exposed along the route is a brownish ledge of the Kasiska Quartzite (upper Brigham Group sandstone) a few tenths of a mile west of milepost 13. Cambrian limestones form the roadcuts and rock fins high above the road at milepost 13.

Farm fields near Mink Creek between mileposts 3 and 4 are planted on sediments deposited in Lake Bonneville. The Bonneville shoreline is indicated at the top of the fields. The bedrock is the Miocene Salt Lake Formation. (42.1988, -111.7708)

Early Cambrian Kasiska Quartzite of the Brigham Group west of milepost 13 on ID 36. The east-dipping brown sandstone contains cross beds that formed in shallow ocean water. (42.3062, -111.6748)

The next roadcuts on the west side of the road, halfway between mileposts 13 and 14, expose the Cambrian Blacksmith Limestone, with oncolites, or algal ball fossils. These spherical stromatolites are common in Cambrian carbonates and became less prevalent once grazing invertebrates such as brachiopods evolved. The fine-grained carbonate nodules formed initially as lime mud patches in a largely siliciclastic muddy seafloor.

Gray-mottled, thin-bedded oncolitic limestone of the upper Bloomington Formation crops out halfway between mileposts 15 and 16. Greenish calcareous shale of the Hodges Shale Member of the Bloomington Formation is exposed near milepost 16. Between mileposts 16 and 18, the Bloomington Formation is mainly limestone with lesser amounts of shale. The layers parallel the road, so not much stratigraphic section is crossed in these couple of miles.

Light- and dark-gray, sugary textured dolostones of the Nounan Formation crop out from a few tenths of a mile west of milepost 20 to milepost 21. These carbonates were altered from original calcium carbonate by magnesium-rich fluids, probably while the sediment was being lithified at the bottom of the basin. Any primary fossils have been recrystallized and are not identifiable.

East of milepost 21, glacial sediment (shown on the map as Qa) is exposed on both sides of the road. Look for prominent white quartzite boulders of the Ordovician Swan Peak Formation, carried here by glaciers that existed higher up in the Bear River Range in Pleistocene time.

Near Strawberry Summit, which is at the Bear Lake county line near milepost 22, the highway crosses the axis of the Bear River Range syncline. The rocks dip eastward west of the summit and westward east of the summit. For this reason, as the highway descends along Emigration Creek, it goes downsection into older rocks,

Oncolites (algal balls) in the Cambrian Blacksmith Limestone from outcrop halfway between mileposts 13 and 14. (42.3145, -111.6624)

Shale of the Hodges Shale Member of the Cambrian Bloomington Formation at milepost 16. (42.3102, -111.6448)

Outcrop of east-dipping, light- and dark-colored dolostone of the Cambrian Nounan Formation halfway between mileposts 20 and 21. (42.3441, -111.5997)

passing in reverse order the formations it crossed while climbing the west side of the Bear River Range. The Nounan Formation is exposed for about a half mile downhill from milepost 22, the Bloomington Formation halfway between mileposts 22 and 23, the Blacksmith Limestone halfway between mileposts 23 and 24, and Brigham Group shale and quartzite that is poorly exposed about half mile east of milepost 24.

The highway enters the Bear Lake Valley and crosses the Paris thrust fault near milepost 33 as the country opens up between Liberty and Ovid. East of the Paris fault are Paleozoic sedimentary rocks of the underlying Meade thrust plate, but they are hidden from view by overlying Miocene to Holocene sediments. East of Ovid, the highway is built on Miocene and younger sediments deposited in the Bear Lake graben. The southern part of the Bear Lake Valley is a true graben, a fault block dropped down between inward-dipping normal faults, one on the east side and one on the west side of the valley. The west side fault runs through Ovid and has been active in the last 15,000 years, but farming has removed any trace of it on the surface.

ID 37 AND ID 38
AMERICAN FALLS—MALAD CITY
71 miles

ID 37, also known as the Rockland Highway, heads south from exit 36 of I-86, the westernmost exit for American Falls. This little-traveled route affords close-up views of rhyolitic tuff, including welded ignimbrite. The highway follows Rock Creek upstream through Rockland Valley, then crosses a low treeless divide into a valley

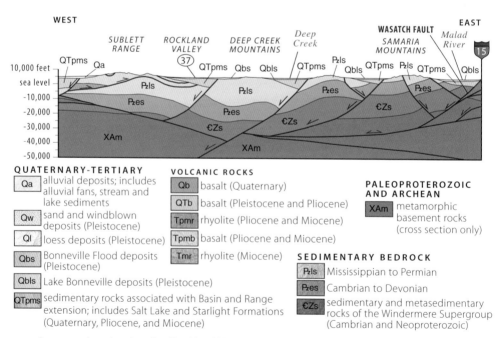

Cross section showing that Rockland Valley is a half graben, dropped down along a fault at the western base of the Deep Creek Mountains.

inundated by Lake Bonneville. Note that mileposts count down in the direction of the road guide.

The route heads uphill from American Falls, through power-generating wind turbines at the hillcrest, rising above mainly dry farms of winter wheat. The hill is young basalt covered with loess. South of the summit, the highway enters Rockland Valley, which contains Quaternary valley fill overlying the Miocene Starlight and Salt Lake Formations. At the north end of the valley is the Late Miocene Massacre Volcanic Complex. The lower northern end of the Rock Creek valley is a narrow canyon cut in the Massacre rocks.

Rockland, a quiet agricultural town, is reached at milepost 56. The river bottom below town is flood irrigated, meaning that creek water, first channeled in irrigation canals, is then allowed to flow across the fields in wide sheets of water.

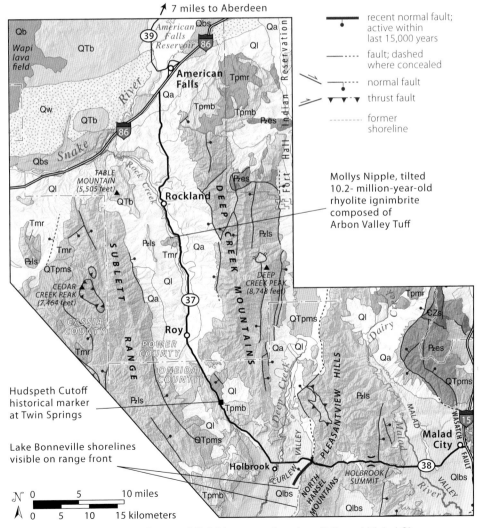

Geology along ID 37 and ID 38 between American Falls and Malad City.

The Deep Creek Mountains to the east and the Sublett Range to the west contain late Paleozoic mixed carbonate and siliciclastic sedimentary rocks. They were deposited in the Oquirrh basin, named for a mountain range in northern Utah. In Pennsylvanian time, this large basin reached from central Utah northward into central Idaho near Sun Valley.

At milepost 55, the highway leaves Rockland and continues south along the Rock Creek valley, here eroded into the Starlight Formation. The gently southeast-dipping Arbon Valley Tuff Member of the middle Starlight Formation forms the caprock of Mollys Nipple, the conical hill south of the highway. The tuff was erupted from the Picabo caldera about 10 million years ago as the Snake River Plain passed over the Yellowstone hot spot. This welded tuff, exposed on the east side of the highway near milepost 50,

contains angular shards of rhyolitic glass that originated as frothy pumice. The shards have been compressed and welded together. Unconformably above the welded tuff is a Quaternary pebble conglomerate. The trough cross beds were formed by small shifting stream channels. The uppermost layer is late Pleistocene windblown loess, which forms the soil of the dry farm fields.

ID 37 crests a low, open pass near milepost 37 and becomes ID 38 at the county line. Twin Springs Campground, which was a much-used campground on the Hudspeth Cutoff in 1849 and 1850, is near milepost 32. The Hudspeth Cutoff left the Oregon Trail west of Soda Springs and took a more direct but mountainous path to connect to the California Trail. South of Twin Springs, ID 38 cuts through a canyon in Pliocene basalt and heads down into the main body of the Curlew National Grassland, established to restore areas damaged by pre–Dust Bowl era ranching. The Curlew Valley

View of Mollys Nipple south of Rockland. The top of the butte is the 10-million-year-old Arbon Valley Tuff Member of the Starlight Formation. (42.5284, -112.8563)

BASIN AND RANGE 175

A roadcut near milepost 50 exposes the Arbon Valley Tuff, a light-colored bedded rhyolite ignimbrite. It is overlain by darker-colored bedded volcanic ash and, at the top, stream channel deposits. (42.5039, -112.8400)

View of the prominent Bonneville shoreline east of Holbrook at the base of the North Hansel Mountains. The irrigated fields in the foreground are on the Provo shoreline at about 4,800 feet. (42.1675, -112.6134)

was once covered by Lake Bonneville.

East of the small hamlet of Holbrook, the highway heads uphill toward the Lake Bonneville shoreline, which is reached between mileposts 18 and 19. The bedrock of the rolling sage- and grass-covered hills consists of late Paleozoic carbonate sedimentary rocks of the Oquirrh basin, but there are only a few low roadcuts on either side of Holbrook Summit (east of milepost 13).

East of Holbrook Summit, ID 38 drops into Malad Valley and crosses the Bonneville shoreline between mileposts 11 and 10. The flat fields between the shoreline and Malad City are former lake bottom. There are big views of the Samaria Mountains to the south and the steep Malad Range to the east. The route ends at I-15 on the east side of Malad City, 500 feet deep in Lake Bonneville.

ID 77
Malta—Declo
31 miles

This short, rural highway begins in the small town of Malta in the Raft River Valley near the boundary between the Basin and Range Province and Snake River Plain. The highway passes three distinct mountain ranges whose rocks range in age from Archean to Miocene. Secondary roads from the highway provide access to City of Rock National Reserve and Castle Rocks State Park, picturesque wonderlands of granite fins and pinnacles.

From Malta, ID 77 heads southwest along the southern edge of the Cotterel Mountains, a north-south range uplifted by steep normal faults during the past 8 million years. Rocks in the Cotterel Mountains are dominated by Miocene rhyolite, erupted as both slow-moving lava and billowing avalanches of ash between 9 and 8 million years ago. South of the highway lie the Jim Sage Mountains, a southward continuation of the Cotterel Mountains with nearly identical rock types and ages. The two highlands were formerly aligned as one unbroken mountain range, but subsequent faulting shifted the northern half of the range eastward, bisecting these highlands. The guilty fault lies beneath the valley, paralleling the highway.

Near milepost 5, westbound travelers have a fine view of the Albion Mountains through the valley ahead. Rocks in these highlands are among the oldest in Idaho, dated to about 2.6 billion years old. Typically, these Archean igneous and metamorphic rocks are found much deeper in the Earth's crust, buried beneath younger rocks, and are often referred to as basement rocks. Here, these ancient rocks have been brought to the surface by extensive faulting over the past 28 or so million years. The Albion Mountains are a fine example of a metamorphic core complex, a region where deep crustal rocks have been exposed due to uplift along low-angle faults called detachment faults. The Albion Mountains core complex was uplifted during three periods of extension: one about 40 million years ago and two at about 20 and 10 million years. Faulting dragged the rocks that once capped the Albions off to the east and west, uplifting and exposing Archean granite and metamorphic rocks along with some slightly metamorphosed Neoproterozoic sandstones correlative with the Brigham Group to the east.

ID 77 abruptly turns north between mileposts 7 and 8. The Elba-Almo Road continues westward then south from here, eventually arriving at the fantastic scenery

Geology along ID 77 between Malta and Declo.

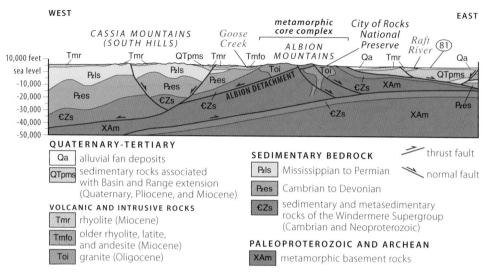

The old basement rocks in the Albion Mountains are exposed as a metamorphic core complex because low-angle detachment faults removed the younger overlying rocks.

of City of Rocks and Castle Rocks. Heading northward, the highway climbs to a pass between the Albion Mountains to the west and the Cotterel Mountains to the east. Outcrops along the base of the Albion Mountains to the west of the highway consist of dark schist with lighter patches of quartzite higher up the slope. The quartzite often splits evenly into slabs along mica-rich layers, forming an ideal flagstone sold as Oakley Stone, named for the town of Oakley, which lies west of the Albion Mountains. The schist near the base of the slope dips eastward more steeply than the hillside, such that the rocks generally get older toward the crest of the range.

The pass at Connor Summit near milepost 11 provides exceptional views in all directions. The most arresting view is to the east, where distinct columnar joints are exposed near the skyline of the Cotterel Mountains. Known to local rock climbers

CITY OF ROCKS NATIONAL RESERVE

A worthwhile diversion—City of Rocks National Reserve and its neighbor, Castle Rocks State Park—lie only 15 to 20 miles southwest of ID 77 and offer awe-inspiring landscapes of granite spires, fins, and monoliths rising above the scenic high desert. Both areas are renowned by rock climbers, who boldly ascend the picturesque faces and cracks. The majority of both parks lie within the Almo pluton, an intrusion of magma that cooled and crystallized about 6 miles below the surface about 28 million years ago during the Oligocene Epoch. This mass of magma intruded a much older collection of Archean basement rocks and Neoproterozoic metasedimentary rocks. Cooling of the magma combined with much later east-west stretching during Basin and Range extension fractured the granite. Uplift of the Albion Mountains ultimately exposed the granite to the effects of weathering, sculpting the rock into a host of unique shapes and forms. The mostly north-trending fins of rock formed as weathering and erosion attacked the granite along the fractures.

Morning Glory Spire in the City of Rocks National Reserve. (42.0822, -113.7256) —Courtesy of Forever Wild Media

The fascinating landscape of eroded Oligocene granite at City of Rocks National Reserve. (42.0709, -113.7068) —Wallace Keck photo

as Connor's Columns, these stately pillars are hexagonal or pentagonal in shape and formed as lava cooled and contracted, similar to mud shrinking and cracking as it dries. While relatively common in basalt flows, the joints at Connor's Columns are in rhyolite lava, a rarer occurrence.

North of Connor Summit, ID 77 descends and enters Marsh Creek Valley and Albion, one of the oldest towns in Idaho. West-dipping layers of rhyolite in the Cotterel Mountains form a dip slope, where the rock is tilted the same direction and angle as the mountain's slope. Streams cutting through the western flank of the mountains nicely expose this relationship. The prominent rhyolite exposures forming the western slope initially erupted as ash-rich pyroclastic flows that were so hot they remelted (became welded) as they came to rest before cooling and crystallizing. About 8 million years old, these volcanic rocks emanated from the Twin Falls volcanic field, some 40 miles away. Volcanic rocks of the same age and origin are exposed just west of the highway, at the north end of Albion, between mileposts 18 and 19. The rhyolite here contains large holes or vesicles formed as gases escaped the hot ash. A layer of black vitrophyre, a glassy rhyolite, lies between two vesicle-rich rhyolite layers.

The most northerly exposures of rock along ID 77 are roadcuts near milepost 22 at the northern terminus of the Albion Mountains. A jumbled and faulted collection of yellow to orange Neoproterozoic schist and white, speckled granite forms much of the roadcut. The age of the granite has not been determined but is likely similar to the 28-million-year-old granite found at City of Rocks National Reserve at the south end of the Albion Mountains. North of these roadcuts, ID 77 drops into the irrigated fields of the Snake River Plain.

Connor's Columns are spectacular columnar joints in Miocene rhyolite within the Cottrell Mountains. (42.3352, -113.4900)

View north, down the Snake River in Hells Canyon, from Dry Diggins Lookout, Idaho. Upper elevations in the canyon above the prominent bench consist of more easily eroded, layered Columbia River Basalts that form a broad upper canyon. Lower elevations consist of basement rocks of the Wallowa terrane, which are more difficult to erode, resulting in a narrower inner canyon.

WESTERN MARGIN

The Western Margin is a land of big rivers—the Snake, Salmon, and Clearwater—that flow through spectacularly deep desert canyons past towering, snow-covered mountain ranges. Both the lowest elevations and some of the highest peaks in the state occur here: Lewiston at 740 feet and She Devil at 9,420 feet in the Seven Devils Mountains. Hells Canyon of the Snake River forms the deepest gorge in North America with nearly 8,000 feet of vertical relief. Despite these elevation differences, remnants of the Columbia Plateau, a level landscape produced by multiple layers of Miocene basalt, exist here. Its flat surface, dotted with farm fields, provides a striking contrast to the rugged canyons.

The Western Margin's large rivers deeply dissected the Columbia Plateau, cutting down through thousands of feet of basalt lava flows to expose ancient accreted basement rocks. The lava flows are part of the Columbia River Basalt Group, erupted in Miocene time from north-trending fissures along the western border of Idaho. Much of the lava flowed west to the Pacific Ocean, transforming the Columbia River drainage basin into an immense volcanic plateau. The lava also flowed eastward to lap up against the Rocky Mountains that rose when a collage of crustal blocks, or terranes, collided with the North American continent in Cretaceous time. The terranes now form the basement to the Columbia Plateau.

A PACIFIC OCEAN CRUSTAL COLLAGE

The Western Margin's basement consists of mostly far-traveled crustal terranes that originated offshore in the Pacific Ocean. These Paleozoic and Mesozoic basement rocks belong to the Blue Mountains Province, which largely lies to the west in northeastern Oregon and southeastern Washington. The province is divided into the Olds Ferry, Baker, and Wallowa terranes, which represent ancient Pacific Ocean tectonic assemblages.

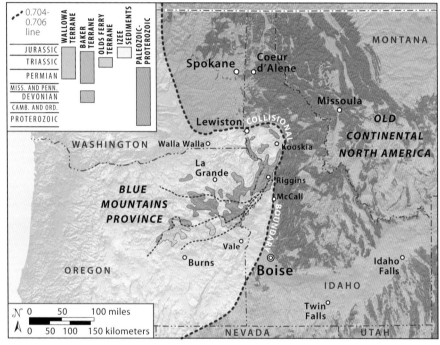

Accreted terranes in Idaho's Western Margin. The collisional boundary, also known as the Salmon River suture zone, is defined by the 0.704-0.706 line, a ratio of strontium isotopes in plutonic rocks.

Olds Ferry and Baker Terranes

In Idaho, rocks of the Olds Ferry and Baker terranes, along with the Weatherby Formation that overlies them, are restricted to a very limited area along the Snake River and in the Cuddy Mountains west of Cambridge. The Olds Ferry terrane consists of Triassic to Jurassic volcanic and sedimentary rocks intruded by Triassic to Cretaceous plutons. This combination of rocks was originally a chain of volcanoes above a subduction zone along the margin of North America. At some point, it peeled off from the continental margin and migrated offshore, even as Pacific Ocean plates subducted beneath it. The volcanic islands of Japan, which were once a volcanic arc on the margin of the Asian continent, are a modern tectonic analog to the Olds Ferry terrane. The arc shifted offshore as rifting opened up the Sea of Japan. We know this because basement rock beneath Japan's volcanic islands consists of ancient Asian crust. Evidence for continental Precambrian granitic rocks in the basement of the Olds Ferry terrane suggests that this terrane spent much of its evolution as a volcanic island arc fringing the North American continent.

The Baker terrane, north and west of the Olds Ferry terrane, represents part of the ancient subduction zone to the Olds Ferry terrane. It comprises mostly shale and chert layers containing blocks of Devonian to Jurassic limestone that were scraped off the subducting seafloor. Subduction zones are messy and consist of mélange, a mixture of oceanic sediments, oceanic crustal fragments, and chunks of deep crust and mantle—figuratively the dog's breakfast. Much of the mélange went down the subduction zone and then slivers of it were shoved by thrust faults back up to the surface, forming a great pile of disrupted crust at the leading edge of the overriding plate. Younger rocks of the Weatherby Formation and related units—Late Triassic to Early Jurassic age—lie on top of the Olds Ferry and Baker terranes. These younger sedimentary rocks are collectively referred to as the Izee overlap assemblage because they were deposited after the terranes were joined together and overlap both. The Weatherby Formation is mostly black shale and siltstone that was deposited in a deep marine basin long after subduction had ceased.

Wallowa Terrane

Most of the Blue Mountains Province exposed in Idaho consists of the Wallowa terrane, a Permian to Jurassic exotic island arc. It's considered exotic because, unlike rocks in the Olds Ferry terrane, it appears to contain nothing remotely relatable to North America. Much of its fossil assemblage belongs to a group of creatures that lived on the other side of the Pacific rather than close to North America.

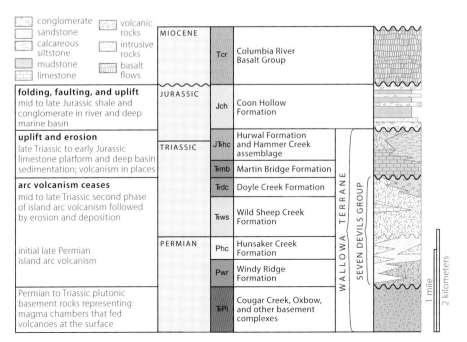

Stratigraphic column showing major units in the Wallowa terrane. —From Lewis et al., 2012

The basement to the Wallowa terrane consists of Permian and Triassic plutons, the magma chambers that fed magma from deep beneath the island arc to the volcanoes at the surface. Above the basement rocks is a thick sequence of Permian and Triassic volcanic and sedimentary rocks of the island arc, including the Permian Windy Ridge and Hunsaker Creek Formations, that belong to the Seven Devils Group. Many of the volcanic rocks in this older sequence are oceanic rhyolites. They differ from normal rhyolites in that they contain very little potassium, consistent with their oceanic crustal source, but they are high in silica just like their continental cousins. These oceanic rhyolites contain many of the copper-gold sulfide ore deposits that occur in upper Hells Canyon in the Homestead and Cuprum mining districts near Oxbow.

After the Permian volcanic and sedimentary rocks had formed, a temporary hiatus in island arc activity was followed by another intense episode in the Middle to Late Triassic Period. Rocks from this later episode, the most prevalent in Hells Canyon and the Seven Devils Mountains, consist of a thick, mostly volcanic sequence called the Wild Sheep Creek Formation followed by conglomerate, sandstone, and shale of the Doyle Creek Formation, the youngest formation in the Seven Devils Group.

The Wallowa island arc went through another period of volcanic quiescence in late Triassic time. A thick limestone platform formed around the eroded volcanic islands and is now the Martin Bridge Formation. Limy shales of the Hurwal Formation were deposited in deeper water along the edge of the platform. The arc was rejuvenated for a relatively brief time around the Triassic-Jurassic time boundary to deposit the Hammer Creek volcanic and sedimentary rocks. A final episode of sedimentation—the conglomerate, sandstone, and shale of the Coon Hollow Formation—took place in deepening marine basins on the eroding islands in late Jurassic time.

The final episode of activity in the Wallowa terrane consisted of extensive intrusion of tonalite, quartz diorite, diorite, and gabbro plutons in late Jurassic and early Cretaceous time. In many regions of the Western Margin, such as along the Clearwater River, these rocks are the most commonly encountered units in the Wallowa terrane. Metamorphosed equivalents to most of the Permian-Jurassic rocks in the Wallowa terrane occur along the Salmon River near Riggins, where the Wallowa terrane collided with the North American margin in Jurassic to Cretaceous time.

TERRANE COLLISION AND THE SALMON RIVER SUTURE ZONE

The Salmon River suture zone, a major highlight of Idaho's geology, was one of the first places in western North America where pioneering geologists in the late 1800s and early 1900s first recognized a fundamental crustal boundary where two very disparate rock assemblages were juxtaposed. The rocks to the east consisted of familiar sequences to these early geologists who knew the geology of the northern Rocky Mountains in the interior of the continent. The sequences to the west were anomalous, and these geologists immediately recognized their oceanic character despite the fact that they are now located more than 370 miles from the nearest coastline. This conundrum took time to work out; after all, it was well over a half century later that the fundamentals of plate tectonics were fully defined in the 1960s. By the late 1970s and early 1980s, the terrane paradigm arrived, which proposed that one mechanism for continental growth through time is by the accretion of far-traveled crustal fragments to continental margins. The suture zone of western Idaho was considered Exhibit A in this once radical idea that has now become a fully accepted tectonic concept.

The suture zone consists of a multitude of units that have been variably metamorphosed, many to such a degree that it's difficult to determine what rock units they were before the collision. The suture zone has also been intruded by younger plutons and covered by younger rocks along much of its length. Even where the suture zone is exposed, it is often difficult to locate the exact boundary between the oceanic Wallowa terrane rocks and the continental North American rocks. This situation is not unusual; suture zones are incredibly messy.

A very useful tool was developed in the 1970s and 1980s to help determine the location of a suture line, and western Idaho was a key location in the tool's development. An analysis of strontium isotopes in plutons can tell you whether the magma rose through continental or oceanic crust. Magma rising through the crust acquires strontium isotopes from its source rocks and from the rocks it passes through as it ascends into the crust. The strontium isotopes are analyzed as the ratio $^{87}Sr/^{86}Sr$, with ^{87}Sr being the decay product from the breakdown of radioactive ^{87}Rb, which is concentrated in older continental crust. Thus, a high $^{87}Sr/^{86}Sr$ ratio (above 0.706) indicates continental basement, while a low $^{87}Sr/^{86}Sr$ ratio (below 0.704) is considered oceanic basement.

The 0.704-0.706 line, which indicates one type of crust versus the other, can be drawn on the map of Idaho (see map on page 182). The line separates plutons of the accreted terranes from those of the ancient North American continent. The continuation of this line runs along the western side of North America from southern Mexico to northern Canada. In most other locations, the "line" is actually a pretty broad zone, on the order of tens of miles across, that encompasses the transition between values of 0.704 and 0.706. In Idaho, the line is quite sharp, less than 1 mile to a couple miles wide at most, and it runs north-south through most of the southern half of western Idaho. However, at the Clearwater River, it bends to the west and disappears beneath the younger basalts of the Columbia Plateau. The simplest explanation for this elbow geometry is that it follows the precollision geometry of the North American margin that was established during rifting of the supercontinent Rodinia at the inception of the Pacific Ocean around 700 million years ago.

Most geologists place the collision of the Wallowa terrane with the North American margin at some time between 150 and 140 million years ago, from the end of Jurassic time to the beginning of Cretaceous time. It wasn't a sudden event, and its aftermath dragged on in pulses of crustal deformation and mountain building along the suture zone that lasted to about 60 million years ago. Regionally, this prolonged episode of deformation and mountain building forms a part of the Sevier and Laramide orogenies. Wallowa terrane rocks in a wide belt on the western side of the Salmon River suture zone were deeply buried and metamorphosed during the collision. Collectively they are called the Riggins Group. The Permian volcanic and sedimentary rocks of the Wallowa terrane were metamorphosed to the Fiddle Creek Schist, Triassic volcanic and sedimentary units became the Lightning Creek Schist, and the numerous marble units were once limestone beds of the Wild Sheep Creek and Martin Bridge Formations. The Hurwal Formation is considered the likely precursor for the Lucile Slate, and the Weatherby Formation of the Izee overlap assemblage is a likely candidate for the Squaw Creek Schist.

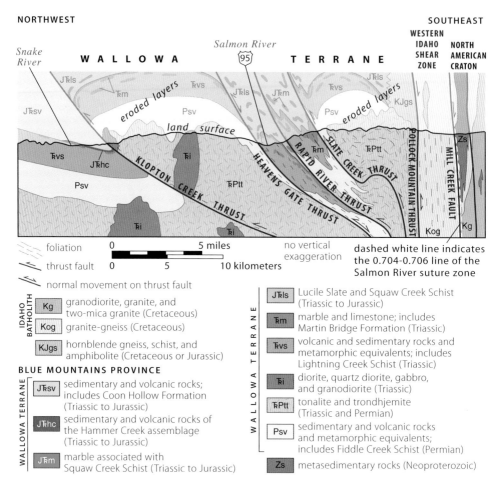

Schematic cross section across the Salmon River suture zone and the eastern side of the Wallowa terrane near Riggins, Idaho.

Metamorphism increases eastward across the Salmon River suture zone to a maximum grade immediately adjacent to the suture (the 0.704-0.706 line). The rocks here, which are too metamorphosed for geologists to know their predecessors with certainty, are high-grade hornblende gneiss and amphibolite with marble and quartzite lenses of the Riggins Group. Equivalent high-grade metamorphic rocks along the east-west oriented segment of the suture zone farther north are called the Orofino Metamorphic Suite. These rocks were buried more than 18 miles deep in the crust by around 120 million years ago; mountain building thickens the crust considerably, causing rocks within the crust to be deeply buried and metamorphosed to high grades.

During and following metamorphism, the metamorphic rocks on the western side of the suture zone were sliced and thrust to the west, pulling higher-grade metamorphic rocks from deeper in the crust in the east and sliding them up and over shallower crustal units to the west. The faults that accommodated this deformation along the Salmon River include the Pollock Mountain, Rapid River, and Heavens Gate thrust faults.

By its very nature, a suture zone requires that the subduction zone that brought together the island arc and the continent is extinguished once accretion occurs. In many cases, however, subduction is rekindled outboard of the accreted terrane following collision. The hallmark of revived subduction is the formation of a new arc that overprints the previous suture zone. In Idaho's Western Margin, the initiation of a new subduction zone is represented by a suite of tonalite and trondhjemite plutons that was emplaced at middle to lower crustal levels in the thickened crust around 120 to 110 million years ago. Well-exposed plutons of this suite include the Six Mile Creek pluton along the Clearwater River, the Blacktail pluton along the South Fork of the Clearwater River, and the Hazard Creek complex along the Little Salmon River.

We know these plutons crystallized about 18 miles deep in the crust because they contain magmatic epidote, a mineral that crystallizes at high pressures. Another form of epidote is a common secondary mineral that crystallizes from hot water circulating through rocks in veins, but magmatic epidote in plutonic rock occurs as well-formed crystals surrounded by other minerals that crystallized from the magma. The presence of magmatic epidote in these plutons at Earth's surface today implies that there has been over 18 miles of uplift and erosion along this belt on the western side of the Salmon River suture zone. Successively younger plutons intruded shallower crustal levels in an eastward progression of continued arc magmatism that culminated with intrusion of the massive Idaho batholith to the east. The oldest plutons of the Idaho batholith intruded along the eastern side of the Salmon River suture zone starting at about 98 million years ago.

Magmatic epidote crystals (elongate minerals indicated by arrows) in a sample of tonalite from a pluton intruded between 120 and 110 million years ago along the western side of the Salmon River suture zone. The epidote signifies magma crystallization at depths of at least 18 miles in the crust. Large black minerals are biotite. —Photo by Reed Lewis

By about 100 million years ago, the dynamics of mountain building along the Salmon River suture zone changed. Rather than accommodating strictly convergence of the accreted Wallowa terrane with the North American margin, the suture zone began to accommodate both convergence and northward sliding of the Wallowa terrane. This type of composite motion is known as transpression (essentially translation and compression together), and the main structure active during this time is known as the Western Idaho shear zone.

The transpressional Western Idaho shear zone, affectionately known as the WISZ to Idaho geologists, is superimposed on the Salmon River suture zone along its north-oriented segment. However, northward slip of the Wallowa terrane into the elbow in the suture zone to the north introduced a geometric problem because the Wallowa terrane was being stuffed into a corner. The northwest-trending, northeast-dipping Ahsahka shear zone along the Clearwater River developed to accommodate the bottleneck. This episode of transpressional tectonics continued to about 90 million years ago and was followed by a series of complex structures that further kinked the elbow into its present geometry; one of these later structures is the Mt. Idaho shear zone near Grangeville. This long episode of faulting and folding along the suture zone had finally dwindled by the time the main phase of the Bitterroot lobe of the Idaho batholith intruded 66 to 53 million years ago, nearly 100 million years following collision. Earth's tectonic collision belts are indeed very long-lived features!

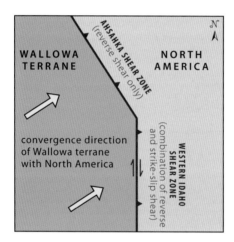

Schematic map showing different styles of shearing on the Ahsahka shear zone and Western Idaho shear zone between about 100 and 90 million years ago due to their orientations relative to the overall plate convergence direction.

MIOCENE FLOOD BASALTS

More than 50 million years of mostly erosion followed the collision of the Blue Mountains terranes and the subsequent mountain building. Starting at about 17 million years ago and continuing until 6 million years ago, huge volumes of basalt lava erupted from north-trending fissures located near the western border of Idaho. The lava flowed across eastern Oregon and Washington and ebbed into western Idaho across the ancient erosional surface to form the Columbia River flood basalt province. Most geologists point to the arrival of a mantle plume as the cause of this episode, the same plume that went on to become the Yellowstone hot spot. The Columbia River flood basalt province is the youngest and smallest flood basalt province on Earth, among scores of continental and oceanic flood basalt provinces that have erupted throughout Earth's history. Despite this lowly distinction, the Columbia River province is impressive with more than 350 individual lava flows, many of which flowed over 300 miles to the Pacific Ocean. All told, this massive pile of lava amounts to 50,000 cubic miles, covering over 80,000 square miles of the Pacific Northwest. The lava is more than 2 miles thick near the center of the plateau but decreases in thickness toward the edges. Idaho, toward the eastern margin, has a relatively thin sequence of these flows—no more than a half mile thick at most—that mostly lie in two major embayments, the Clearwater and Weiser embayments. This collection of lava flows is known as the Columbia River Basalt Group.

Basalt lava is hot and very fluid, and the fact that basalt lava travels long distances isn't surprising to most geologists. The main obstacle lava must overcome if it is to travel far is that it can't lose much heat or it will solidify. Lava flows in volcanically active parts of the Earth today such as Hawaii travel like rivers down drainages away from vents, and the tops, sides, and bases of these flows quickly lose heat to their surroundings and solidify. The solid exterior insulates the molten lava within, which continues flowing through the tube-shaped conduit. These types of flows can advance miles to tens of miles in the span of hours to days. Typical Hawaiian eruptions emplace around 0.1 cubic miles of lava over the course of a year.

Columbia River basalt flows operated in a similar fashion to their Hawaiian counterparts, except at an astoundingly different scale. During the effusive early stages of eruption of the Columbia River Basalt Group, in which more than 80 percent of the total volume of basalt erupted in less than 1 million years, individual lava flow volumes topped 480 cubic miles. Rather than in narrow lava tubes, these humungous flows advanced as lobes, which later coalesced into sheets that inflated to permit large volumes of magma to feed the lava fronts as they advanced hundreds of miles westward across the Columbia Plateau.

The pre-eruption topography was a more subdued version of today's dramatic topography, but even so, it included deep canyons and high mountain peaks. The early flows quickly filled the deep canyons and then spilled out across the canyon divides. When enough time passed for new drainages to establish themselves on the young plateau, new lava flows dammed the drainages to produce lakes and wetlands. Lakes and stream sediments ranging from sand to mud are preserved as sedimentary interbeds between basalt flows. The interbeds are collectively referred to as the Latah Formation in the north and the Payette Formation in the south, and many interbeds preserve a spectacular Miocene assemblage of fossil plants, fish, and insects. The Miocene climate was tropical in the Pacific Northwest, and many tree species of the southeastern United States today have ancestors that grew on the Columbia Plateau in the Miocene Epoch.

This lava invariably flowed into lakes and wetlands, producing pillow lavas where the water was deep enough to prevent the molten lava from exploding. Basaltic tuff formed where the lava interacted violently with shallow water, exploding the lava into glassy fragments. By the time the entire sequence of lava flows had been emplaced around 6 million years ago, the Columbia River basin had been completely transformed to a lava plateau. The only features remaining from its previous history are some of the higher peaks, which poke out above the lava plateau. These ancient mountaintops are known as steptoe buttes from their namesake just across the border from Idaho in Washington. Some rose more than 4,000 feet above the Miocene river bottoms.

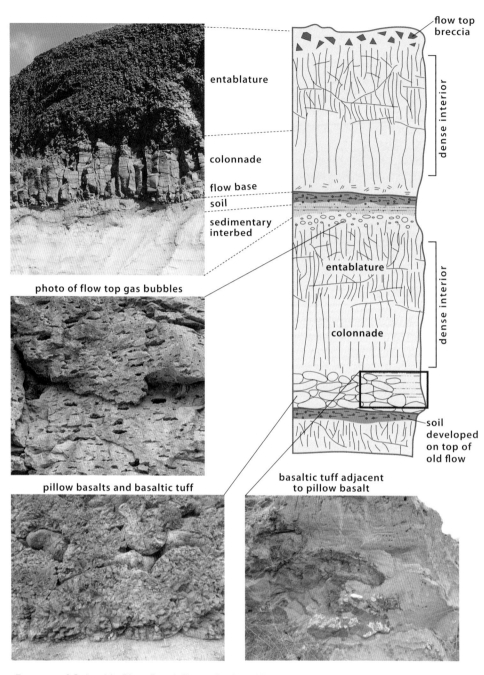

Features of Columbia River Basalt flows displayed in an idealized cross section through two flows stacked one on top of the other. A typical lava flow includes fractured basalt that forms as lava solidifies and continues to cool, forcing the new rock to shrink by about 10 percent by volume. This process is accommodated by fracturing, with vertical fractures propagating from the colder base and top inward into the warmer interior of the flow. The base of typical flows fractures more slowly and consistently because it cools slower, forming vertical columns in the part of the flow called the colonnade. The upper parts of lava flows cool more quickly because they are in contact with air and the fractures are much more randomly oriented, forming the part of a flow called the entablature. Rapidly solidifying lava flow tops trap gas bubbles rising through the lava and may become fragmented or brecciated as the solidified carapace to a flow is broken up by still-molten lava moving in the flow interior. —Column sketch from Miller, 2014

One interesting question that has occupied volcanologists is how long it took for one of these massive lava sheets to flow all the way from western Idaho to the Pacific coast. The short answer to this problem is a matter of years. One way to determine the rate of flow is to pay attention to what was engulfed by the advancing lava, and equally important, what was not. We can safely say that these lava sheets advanced fast enough that entire tropical Miocene forests were overrun by lava, leaving holes in the rock, called tree molds, where the hot lava instantly solidified against the much cooler trunks and branches. Once encased in hot lava, the woody material would largely burn out to leave the hole. The lava was moving fast enough that the forests were not completely destroyed by fire before the advancing lava engulfed the trees. On the other hand, it is extremely rare to find an animal cast in the basalt. One famous example from Washington state is a Miocene rhinoceros speculated to have been an engulfed carcass, or an individual that was too weak to escape, rather than a healthy animal. Apparently, the advancing lava was slow enough that larger animals were generally able to avoid being overrun even though the lava flows covered thousands of square miles.

Distinguishing the different basalt lava flows in the Columbia River Basalt Group from each other is quite difficult in outcrops because most flows look the same. In fact, most of the basalts show no visible crystals. Stratigraphers who study the basalt sequence use lava rock chemistry and magnetic polarity to separate the main sequence into four major formations, and each formation contains between tens and hundreds of individual lava flows. Eruption of the Columbia River Basalt Group and related units initiated with the Steens Basalt between about 16.8 and 16.6 million years ago

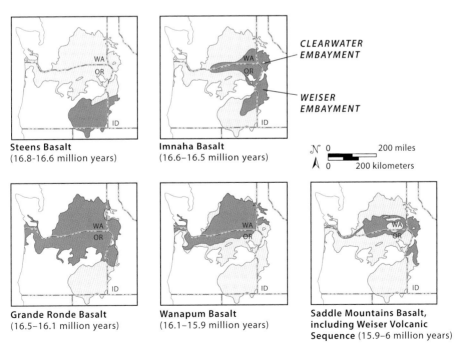

Distribution of major units of the Columbia River Basalt Group in the Pacific Northwest. Note that ages of the Columbia River Basalt Group were significantly revised in the 2010s. These numbers reflect results of Kasbohm and Schoene (2018). —Modified from Miller, 2014

in the southern part of the volcanic province that laps into the southwestern corner of Idaho. Eruption of the main sequence of the Columbia River Basalt Group began with the Imnaha Basalt that erupted between about 16.6 and 16.5 million years ago, followed by the Grande Ronde Basalt between about 16.5 and 16.1 million years ago. A whopping 93 percent of the entire volume of the Columbia River basalts erupted in less than 700,000 years to form these lower basalts.

This massive flux of lava to the Earth's surface occurred right at about the beginning of the mid-Miocene climatic optimum, a major global warming event in Earth's climate history. Basalt volcanism releases huge amounts of the greenhouse gas CO_2 and other gases that affect climate. Determining whether the main eruption stage of the Columbia River Basalt Group caused the mid-Miocene climatic optimum is currently an area of active research. The remaining volume of the Columbia River Basalt Group erupted as the Wanapum Basalt between about 16.1 and 15.9 million years ago and the Saddle Mountains Basalt that erupted intermittently up to about 6 million years ago. A distinctive group of andesite and basalt lava flows and some tuffs of the Weiser volcanic sequence also erupted during Saddle Mountains Basalt time between 15.0 and 14.8 million years ago.

CANYON CUTTING, UPLIFT, AND ICE-AGE FLOODS

As eruptions of the Columbia River basalts waned in the Late Miocene, the region's streams were able to start cutting canyons into the new plateau, although many of the late eruptions filled nascent canyons. Only after the last eruption 6 million years ago were the modern canyons able to fully establish themselves on the plateau.

Much of the crust beneath the plateau had been underplated by the dense, crystallized remains of magma systems that fed the basalt. In places, these dense layers peeled-off and sank into the mantle below. The freshly peeled areas of the plateau rebounded with the loss of weight, resulting in uplifting and mountain building. In the taller mountain ranges of the plateau, Wallowa terrane rocks that were once deeply buried by the thick sequence of Columbia River Basalts now form peaks that rise to more than 9,000 feet in elevation. The peaks in the Seven Devils Mountains, for example, were once covered with Imnaha and Grande Ronde lava flows, and there are many remnants of these lava flows at high elevations. Erosion accompanied the uplift, and downcutting began in the section of Hells Canyon adjacent to the Seven Devils Mountains, producing the deepest part of the canyon today. The highest parts of the Seven Devils, where uplift and erosion were the most pronounced, now expose basement rocks of the Wallowa terrane. Basin and Range faults cutting northward across the plateau also uplifted areas.

As the region's streams matured, they expanded their drainage basins. The stream that started cutting Hells Canyon was originally part of the Salmon River drainage basin. With time, it eroded higher elevations to the south, cutting Hells Canyon slowly over millions of years. This headward erosion eventually migrated far enough south that by 3 to 2 million years ago, it managed to cut through the original divide between the Salmon and Snake River drainage basins that existed at the northwestern end of the western Snake River Plain. This event spelled the demise of Pliocene Lake Idaho as its water began to escape north into Hells Canyon.

By the time the Pleistocene ice age began about 2 million years ago, Hells Canyon and other canyons in the Western Margin had been fully cut. Alpine glaciers covered

View west from the summit of He Devil (9,420 feet) in the Seven Devils Mountains across Hells Canyon of the Snake River (1,275 feet at river-level here) to the Wallowa Mountains in Oregon (9,838 feet), representing more than 8,000 feet of topographic relief. By this measure, Hells Canyon is the deepest gorge in North America, deeper than the Grand Canyon.

the mountains of the region and extended down their valleys high above these canyons. Glacially derived silt and fine sand was blown across the dissected Columbia Plateau to form thick deposits called loess. The upland farmland in the region today utilizes the fertile loess soil. Saber-toothed tigers, Columbian mammoths, and other Pleistocene fauna roamed the frozen tundra on the plateau.

Two major glacial flooding episodes affected rivers in the Western Margin. The Bonneville Flood spilled down the Snake River about 17,500 years ago when Lake Bonneville, the ancestral Great Salt Lake in Utah, catastrophically overtopped its northern shore. At peak flood, the narrower parts of Hells Canyon were nearly completely filled with raging floodwater. The flood left huge gravel bars in the more open areas such as Pittsburg Landing and Lewiston. The other flooding episode involved the catastrophic collapses of the Purcell Trench ice dam in northern Idaho that repeatedly backed up Glacial Lake Missoula between 19,000 and 14,000 years ago. The Western Margin of Idaho lay well upstream from the massive floods that swept across the Columbia Plateau, but the narrows such as at Wallula Gap on the Columbia River impeded the water. Floodwaters backed up repeatedly on the Snake River, drowning Lewiston, 139 river miles upstream from the Columbia, in over 500 feet of water during the biggest floods, and reaching as far as Pittsburg Landing, 215 miles from the Columbia River.

Liam's Rock, a granite ice-rafted boulder, melted out from a stranded iceberg about 500 feet above the modern water level during one of the larger Missoula floods. View east from Washington up the Snake River toward Lewiston in background. (46.4306, -117.1717)

US 12
Lewiston—Kooskia
74 miles

US 12 crosses north-central Idaho, following the Clearwater River from its confluence with the Snake River at Lewiston east to its headwaters in the Bitterroot Mountains. Meriwether Lewis and William Clark followed parts of this route in their exploration of the west in 1805–1806. The section of US 12 in the Western Margin traverses the Clearwater embayment in basalt on the eastern margin of the Columbia Plateau. The Clearwater River and its tributaries cut deep valleys, up to several thousand feet deep, into the stacked lava flows of the Columbia River Basalt Group and in many places into the underlying basement rocks of the accreted Wallowa terrane.

US 12 enters Lewiston, Idaho, from Clarkston, Washington, across a bridge that spans the north-flowing Snake River just upstream from its confluence with the west-flowing Clearwater River. Much of Lewiston lies hidden from view on a huge Bonneville flood bar that sits more than 80 feet above the bridge to the south. The modern Snake River cut down into the bar, revealing gravels of the Bonneville Flood that are clearly visible on the east (Lewiston) side of the bridge. The large bar, along with one underlying Clarkston, was deposited as floodwaters slowed as they exited Hells Canyon. Note the north-slanting beds, called foresets, in these gravels. Modern Snake River sandbars have similar foresets, which form during normal river flow

12 WESTERN MARGIN   195

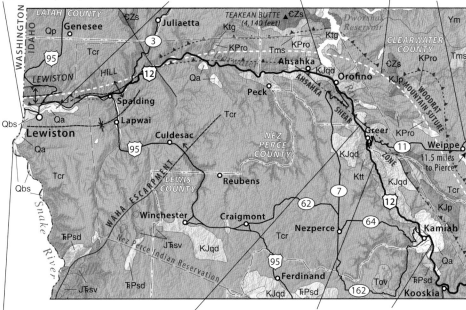

- Lewiston basin forms the lowest point in Idaho at the confluence of the Clearwater and Snake Rivers; it's also Idaho's only seaport
- large landslide complex on south side of Clearwater River at milepost 308
- spectacular exposures of 190- to 116-million-year-old mylonite in Ahsahka shear zone on both sides of Dworshak Dam
- dashed white line indicates the 0.704-0.706 line, demarking the edge of cratonic North America
- site of first gold strike in Idaho is 11.5 miles from Weippe in Pierce
- Missoula Flood deposits lying on Bonneville Flood deposits at Hells Gate State Park
- exfoliation domes developed in quartz diorite in Clearwater River Canyon between milepost 49 and 58
- 120- to 115-million-year-old epidote-bearing tonalite of the Six Mile Creek pluton at milepost 56 indicates crystallization at more than 18 miles depth in the crust
- Heart of the Monster, a basalt slide block south of milepost 68 that figures prominently in the Nez Perce creation story

QUATERNARY-TERTIARY
- Qa — alluvial deposits; includes landslides
- Qbs — Bonneville Flood deposits (Pleistocene)
- Tms — sedimentary rocks associated with flood basalts (Miocene)

VOLCANIC ROCKS
- Tcr — Columbia River Basalt Group (Miocene)
- Tov — volcanic rocks (Oligocene)

INTRUSIVE ROCKS
IDAHO BATHOLITH
- Ktg — tonalite, granodiorite and quartz diorite (Cretaceous)

BLUE MOUNTAINS PROVINCE
- Ktt — tonalite and trondhjemite (Cretaceous)
- KJqd — quartz diorite (Cretaceous and Jurassic)
- KJp — plutonic rocks along the Western Idaho shear zone (Cretaceous and Jurassic)

BLUE MOUNTAINS PROVINCE
WALLOWA TERRANE
- KPro — Riggins Group, Orofino Metamorphic Suite, and related rocks (Permian to Cretaceous)
- JŦsv — sedimentary and volcanic rocks (Jurassic and Triassic)
- ŦPsd — Seven Devils Group (Triassic and Permian)

SEDIMENTARY AND METASEDIMENTARY BEDROCK
- €Zs — Windermere Supergroup (Cambrian and Neoproterozoic)

LEMHI SUBBASIN OF BELT BASIN
- Ym — gneissic and schistose metasedimentary rocks (Mesoproterozoic)

Geology along US 12 between Lewiston and Kooskia.

conditions by the addition of sand to the leeward end of bars, causing them to build progressively downstream. However, in the case of the Bonneville Flood 17,500 years ago, the foresets are gigantic because the flow was colossal (on the order of 18 million cubic feet per second as compared to modern flood discharges up to 150,000 cubic feet per second). Instead of the river's normal sedimentary load of sand, the flood carried large cobbles and boulders.

Although no evidence is visible along the road here, the Missoula floods also left their mark on the Lewiston-Clarkston Valley. As floodwater backed up behind Wallula Gap on the Columbia River—150 river miles downstream—during repeated glacial outburst floods emanating from northern Idaho, the valley filled with flood-water. The present elevation of the confluence of the Clearwater and Snake Rivers is 740 feet. During the largest floods, the water rose above 1,250 feet elevation, 500 feet above this level. Icebergs carrying cobbles and boulders originally entrained in the great Cordilleran ice sheet to the north were rafted into the valley by floodwaters and then stranded on high-stand shorelines, leaving a "bathtub ring" of ice-rafted sediment on the valley sides as they melted. At lower elevations, the floods left telltale sand-silt sequences, called rhythmites, each of which consists of a sand layer overlain by a silt layer, the two together representing a single flood event. The sand layer was deposited as a flood flowed rapidly *up* the valley, and the overlying silt layer was deposited as the water slowly drained back *down* the valley through Wallula Gap. A good location to see both Bonneville gravel and Missoula sand/silt rhythmites is at

Bonneville (gray gravels containing boulders at bottom) and Missoula flood deposits (tan-brown silt and sand toward top) at Hells Gate, south of Lewiston. (46.3645, -117.0537)

the parking lot for the Army Corps of Engineers Hells Gate Habitat Management Unit, 5.3 miles south of US 12 on Snake River Avenue–Tammany Creek Road. The Corps of Engineers land lies adjacent to Hells Gate State Park.

The highway follows Main Street through historic downtown Lewiston. Established in 1862 following gold strikes upriver in the Clearwater River basin, this river port town was the original capitol of the Idaho Territory. The capitol was relocated near to its present seat in Boise in 1864, following new gold strikes in the Boise Basin.

Sand-silt rhythmite in Missoula flood deposits at Hells Gate, south of Lewiston. Current directions in sand sequences are upriver, deposited as floodwaters swept up the Snake River; current directions in silt sequences are downriver, representing the slow drainage back down the Snake River as floodwaters slowly drained through Wallula Gap. (46.3573, -117.0516)

The original location of Lewiston below the high Bonneville bar and at the confluence of two of the Northwest's great rivers established Lewiston as an important economic center for mining and timber, but it also subjected it to flooding on many occasions. Lewiston holds its place as Idaho's only "seaport" because a system of seven dams downstream on the Snake and Columbia Rivers allow barges to reach this far upstream.

East of Lewiston, the combined US 12 and US 95 (mileposts along this initial stretch are for US 95) continue up the Clearwater River parallel to the east-trending axis of the Lewiston basin, a broad syncline developed in lava flows in the Columbia River Basalt Group. The basin is bounded by two structural zones of faults and folds, the Waha escarpment to the south and Lewiston Hill structure to the north, that drop the basin relative to the plateaus on either side.

At milepost 308, the highway passes a large landslide complex across the Clearwater River to the south. The prominent continuous bench in the hillside marks the location of a thick sedimentary interbed consisting of sand, silt, and clay, known here as the Sweetwater Creek interbed, between the Wanapum Basalt and Saddle Mountains Basalt. The interbed thickens dramatically into the Lewiston basin, indicating that the Waha and Lewiston Hill structural zones were actively dropping the Lewiston basin during the interbed's deposition around 16 million years ago. The landslide occurred in the interbed's weak sedimentary material, taking large blocks of the overlying Saddle Mountains Basalt with it. This particular interbed is responsible for numerous landslide problems in the region.

Between mileposts 305 and 304, US 12 splits off from US 95 following the east-northeast trend of the Clearwater River. Mileposts from here onward are for US 12. The Lewiston Hill structure forms an anticline that follows the river northeast to where the river bends to the east at about milepost 21, just east of Cherry Lane Bridge. As is typical for the very hard and brittle basalts, which do not fold easily, the river has preferentially eroded the shattered hinge of the anticline, leaving remnants of the fold limbs

Landslide complex across Clearwater River at milepost 308. The prominent high bench marks the location of the Sweetwater sedimentary interbed, the weak unit that slides. It is sandwiched between Wanapum Basalt below and Saddle Mountains Basalt above. (46.4280, -116.9181)

in the canyon walls on both sides of the river. Basalt and sedimentary interbeds in the limbs of this fold dip from 10 to 60 degrees, and the dipping layers are more obvious where sedimentary interbeds occur between the basalt layers. Notable examples include roadcuts on the north side of the road between mileposts 14 and 15 that expose rocks that dip about 60 degrees to the south and on the south side of the highway at milepost 19 (just west of the junction with Gifford-Reubens Road), which dip 30 degrees to the south. Basalt layers along this stretch belong to the Imnaha and Grande Ronde Basalts.

Just east of milepost 31, Mesozoic basement rocks appear in roadcuts beneath the basalt. These outcrops consist of Jurassic to Cretaceous diorite and gabbro plutons that were intruded into the Wallowa terrane. At milepost 35, just east of the turnoff to Peck, a large roadcut on the south shows some of the wall rocks that these plutons intrude: highly metamorphosed hornblende-bearing gneiss and amphibolite and minor calc-silicate gneiss that were likely Permian-Jurassic marine sediments and island arc volcanics of the Wallowa arc. The term *calc-silicate* refers to metamorphic minerals consisting of calcium and silica that formed from metamorphism of calcite, a calcium carbonate mineral.

A little west of milepost 39, near Pink House Recreation Site, the highway crosses a major structure, the Ahsahka shear zone. This fault zone trends northwest and dips steeply to the northeast. The rocks in the shear zone are mylonite, ductily deformed crystalline rocks that show stretching and even ribboning of minerals during shear and metamorphism. As one crustal block moves past another across a shear zone, asymmetry in the shapes of stretched minerals gives clues to the direction of shear. At this location, they show that the more deeply seated rocks of the Orofino Metamorphic Suite on the northeast side of the shear zone moved up and to the southwest over the top of the more shallowly seated rocks of the Wallowa terrane on the southwest side of the shear zone.

Spectacular exposures of mylonite in the Ahsahka shear zone occur on both sides of Dworshak Dam, which comes into view for eastbound travelers at milepost 40. Construction of the dam in the 1970s on the North Fork of the Clearwater River ended the annual threat of flooding to Lewiston. To get to the dam, cross the Clearwater River at the Orofino Bridge at milepost 44 and head 5 miles west on ID 7 to Ahsahka, where turnoffs on either side of the North Fork of the Clearwater River lead to the northwest and southeast sides of the dam. The shear zone, in 116-million-year-old quartz diorite, is nicely exposed in outcrops next to the southeast abutment of the dam along the road at Bruce's Eddy launch area. On the northwest side of the dam, drive to Big Eddy marina and walk down the small boat ramp on the north side of the marina to view 96-million-year-old granodiorite and metasedimentary rocks of the Orofino Metamorphic Suite that have been intensely mylonitized. Note that the outcrops at Big Eddy are only accessible when the level of Dworshak Reservoir is drawn down, usually late August through early April.

Southeast of Orofino, US 12 recrosses the Ahsahka shear zone into metamorphosed Wallowa terrane rocks to the southwest of the shear zone. The Clearwater Canyon becomes narrow and rocky between mileposts 49 and 58. The rocks in the canyon consist of Jurassic and Cretaceous quartz diorite that intruded the Orofino Metamorphic Suite before the Ahsahka shear zone became active. The intrusive rocks make tall, rounded cliffs formed by slabs of rock exfoliating as the rocks weather, much like skin peeling off an onion.

Ahsahka shear zone at Bruce's Eddy. View in both photographs is to the northwest showing northeast-over-southwest movement across the shear zone. (46.5117, -116.2923)

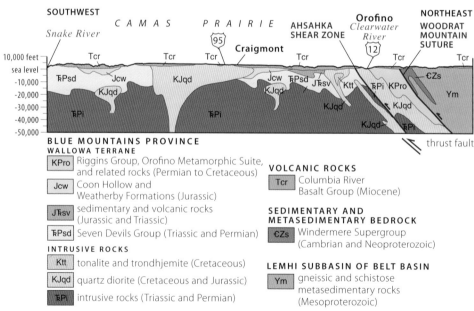

Cross section from the Snake River northwest across the Camas Prairie to Orofino and the Woodrat Mountain suture.

BLUE MOUNTAINS PROVINCE
WALLOWA TERRANE

- **KPro** — Riggins Group, Orofino Metamorphic Suite, and related rocks (Permian to Cretaceous)
- **Jcw** — Coon Hollow and Weatherby Formations (Jurassic)
- **JTRsv** — sedimentary and volcanic rocks (Jurassic and Triassic)
- **TRPsd** — Seven Devils Group (Triassic and Permian)

INTRUSIVE ROCKS

- **Ktt** — tonalite and trondhjemite (Cretaceous)
- **KJqd** — quartz diorite (Cretaceous and Jurassic)
- **TRPi** — intrusive rocks (Triassic and Permian)

VOLCANIC ROCKS

- **Tcr** — Columbia River Basalt Group (Miocene)

SEDIMENTARY AND METASEDIMENTARY BEDROCK

- **€Zs** — Windermere Supergroup (Cambrian and Neoproterozoic)

LEMHI SUBBASIN OF BELT BASIN

- **Ym** — gneissic and schistose metasedimentary rocks (Mesoproterozoic)

South of milepost 51, the highway passes the bridge to Greer. ID 11 continues from Greer to Pierce, the site of the first gold strike in Idaho in 1860. Pierce lies on the North American continent proper, in an ancient Tertiary basin. Gold-bearing gravels in the basin were placer mined, a process involving washing the gravels through a flume containing riffles that capture the heavy gold particles. Later mining activity included hard-rock mining of ore veins in Precambrian metamorphic rocks and the Cretaceous and Tertiary igneous rocks that intrude them. The initial gold discovery brought a flood of new settlers into north-central Idaho that upset the delicate balance between Nez Perce tribal society and the first settlers, as established in an initial treaty in 1855. A new treaty, drawn in 1863 between the US government and a few of the Nez Perce bands, ceded much of the tribal land promised by the first treaty. Mounting conflicts between "non-treaty" tribal members and ever-increasing numbers of new settlers in the region culminated in the 1877 Nez Perce War.

Another important rock that intrudes the Wallowa terrane adjacent to the suture zone appears along the highway at milepost 56 (pullout on the river side of the road). These rocks are 120- to 115-million-year-old, epidote-bearing trondhjemite of the Six Mile Creek pluton. Trondhjemite, a granitic rock, is very light colored and contains considerable quartz and sodium-rich feldspar but lacks the potassium-rich feldspar that makes a true granite. The reason this important mineral is missing is that the magma for these rocks was sourced from oceanic crustal basement of the Wallowa terrane. The rock contains the type of epidote that crystallized from magma, indicating crystallization more than 18 miles deep in the crust. The highway crosses back into hornblende-bearing quartz diorite rocks that persist in roadcuts to Kamiah. A particularly nice roadcut is south of Six Mile Creek, just south of milepost 59 (large pullout on river side of road).

View of Kamiah and the Clearwater River valley looking south. Rocks in the foreground and flat-topped ridges are Miocene Columbia River Basalt Group. —Photo by Reed Lewis

At Kamiah, the Clearwater River valley broadens dramatically. Between Kamiah and Kooskia, US 12 crosses rocks of the Columbia River Basalt Group that filled an ancestral canyon. Large landslides contributed to broadening the valley here, and numerous slide blocks are preserved on the valley sides and floor. An impressive block in a field between the highway and river can be seen at the Nez Perce National Historical Park between mileposts 68 and 69. The basalt block is known as Heart of the Monster and is the creation site for the Nez Perce people.

US 95
Weiser—New Meadows
76 miles

US 95 between Weiser and New Meadows generally follows the Weiser River across the northern part of the Weiser embayment of the Columbia Plateau. Beneath the Miocene volcanic rocks and sedimentary deposits of the embayment are rocks of the Olds Ferry terrane, which likely began as a volcanic arc along the western side of North America. Hells Canyon lies just west of this route, and the Salmon River suture zone lies to the east.

Weiser is located on the Snake River at the northern margin of the western Snake River Plain, a northwest-trending graben system filled with Miocene and Pliocene volcanic rocks and sediments. Many of the sediments are from Lake Idaho, which existed between 10 and 3 million years ago. The lake drained into Hells Canyon when a north-flowing stream breached its drainage divide, less than 20 miles downriver from Weiser, between 3 and 2 million years ago.

North of Weiser, US 95 leaves the western Snake River Plain and enters the Weiser embayment as it winds its way northward through low, easily eroded hills in sedimentary deposits of the Payette Formation. These sediments are older than the Lake Idaho sediments and consist of mostly sand deposited by ancient rivers, some muddy sediments, and layers of volcanic ash deposited during explosive eruptions of nearby volcanoes. The volcanic ash layers are easy to spot in roadcuts and outcrops by their distinct white color.

North of milepost 91, craggy outcrops on hilltops and ridges belong to the Weiser volcanics, coeval with Saddle Mountains Basalt, that generally lie above and are interlayered with the Payette Formation. North of Mann Creek, the highway climbs Midvale Hill. At milepost 96, a big roadcut on the east side of the road exposes several Weiser basalt lava flows interlayered with muddy sediments of the Payette Formation that were deposited in an ancient lava-dammed lake. The light-colored muddy sediments were baked to reddish brown along their contact with the once-scorching basalt lava. Look for basaltic tuff and chunks of volcanic glass where lava flows reacted explosively with water as the flows advanced into the lake. The prominent white layer under the uppermost lava flow is diatomite, a rock composed of microscopic shells of silica from plankton that were living in the lake. As the organisms died, their shells accumulated on the floor of the lake to form the deposit. Exposures of basalt and the Payette Formation continue along the highway across Midvale Hill Summit.

Midvale sits along the southwest-flowing Weiser River, which drains most of the Weiser embayment between the West Mountains on the skyline to the east and

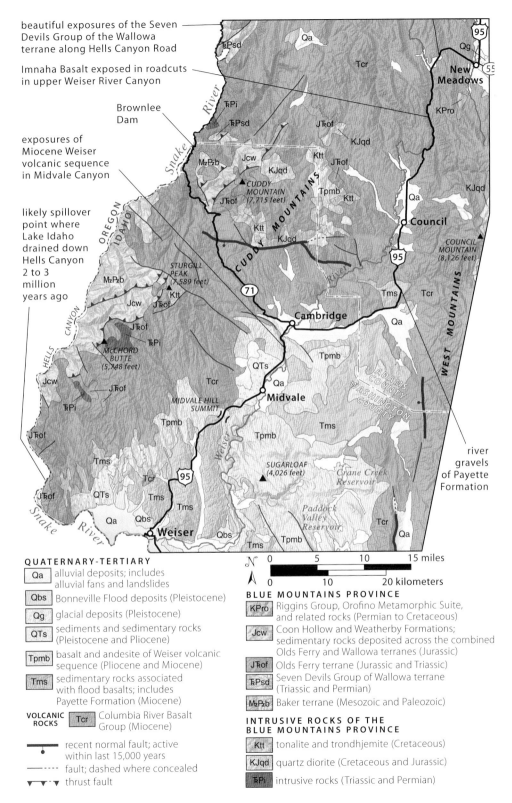

Geology along US 95 between Weiser and New Meadows.

the Cuddy Mountains to the west. Between the broad valleys that host Midvale and Cambridge lies Midvale Canyon, which the Weiser River cut through a ridge of the Weiser volcanic sequence. Nearly continuous roadcuts display flow-banded basalt with vesicles and rubbly zones.

Roadcut showing interlayered Weiser basalt and Payette Formation at milepost 96. (44.4007, -116.8698)

View looking downstream in Midvale Canyon.

HELLS CANYON ROAD

More adventurous travelers with plenty of time can explore Hells Canyon on Hells Canyon Road. After driving to Brownlee Dam on ID 71, follow OR 86 to Oxbow, then cross the Snake River back to the Idaho side to drive the 21.6 miles of Hells Canyon Road along the east shore of Hells Canyon Reservoir to Hells Canyon Dam. Spectacularly exposed along this gravel road is the entire type section of the Seven Devils Group of the Wallowa terrane as defined by Tracy Vallier when he was a graduate student at Oregon State University in the mid-1960s, before Hells Canyon Dam was constructed. He went on to work for the US Geological Survey Marine Program, taking part in cruises to modern island arcs all over the world. It quickly dawned on him that what he had mapped in Hells Canyon of Idaho during his early career was an ancient island arc that was now part of North America.

The first 8 miles of Hells Canyon Road (set odometer at the bridge that crosses the Snake River from Oregon) cross through Wallowa terrane rocks that were deposited and erupted when the island arc first began forming in early Permian time. Most of the igneous rocks are unusually felsic for an island arc complex; they are light colored and contain quartz and feldspar crystals. We'd call these volcanic rocks *rhyolites* if they erupted on a continent. The problem with calling them rhyolite in the Wallowa terrane is that the feldspar is decidedly rich in sodium and poor in potassium. A common informal name for them is *oceanic rhyolite*.

Outcrops in the first 1.7 miles are mostly light-colored quartz and feldspar crystal tuffs and volcanic breccias of the Windy Ridge Formation. Between approximately 1.7 to 8 miles, outcrops along the road belong to the Hunsaker Creek Formation. Most of these rocks are volcanic breccia and sandstone derived from erosion of the volcanic rocks. Like its older Windy Ridge cousin, the Hunsaker Creek Formation includes largely felsic volcanic rocks, including some oceanic rhyolite tuff, although along Hells Canyon Road it is uncharacteristically mafic in places. These rocks were mineralized by magma-heated water that coursed through fractures in the rock and deposited veins rich in copper, gold, and silver. Where the hot water discharged on the seafloor, hot spring deposits rich in these metals built up on the flanks of undersea arc volcanoes that subsequently collapsed to form mineralized breccias in the Hunsaker Creek Formation.

View down Hells Canyon Reservoir. Limestone of the Triassic Martin Bridge Formation in foreground; volcanic and sedimentary greenstone of the Triassic Wild Sheep Creek Formation in distance.
—Photo by Adam Stocks

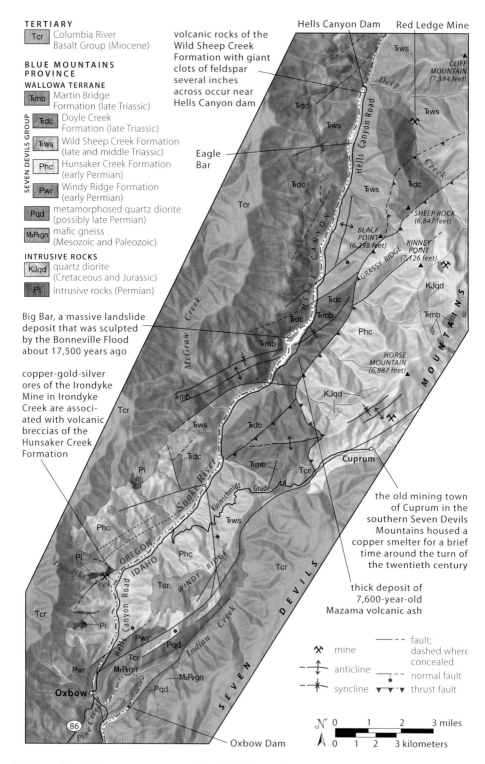

Geology of the Wallowa terrane surrounding Hells Canyon Reservoir. —Modified from Vallier, 1998; Lund, 2004

Prospector pits and mine dumps dot the hillsides along this part of the canyon; Irondyke Creek hosts one of the more successful copper-gold-silver mines that was in operation in the first third of the 1900s.

The remaining 13.6 miles of Hells Canyon Road pass through the Triassic section of the Wallowa terrane. After a hiatus of more than 20 million years following deposition of the Permian section, the Wallowa island arc entered a second phase of arc magmatism. The Middle to Late Triassic Wild Sheep Creek Formation consists of decidedly mafic volcanic rocks, mostly basalt and andesite, compositions much more common in oceanic island arc settings. Many lavas preserve pillow structures indicating that they erupted underwater. Sedimentary sections in the Wild Sheep Creek Formation consist of sandstone, shale, and some limestone. As the main Triassic volcanic phase waned, the arc eroded and the Doyle Creek Formation was deposited in adjacent basins. Distinctively reddish coarse sandstone and conglomerate are present in the Doyle Creek, the red color indicating an oxidizing environment either above sea level or in shallow water. Rocks of the Wild Sheep Creek and Doyle Creek Formations are present in outcrops from mile 8 to 12.2 and from mile 14.8 to the dam at mile 21.6.

Between mile 12.2 and 14.8, the road passes through a thick section of limestone that belongs to the Martin Bridge Formation, the next unit above the Doyle Creek Formation. This unit was deposited in a broad shallow carbonate bank built on top of the eroding arc, an environment that teemed with Triassic marine life. Although few fossils are found in the limestone along Hells Canyon Road, the Martin Bridge Formation elsewhere contains abundant corals, clams, brachiopods, and even marine vertebrates such as ichthyosaurs.

Just south of Allison Creek is a large deposit of young volcanic ash from the explosive eruption of Mt. Mazama in the Oregon Cascades volcanic arc that created Crater Lake approximately 7,600 years ago. The original air-fall ash deposit that blanketed the landscape wasn't very thick here, maybe an inch or two in total thickness. As rains washed the ash off the hillsides and into the valleys, however, it piled up in deposits that reach many feet or even yards in thickness on the valley floors.

Big Bar at about mile 13 is mostly submerged by the reservoir, but the sculpted island consists of remnants of a large landslide deposit that must have temporarily dammed the Snake River. Approximately 17,500 years ago, the Bonneville Flood surged through the canyon, sweeping away landslide debris finer than boulder-size material, then leaving a streamlined gravel deposit over the top of the original landslide deposit.

Big Bar near mile 13, an ancient landslide deposit sculpted by the Bonneville Flood 17,500 years ago. View to west. —Photo by Adam Stocks

At Cambridge, travelers can head 28 miles northwest on ID 71 to Brownlee Dam, the gateway to Hells Canyon. ID 71 crosses a broad pass between Cuddy Mountain to the north and Sturgill Peak to the south before dropping down to the Snake River at Brownlee Reservoir. The highway travels across lower basalts (Steens, Imnaha, and Grande Ronde Basalts) of the Columbia River Basalt Group almost the entire way. The mountains to either side of the pass expose the deep basement rocks of the Blue Mountains terranes, and these exposures are particularly important to our understanding of these various terranes. From northwest to southeast across this range are tectonic slices of the Wallowa, Baker, and Olds Ferry terranes that represent, respectively, pieces of an exotic island arc, the subduction zone complex for an ancient oceanic subduction zone, and a continental arc that was pulled offshore along the fringe of the North American continent.

North of Cambridge, US 95 heads east, away from the Weiser River and toward the West Mountains that lie straight ahead on the skyline. This mountain range is bounded by north-trending Basin and Range normal faults that raise the range and drop the Weiser embayment down to the west. Low hills to either side of the highway are Payette Formation sedimentary layers. At milepost 116, north-dipping Weiser basalt lava flows that lie above the Payette Formation can be seen in hills to the south. East of milepost 120, the highway enters a small stream valley that has eroded below the Payette sedimentary rocks and into the underlying basalt lava flows of the Columbia River Basalt Group. These rocks appear in roadcuts for the next few miles. North of milepost 126, the highway turns north and climbs out of the valley and back up into Payette Formation strata that here consist of mostly cobble-sized gravel deposited by a Miocene river system. North of milepost 130, the road again dips back down into the lower Columbia River Basalt Group, which persists all the way to Council.

North of Council, the open valley of the south-flowing Weiser River lies to the west. The broad peak to the east is Council Mountain, mostly metamorphosed igneous rocks of the Hazard Creek complex that intruded the Blue Mountains terranes along and west of the suture zone after their collision with North America. These rocks were deeply buried and metamorphosed prior to 110 million years ago. Lava flows of the Columbia River Basalt Group cover the western flank of Council Mountain and are tilted down toward the west, indicating that uplift of the range occurred after the lava flows erupted. The main Basin and Range normal fault that accommodated uplift of the range is in Long Valley on the east side of the range. The massif in the distance to the west is the Cuddy Mountains, also basement terranes of the Blue Mountains Province.

Between mileposts 143 and 144, the highway drops into the beautiful forested upper canyon of the Weiser River. Higher roadcuts are in the Grande Ronde Basalt. Along the canyon floor, between mileposts 144 and 153, roadcuts expose the underlying Imnaha Basalt, distinctive in the Columbia River Basalt Group for its large feldspar crystals, typically reaching an inch or more in size.

In the upper reaches of the canyon, the valley opens and the Weiser River meanders through meadows. The highway turns east to cross a drainage divide and descends to the north-flowing Little Salmon River in Meadows Valley, another north-trending valley bounded by normal faults on either side.

US 95
NEW MEADOWS—GRANGEVILLE
80 miles

US 95 between New Meadows and Grangeville follows the Little Salmon and Salmon Rivers that are entrenched into the eastern side of the Columbia Plateau. These deep river valleys afford views into the older Paleozoic and Mesozoic basement rocks of the Wallowa accreted terrane that lie below the much younger Columbia River Basalt Group. This route has so many spectacular rock exposures that you might consider slowing down on this section to fully appreciate the delightful array of geologic treasures that lie in store. Traversing the western side of the Salmon River suture zone, the collisional boundary between the Wallowa terrane and North America, US 95 crosses rocks that have been extensively folded and faulted along a world-renowned showcase for tectonic collision belts. The highway is in the deepest crustal levels at New Meadows and traverses one east-dipping thrust fault after another as it travels northward into progressively less-metamorphosed rocks from more shallow regions of the crust.

New Meadows lies in Meadows Valley, a relatively young graben, a down-dropped fault block that formed during Basin and Range extension. It filled with Columbia River Basalt flows in Miocene time and later with glacial and river sediments. The

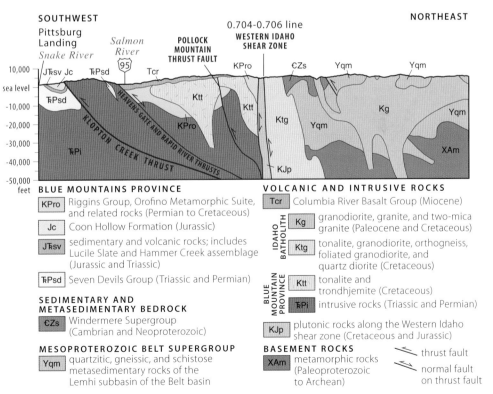

Cross section between Pittsburg Landing on the Snake River northeast across the 0.704-0.706 line to the Idaho batholith.

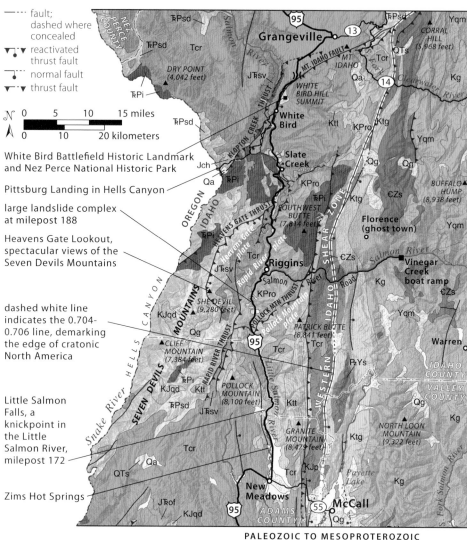

Geology along US 95 between New Meadows and Grangeville.

— — — fault; dashed where concealed
▼ ▼ reactivated thrust fault
● ● normal fault
▼ ▼ thrust fault

White Bird Battlefield Historic Landmark and Nez Perce National Historic Park

Pittsburg Landing in Hells Canyon

large landslide complex at milepost 188

Heavens Gate Lookout, spectacular views of the Seven Devils Mountains

dashed white line indicates the 0.704-0.706 line, demarking the edge of cratonic North America

Little Salmon Falls, a knickpoint in the Little Salmon River, milepost 172

Zims Hot Springs

QUATERNARY-TERTIARY
- Qa alluvial deposits; includes alluvial fans, and landslides
- Qg glacial deposits (Pleistocene)
- QTs sediments and sedimentary rocks (Pleistocene, Pliocene, and Miocene)

VOLCANIC ROCKS Tcr Columbia River Basalt (Miocene)

BLUE MOUNTAINS ISLAND-ARC COMPLEX
- KPro Riggins Group (Cretaceous to Permian): Rapid River plate (west) Pollock Mountain plate (east)
- Jch Coon Hollow Formation (Jurassic)
- JTrsv sedimentary and volcanic rocks; including Lucile Slate and Hammer Creek assemblage (Jurassic and Triassic)
- JTrof Olds Ferry terrane (Jurassic and Triassic)
- TrPsd Seven Devils Group of the Wallowa terrane (Triassic and Permian)

PALEOZOIC TO MESOPROTEROZOIC
- PzYs metasedimentary rocks of uncertain age (Paleozoic to Mesoproterozoic)
- €Zs sedimentary and metasedimentary rocks of the Windermere Supergroup (Cambrian and Neoproterozoic)
- Yqm quartzitic, gneissic, and schistose metasedimentary rocks of the Lemhi subbasin of the Belt basin (Mesoproterozoic)

INTRUSIVE ROCKS

IDAHO BATHOLITH
- Kg granodiorite, granite, and two-mica granite (Cretaceous)
- Ktg tonalite, granodiorite, and quartz diorite; includes deformed and metamorphosed units (Cretaceous)

BLUE MOUNTAINS COMPLEX
- Ktt tonalite and trondhjemite (Cretaceous)
- KJqd quartz diorite (Cretaceous and Jurassic)
- TrPi intrusive rocks (Triassic and Permian)
- KJp plutonic rocks along the Western Idaho shear zone (Cretaceous and Jurassic)

basalts are visible in roadcuts on the north end of town. US 95 follows the Little Salmon River, which flows northward through the graben. Between mileposts 163 and 164, the highway crosses the 45th parallel, the latitude halfway between the equator and the North Pole. Zims Hot Springs, on the left between mileposts 165 and 166, is fed by hot water that flows from relatively deep levels in the crust along the graben's normal faults, permeable pathways of pulverized rock.

Granite Mountain, the high point to the east of the valley, and Pollock Mountain, which becomes visible straight ahead in the distance to the west of the road north of milepost 166, consist of gneiss of the Hazard Creek complex, which is present west of the 0.704-0.706 line, the tectonic suture between North America and the Wallowa terrane. These gneissic rocks are part of a continuous belt of tonalite and trondhjemite plutons that intruded deeply buried Wallowa terrane rocks at lower crustal depths about 120 to 110 million years ago during continued convergence between the Wallowa terrane and North America following the initial collision. North of milepost 170, whitish weathering outcrops of the Hazard Creek complex appear on the east. A distinctive orange-weathering roadcut on the east between mileposts 171 and 172 consists of basalt breccia of the Columbia River Basalt Group. The lava flow encountered shallow water as it flowed into the Meadows Valley graben from the west, producing an explosive reaction that fragmented the lava into a glassy basaltic tuff.

The highway crosses the Little Salmon River at the northern end of Meadows Valley near milepost 172. Below the bridge is Little Salmon Falls, where the stream gradient dramatically changes from the nearly flat meadows southward to the narrow, steep-gradient canyon to the north. Faulting, which uplifted the region to the south and down-dropped areas to the north, disrupted the natural gradient of the river. The transition between the shallow and steep gradient stretches of the stream, known as a knickpoint, is slowly migrating upstream as the Little Salmon River attempts to smooth its gradient along the entire length of the river.

North of Little Salmon Falls is a steep canyon where roadcuts expose high-grade gneiss of the Riggins Group in the Salmon River suture zone. These rocks, which originated as island arc rocks of the Wallowa terrane, were metamorphosed to gneiss when they were buried more than 18 miles deep in the crust between 141 and 124 million years ago. A typical outcrop of well-layered hornblende gneiss appears across from the first big pullout on the west side of the road north of milepost 174.

Between mileposts 176 and 181, the highway traverses flows of the Imnaha Basalt, the lowest unit of the Columbia River Basalt Group, that partially filled a canyon cut into the gneiss. Many of the basalt lava outcrops contain vesicles encrusted with zeolite minerals (a hydrous, aluminum-rich silicate mineral group). These minerals formed after the lava had solidified but was still cooling, with lots of steam and hot water circulating through the hot rock.

Just north of milepost 184, the highway crosses the trace of the northeast-trending Pollock Mountain thrust fault. Like most major fault zones that extensively fracture the rocks they cut, this one is covered by grassy slopes that arise from intense weathering of the fractured rock. The deep-seated gneisses that make up the steep, narrow canyon in the Pollock Mountain plate to the southeast have been thrust northwestward over shallower-level rocks of the Squaw Creek Schist that lie within the Rapid River plate. The Squaw Creek Schist consists of well-layered, fine-grained, quartz-rich mica schists that commonly contain the mineral graphite, a mineral form of carbon.

Before metamorphism, they were probably Jurassic sedimentary rocks of the Wallowa terrane or an adjacent sedimentary basin. They were metamorphosed during burial between 124 and 113 million years ago as the Pollock Mountain plate was thrust over them from the southeast around 117 million years ago. Most outcrops between the fault and Riggins are Squaw Creek Schist, and good spots to look at them are across from the rest area just south of milepost 189 and along the bank of the Salmon River, accessed from a large pullout on the east side of the road at milepost 197, just south of Time Zone Bridge.

Zeolite minerals filling a vesicle in the Imnaha Basalt. The vesicle is about a half inch across.

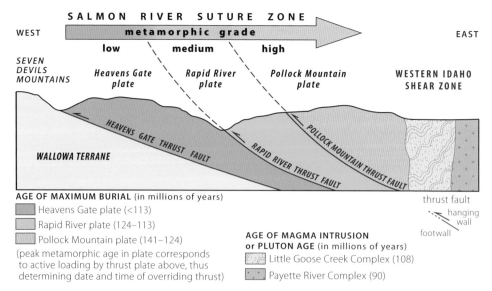

Cross section showing major thrust plates along the western side of the Salmon River suture zone of the Riggins area. —Modified from McKay et al, 2017

A huge landslide complex that extends more than 1 mile across on either side of milepost 188 occurs across the river to the west of the highway. Heavily weathered Squaw Creek Schist in the canyon sides above, just below the Columbia River basalts, are responsible for this deposit. Note the large scattered blocks of rock in the debris. High cutbanks along the Little Salmon River here indicate that past landslides dammed the river temporarily, and the deposits were subsequently breached to make the modern channel. The slope above the highway at milepost 188 has also been a problem, producing rockfall and rock slides that closed US 95 for several weeks in 2020.

North of milepost 194, the highway passes the junction with Seven Devils Road, which climbs west for 16.5 miles to Windy Saddle on the divide between the Salmon River Canyon and Hells Canyon of the Snake River, just northeast of the main peaks of the Seven Devils Mountains. The saddle provides some views, but better views can be had by driving another 1.5 miles north to the trailhead for Heavens Gate Lookout, a short walk from the parking area. The rocks of the Seven Devils consist of mostly Triassic volcanic and sedimentary rocks of the Wild Sheep Creek Formation of the Wallowa terrane. A major fault, the Heavens Gate thrust, runs north-northwest through Windy Saddle and separates more deformed rocks at the lookout to the east from less deformed rocks that make up the high peaks of the Seven Devils Mountains and Hells Canyon to the west.

View to the west of the Seven Devils Mountains from Heavens Gate Lookout. Rocks in foreground are chlorite schist and greenstone of the Triassic Wild Sheep Creek Formation of the Wallowa terrane deformed along the Heavens Gate thrust.

SALMON RIVER ROAD FROM RIGGINS

Salmon River Road, which intersects US 95 just south of milepost 195 on the south side of Riggins, heads 26 miles east (upstream) to Vinegar Creek boat ramp on the Salmon River and provides a spectacular transect across the Salmon River suture zone and Western Idaho shear zone. The first half of the road is paved, the rest is gravel. The road begins in the Rapid River plate along the western side of the Salmon River suture zone, in the Squaw Creek Schist. This formation was once limy siltstone and sandstone, and limestone of probable Jurassic age, now metamorphosed to schist, amphibolite, quartzite, and marble. These rocks continue for about 4 miles upriver. Between mileposts 4 and 5, the road makes a sharp righthand bend across from the Berg Ranch on the north side of the river. Roadcuts in the bend expose the Berg Creek amphibolite and

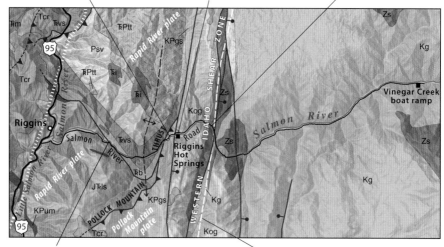

Pollock Mountain amphibolite between mileposts 8 and 9 contains garnets that indicate burial of these rocks deep in the crust following collision of the Wallowa terrane after 140 million years ago

rocks of the Hazard Creek complex have been sheared in the Western Idaho shear zone just west of milepost 9 near Riggins Hot Springs

metasedimentary rocks of the Neoproterozoic to early Paleozoic Windermere Supergroup appear east of milepost 11

Berg Creek amphibolite in the Rapid River plate between milepost 4 and 5 was deeply buried in the crust after 117 million years ago as the Pollock Mountain plate was thrust over it

dashed white line indicates the 0.704-0.706 line, demarking the edge of cratonic North America

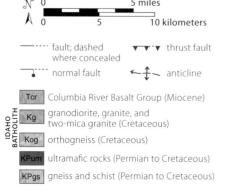

Simplified map of the geology along the Salmon River Road. —Modified from Gray, 2013

marble that record burial to a depth of over 18 miles in the crust, as a result of being overthrust about 117 million years ago by the Pollock Mountain thrust plate. In the amphibolite, look for zoisite, a calcium-rich silicate that forms tiny, clear-white needles; garnet, a deep-red, spherical mineral; and amphibole, a shiny, black mineral.

Roadcuts east (upstream) of the Berg Creek amphibolite expose rocks of the Lightning Creek Schist that include mostly biotite (black) and chlorite (green) schist. These rocks are interpreted to be metamorphosed andesite and basalt-rich Triassic Wallowa terrane rocks. Rocks of the Fiddle Creek Schist are exposed east of milepost 6, on the south side of Lake Creek Bridge. This schist's distinctive light-silver-green color results from the combination of chlorite (green) and muscovite (silver) micas. The Fiddle Creek Schist was likely part of the Permian section of the Wallowa terrane.

East of Lake Creek Bridge, the road continues along the north side of the Salmon River through schist and gneiss of the Fiddle Creek Schist. East of milepost 8, it crosses the Pollock Mountain thrust fault into garnet- and biotite-bearing Pollock Mountain amphibolite, which lies in the Pollock Mountain thrust plate. These rocks represent a higher metamorphic grade than rocks in the underlying Rapid River plate because they were thrust up from deeper crustal levels; they likely were derived from the same original sedimentary rock package in the Riggins Group. The rocks contain a strong foliation and steep lineation. The small garnets in the amphibolite slowly crystallized over the course of two metamorphic events. The garnet cores began growing about 140 million years ago when the rock was buried about 15 miles deep during the original collision of the Wallowa terrane with North America. The garnet rims record continued burial to depths of about 19 miles around 124 million years ago as convergence continued, and the crust was further thickened.

After passing Riggins Hot Springs commercial resort to the south, just west of milepost 9, outcrops along the north side of the road show a bewildering array of vertical sheets of metamorphosed, folded, and sheared plutonic rocks. These are likely part of the 120- to 110-million-year-old Hazard Creek complex that intruded the Pollock Mountain amphibolite when it was still deeply buried. The gneissic rocks, originally mostly

Gneiss and amphibolite in the Pollock Mountain thrust plate east of milepost 8 along the Salmon River Road. (45.4131, -116.1859)

tonalite and trondhjemite, were folded and sheared in the Western Idaho shear zone, which sliced along the already established Salmon River suture zone at 105 to 90 million years ago. They developed a strong, nearly vertical, north-trending foliation and steep lineation.

Tonalite-trondhjemite gneiss of the Hazard Creek complex in the western margin of the Western Idaho shear zone along the Salmon River Road. (45.4188, -116.1713)

The road continues upriver, passing more outcrops of metamorphosed Hazard Creek rocks in the Western Idaho shear zone. At approximately milepost 11, the road crosses the actual suture between Cretaceous Hazard Creek igneous rocks of the Wallowa terrane to the west and schist containing visible crystals of sillimanite, garnet, and biotite of the North American continent to the east. Before metamorphism, the schist was mudstone and sandstone of the Windermere Supergroup, deposited in Neoproterozoic to early Paleozoic time along the initial coastline of western North America as the Pacific Ocean successfully opened during rifting of the supercontinent Rodinia about 700 million years ago.

Between the Western Idaho shear zone and Vinegar Creek at the end of the road, a variety of plutons of the suture zone and Idaho batholith intrude schist of the Windermere Supergroup. Most of the rock in outcrops along the road is hornblende- and biotite-bearing tonalite and granodiorite of the Idaho batholith.

At Riggins, the north-flowing Little Salmon River joins the Salmon River flowing from the east. The Salmon, known as the River of No Return, has run 338 miles of its 425-mile total to this point, 79 miles of which pass through designated Wild and Scenic River sections of the Frank Church–River of No Return Wilderness just upstream. Riggins is situated in the hinge of the broad, southeast-plunging Riggins syncline, part of which is visible from town looking north to where the river takes a sharp left-hand bend. The Salmon River north of Riggins follows the trend of this fold in rocks of the Squaw Creek Schist.

Between mileposts 197 and 198, the highway crosses Time Zone Bridge (officially Goff Bridge) over the Salmon River. The time zone boundary follows the Salmon River across Idaho, separating northern Idaho, which is on Pacific Time, from southern Idaho on Mountain Time. In the north-trending Salmon River Canyon, which US 95 follows for approximately 25 miles between Riggins and White Bird, this creates an interesting, although very localized, conundrum. The east side of the river is on Pacific Time, while the west side of the river is on Mountain Time.

Just north of Time Zone Bridge, a large roadcut on the east side of US 95 exposes ultramafic rocks that occur in a pod between Squaw Creek Schist to the south and Lightning Creek Schist to the north. A huge pullout on the west side of the highway provides access to this impressive outcrop. The ultramafic rocks—consisting of mostly dark green, soft serpentinite with white talc and calcite—were likely peridotite before being metamorphosed in the presence of copious amounts of hot water. Peridotite, a rock consisting of mostly the magnesium-iron silicate mineral olivine, is the main rock type of the upper mantle. Alteration of olivine-rich rocks in water-charged environments produces serpentine. How rocks of this apparent origin came to be here, and in many other locations along and west of the Salmon River suture zone, is a topic of some deliberation. Most geologists who worked on these fascinating units consider them to be mantle slices that were faulted into their present locations within the suture zone during collision, but other ideas include their formation during rifting in a basin adjacent to the Blue Mountains terrane complex prior to collision. The serpentine is actually a mix of minerals consisting here of mostly antigorite (green) and magnesite (brown), with tiny chrome spinel crystals that form little black spots. Other minerals that occur in this ultramafic package include actinolite, biotite, and chlorite.

The highway traverses exposures of Lightning Creek Schist for the next few miles north of Time Zone Bridge. This unit consists of schist, gneiss, and marble that

Ultramafic outcrop on north side of Time Zone Bridge. The fault in the middle of the outcrop duplicates the ultramafic package and includes a slice of Squaw Creek Schist in the exposure. (45.4470, -116.3108)

originated as basalt and andesite volcanic rocks, and sedimentary rocks that include shale, sandstone, conglomerate, and limestone. Many geologists suspect that this unit is the metamorphosed Triassic section of the Wallowa terrane within the suture zone, particularly the Wild Sheep Creek Formation. A beautiful stretched-cobble metaconglomerate is exposed in roadcuts between mileposts 199 and 200 (pullout on west side). North of Fiddle Creek at milepost 201, roadcuts along the highway are dominated by the Fiddle Creek Schist, another metamorphic unit consisting of rock types similar to the Lightning Creek Schist, with notable differences: abundant schist and gneiss originated as oceanic rhyolite volcanic rocks (distinctively lighter in color than Lightning Creek metavolcanic rocks) and the absence of marble (metamorphosed from limestone). The Fiddle Creek Schist meets the compositional criteria for being metamorphosed Permian Hunsaker Creek Formation of the Wallowa terrane.

Just north of milepost 203, the river and highway cross a thick marble unit, still within the Rapid River plate, that forms prominent light-colored cliffs on both sides of the river. This marble is likely metamorphosed Triassic limestone of the Wallowa terrane. Rocks beneath the marble to the northwest (gray, crumbly rocks best exposed across the river) are mostly limy, graphite-bearing phyllite of the Lucile Slate, part of the Heavens Gate plate. The marble in the Rapid River plate was thrust northwestward over the Lucile Slate along the Rapid River thrust fault 113 million years ago.

Between mileposts 204 and 205, a prominent Salmon River gravel terrace stands out 400 feet above the river on both sides of the canyon. The terrace represents a time in the Pleistocene when the bed of the Salmon River was at a much higher level. The river subsequently eroded its canyon to the present level, leaving older riverbed gravels stranded high above on the valley side. Lucile Caves occur in the gravel terrace where calcite precipitated by springs cemented the gravel together. The caves are accessible from a rough trail at the base. Easy parking can be found at the Old Lucile boat ramp on the west side of the highway. From here to White Bird, numerous Pleistocene gravel river terraces occur along the river, and the highway makes impressive cuts through many of them.

Just south of milepost 207, the road crosses the Heavens Gate thrust fault. A large pullout on the east side of the road gives access to outcrops of Triassic metavolcanic rocks located at the northern end of the pullout. These rocks are intensely sheared in the Heavens Gate fault zone, so much so that individual mineral grains are stretched out into ribbons. The geometry of small shear zones in the outcrop show that the southeast side of the outcrop moved up relative to the northwest side. Deformed and metamorphosed Wallowa terrane rocks within the Heavens Gate plate southeast of here were thrust to the northwest over relatively unmetamorphosed Wallowa terrane rocks sometime after 113 million years ago. The Heavens Gate fault follows the highway for several miles north of here.

Between mileposts 208 and 209, US 95 crosses John Day Creek and old placer workings in Pleistocene gravel with an abandoned trommel on the east. The trommel, consisting of a cylindrical screened drum that rotates when in operation, was used to separate the sand-sized material that contained gold from coarser-sized gravels. These workings and others across the river and throughout the canyon were gold mining operations that continued as late as the 1980s.

Just south of the highway ponds between mileposts 210 and 211, the highway passes a large slump scarp developed in the limy phyllite of the Lucile Slate that lies

Small, nearly horizontal thrust faults in Permian tonalite of the Wallowa terrane. The Heavens Gate fault is 100 yards to the right of photo. (45.6126, -116.2786)

above the Heavens Gate fault in the cliff above the east side of the road. This weak unit has caused numerous landslide problems in the region, as displayed here by the extensive reconstruction of the highway. Prominent outcrops northwest of the highway ponds east of the highway consist of Permian tonalite and metamorphosed volcanic and sedimentary rocks in the Wallowa terrane that lie just beneath the Heavens Gate thrust. Note the small, nearly horizontal thrust faults in this outcrop.

Just south of the town of Slate Creek, between mileposts 213 and 214, impressive roadcuts east of the highway expose a broad shear zone in Permian tonalite and metamorphosed volcanic and sedimentary rocks. The shear zone is developed in the combined Rapid River and Heavens Gate faults that are telescoped together. Slate Creek grew rapidly in the early 1860s as a supply point for the Florence mining district, located about 30 miles up Slate Creek and to the southeast. The Florence gold placer deposit is in Tertiary and Quaternary gravels that formed in a small perched basin high above the Salmon River Canyon.

North of milepost 214, basalt of the Columbia River Basalt Group dominates the geology along the Salmon River Canyon. Lava flows of the Imnaha and Grande Ronde Basalts filled a canyon here between 17 and 15 million years ago, and north-trending normal faults cut the lava flows. Note the tilted layers of basalt that constitute many of these faulted blocks within the modern canyon. Uplifting and down-dropping by the faults disrupted stream gradients, causing tributaries to cut downward faster in places. The evidence for this dynamic process is very apparent in a dissected alluvial fan that is visible across the river between mileposts 217 and 218. The stream draining the side canyon cut a deep gully through its uplifted fan.

Dissected alluvial fan developed on faulted section of Columbia River Basalt Group lava flows between mileposts 217 and 218. The main fault responsible for uplifting the ridge in the middle ground lies just beyond the right-hand side of the photograph. (45.6585, -116.3060)

The road to Pittsburg Landing (Old US 95 to Deer Creek Road/National Forest 493) turns west from the highway at milepost 222. Pittsburg Landing, one of the few road access points to the bottom of Hells Canyon, provides closeup views of Wallowa terrane rocks and deposits of the Bonneville Flood.

North of White Bird, US 95 follows the White Bird fault, a major northeast-trending, southeast-dipping normal fault that drops the crustal block on the southeast side relative to the uplifted block that forms the ridge above the highway to the northwest. Normal faults are typically shaped like the runner of a sled—steeply dipping near the surface, like the leading edge of a runner, and then curving to shallower angles at depth. This geometry causes the down-dropped blocks to rotate as they slip along the fault and tilt back toward the fault. The White Bird fault, viewed from the White Bird Battlefield Memorial overlook at milepost 227, provides a nice example of this rotation. Look across White Bird Creek to the southeast to see basalt layers that dip parallel to the slope, gently toward the White Bird fault below the highway. The roadcuts along the highway expose basalts in the uplifted block of the White Bird fault. The highway follows the contact between the Imnaha Basalt and overlying Grande Ronde Basalt for much of the grade.

The much older thrust faults in this area exerted a strong control on the location of the more recent normal faults such as the White Bird fault along the Salmon River Canyon. At depth here is the westernmost thrust fault, the Klopton Creek thrust, in the series of thrusts along the western side of the suture zone. The Klopton Creek fault thrusts Triassic diorite of the deep crustal basement northwestward over Triassic and Jurassic sedimentary and volcanic rocks of the Wallowa terrane. It is likely that this moderately dipping thrust fault was reactivated by the White Bird fault, whose dip likely shallows with depth to the point that it merges with the older thrust fault.

The White Bird Battlefield below US 95 to the southeast consists of landslide blocks and debris that slid into the canyon. This geology played a critical role in the White Bird battle that took place on June 17, 1877. Nez Perce warriors led by Chief Joseph held the higher ground in the hummocky terrain and blocks of basalt in the landslide deposit, which ultimately proved to be a decisive factor in their rout of US forces.

The highway passes pillow basalts in a Grande Ronde lava flow in roadcuts just north of milepost 228. The pillows indicate that the interaction between lava and

View to the southeast from White Bird Hill looking across the White Bird Battlefield at down-dropped basalt layers tilted toward White Bird normal fault.

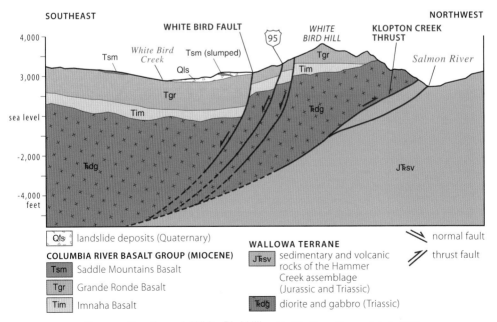

Cross section through White Bird Hill. —Modified from Garwood et al., 2008

HELLS CANYON AT PITTSBURG LANDING

To reach Pittsburg Landing in Hells Canyon, northbound travelers can follow the exit for Hammer Creek at milepost 222 and continue on Old US 95 (River Road), which links to Doumecq Road. Southbound travelers can head south on Old US 95 from White Bird. After crossing the Salmon River on Doumecq Bridge, turn south onto Deer Creek Road (National Forest 493), a steep, winding, all-season gravel road that continues 17.5 miles over the high ridge that separates the Salmon and Snake Rivers. On the Salmon River side of this high ridge, the road crosses through mostly rocks of the Columbia River Basalt Group of Miocene age and distinctively light-colored sedimentary interbeds that lie between individual basalt lava flows and have caused numerous landslides. The road crosses into rocks of the underlying Wallowa terrane by the first of many switchbacks in the final climb over the main ridge. These are mostly Permian and Triassic crystalline plutonic and metamorphic rocks of the Cougar Creek complex. Plutons within the Cougar Creek complex were essentially magma chambers deep in the crust that fed volcanoes of the Wallowa island arc. Columbia River basalts cap Wallowa terrane rocks at the top of the ridge.

A pullout within the first half mile down the Hells Canyon side of the ridge gives a great overview of Pittsburg Landing, which forms an obvious topographic bowl within the otherwise dramatically steep and rugged topography. The subdued topography, which permits road access to the bottom of Hells Canyon, is eroded in relatively weak Triassic and Jurassic sedimentary units. Rocks in craggy outcrops in steep hillsides to the south and below this vantage point are the Cougar Creek complex and the Klopton Creek pluton that intrudes the complex. They have been thrust to the northwest on the Klopton Creek thrust fault over the Triassic and Jurassic sedimentary and volcanic rocks.

The Triassic-Jurassic sedimentary sequence consists of the following units from oldest to youngest. The craggy-looking, steep hills to the north of Pittsburg Landing are lava flows of the Triassic Wild Sheep Creek Formation, many containing pillow lavas that formed as the molten rock flowing down the undersea flanks of arc volcanoes solidified quickly in deep seawater. Softer sedimentary rocks of the Triassic Doyle Creek

View of Pittsburg Landing to the west from Pittsburg Saddle. (45.6607, -116.3938)

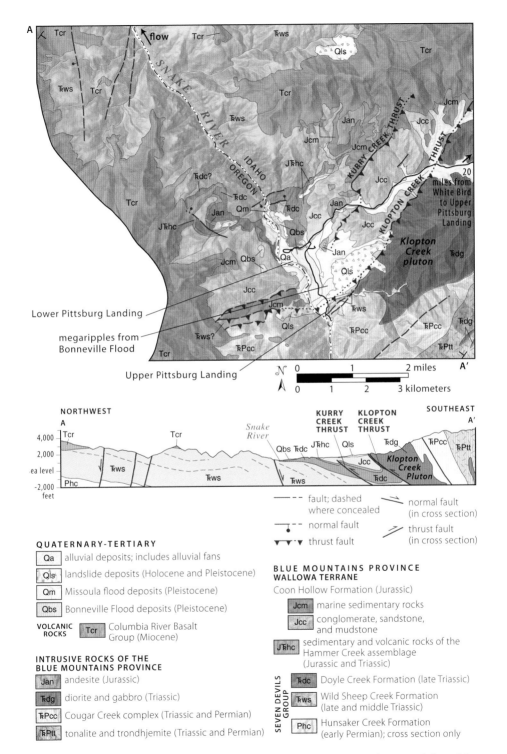

Map and cross section of Pittsburg Landing area. The line of the cross section extends from A in upper left of map to A' at bottom right. —Modified from Vallier et al., 2016

View looking downriver from Pittsburg Landing to the northwest of the Triassic Wild Sheep Creek Formation (craggy outcrops) and the overlying Triassic Doyle Creek Formation (tan hills in lower right). Inset shows fossil clam (oval-shaped object) and coral from a Triassic pocket-reef that developed on the flank of an arc volcano. (45.6423, -116.4770)

Formation lie in the very subdued topography to the south of the Wild Sheep Creek rocks. This tan-colored limy sandstone, limestone, and varicolored shale contains marine fossils of clams and ammonites, extinct relatives of the modern nautilus. Sediment in the Doyle Creek was eroded from the Wild Sheep Creek Formation.

After deposition of the Doyle Creek Formation, a hiatus of nondeposition or erosion occurred, after which the overlying layers of reddish tuff, conglomerate, and sandstone of the Late Triassic to Early Jurassic Hammer Creek assemblage were deposited. The tuffs are 197 million years old, erupted early in the Jurassic. Another hiatus followed deposition of the Hammer Creek rocks, and then the Coon Hollow Formation was deposited between 160 and 150 million years ago in Late Jurassic time. The oldest unit in the Coon Hollow Formation consists of river-deposited conglomerate, sandstone, and siltstone. Many of the conglomerate layers form lenses within channels cut into underlying sandstone and siltstone layers. Plant fossils that include ferns are common in the sandstone and siltstone, attesting to a time when the Wallowa volcanic arc was above sea level. Both the Hammer Creek assemblage and lower Coon Hollow Formation rocks form the hilly areas within Pittsburg Landing because the conglomerates and sandstones are more resistant to weathering. The upper unit of the Coon Hollow Formation consists of black shale and siltstone, soft rocks that form more subdued topography. This unit formed in a deep marine basin and contains fossils of clams and ammonites. Andesite dikes and stocks, which intruded these sedimentary rocks 152 million years ago, stand out from the easily eroded sedimentary rocks. The southeast-dipping Klopton Creek thrust fault was active sometime later, probably during the post-collisional period of crustal shortening that formed the other thrust faults along the Salmon River to the east.

Long after the Columbia River basalts erupted and covered the Wallowa terrane, rivers cut new canyons deep enough to expose the island arc rocks. Then, a final notable event occurred at Pittsburg Landing. Approximately 17,500 years ago, the Bonneville Flood came close to filling the narrow confines of Hells Canyon with a furious torrent that scoured and stripped away gravel from the canyon sides and bottom. In the uncommon broader stretches, such as at Pittsburg Landing, the slowing floodwaters deposited huge gravel bars.

From the saddle, the road descends steeply through many initial switchbacks cut into metamorphosed and sheared rocks of the Cougar Creek complex. Landslide deposits are common in these roadcuts. The road then straightens and follows Kurry Creek in the more subdued topography of the bowl, crossing the Klopton Creek thrust, which is buried under Kurry Creek stream sediments, followed by, in succession and visible in the hills on either side of the road, shale of the upper marine Coon Hollow Formation, sandstone and conglomerate of the lower river-deposited Coon Hollow Formation, and tall walls of intrusive andesite that Kurry Creek detours around. The road then crosses an alluvial fan formed by Bonneville Flood deposits and modern alluvium from Kurry Creek. Farther down the alluvial fan, the road crosses an enormous Bonneville Flood gravel bar that lies 200 feet above the modern Snake River. Bonneville gravels continue upward on both sides of the canyon here to an elevation of 1,700 feet, almost 550 feet above the modern river level.

Below the Bonneville bar, on the road's final drop to the boat ramp at Lower Pittsburg Landing, the road cuts through conglomerate outcrops of the Hammer Creek assemblage. The low hills north of the road are made of mudstone and sandstone of the Doyle Creek Formation. On the upstream side of the boat ramp, sandstone and conglomerate of the river-deposited Jurassic Coon Hollow Formation can be examined.

Fossils in Jurassic sedimentary rocks of the Coon Hollow Formation. Fern leaves and immature fern whorls are from mudstone in the river-deposited lower unit; ammonite is from the marine upper unit. —Photos by Tracy Vallier

Take the Upper Pittsburg Landing road to the south to cross over the Bonneville gravel bar. The top of the bar is littered with huge, rounded boulders, the remains of an older landslide deposit that was mostly removed by Bonneville floodwaters. Only the largest boulders in the original deposit were left behind, unable to be moved by the floodwaters. Bonneville Flood megaripples can be seen on the southern end of the Bonneville bar. These giant dunes of coarse gravel form in the same manner as their small-scale sandy ripple counterparts, except on a much grander scale.

Sombrero Rock, a flood-carved boulder in the shape of a ship's propeller, was carried here from an earlier landslide and now lies within the Bonneville gravel bar and was smoothed by the floodwater. Boulder is about 4.5 feet high. (45.6264, -116.4681)

View of Upper Pittsburg Landing from Oregon side of the Snake River showing Bonneville Flood megaripples.

water wasn't explosive; either the lake was deep enough to prevent the lava lobes at the advancing flow front from exploding or the lava had largely degassed before it flowed into shallow water. A sedimentary layer sandwiched between flows is visible between mileposts 230 and 231. These sedimentary interlayers formed when there was enough time between lava eruptions to allow sediments to accumulate in ponds and lakes. Many of these water bodies formed behind basalt lava flows that dammed streams.

From White Bird Hill Summit, the highway descends to the Camas Prairie, a relatively high plateau of nearly horizontal lava flows of the Columbia River Basalt Group. The road follows the northern flank of Mt. Idaho, a ridge uplifted by the Mt. Idaho fault. As the north-trending White Bird fault curves to the northeast and crosses to the northwest side of Mt. Idaho, the sense of displacement on the fault reverses from normal faulting that drops the east side to reverse faulting that uplifts on the southeast, where it becomes the northeast-trending Mt. Idaho fault. These faults are younger than 15 million years old. Below the basalt lava flows is the much older though similarly named Mt. Idaho shear zone (see ID 14: ID 13 Junction— Elk City section).

US 95
Grangeville—Lewiston Hill
78 miles

Between Grangeville and Winchester, US 95 crosses the Camas Prairie, an approximately 3,500-foot-high plateau of Miocene lava flows of the Columbia River Basalt Group. The canyon of the Salmon River lies south of the Camas Prairie. A veneer of loess, Pleistocene windblown silt deposits derived from glacial outwash streams farther north, covers the flat plateau and provides fertile soil for crops. During ice ages of the past 2 million years, the Camas Prairie was home to a variety of large cold-adapted mammals. Grangeville hosts an impressive Pleistocene Columbian mammoth skeleton at Eimers-Soltman Park behind the Chamber of Commerce building (north of milepost 240 on US 95, turn right on Pine Street and follow signs to the parking lot). This skeleton was discovered at the nearby Tolo Lake recreation site, where the bones from a number of other mammoths and bison were uncovered from the loess during a dredging project in 1994. Tolo Lake, about 3 miles east of Grangeville, can be accessed by various routes, the most direct by turning south at milepost 246 from US 95 onto Lake Road and driving 3 miles.

Poking out above the plateau are steptoes that were the tops of mountains sticking out several thousand feet above the plateau at the time the first lavas of the Columbia River basalts poured out onto the landscape, filling the low areas. The lava surrounded the base of the mountains but left their tops uncovered. Cottonwood Butte, the prominent butte west of Cottonwood, consists of Permian volcanic and sedimentary rocks of the Wallowa terrane. Steptoes visible along the highway between Cottonwood and Ferdinand are Jurassic to Cretaceous diorite and gabbro that intruded the volcanic and sedimentary rocks of the Wallowa Terrane. The low buttes in the distance to the northeast are Kamiah Buttes, which consist of mostly andesite and rhyolite volcanic deposits of likely Oligocene age.

Just south of milepost 267, a bright-red ancient soil, called a paleosol, and pillow basalts are visible in the Grande Ronde Basalt in the roadcut on the east side of the

228 WESTERN MARGIN (95)

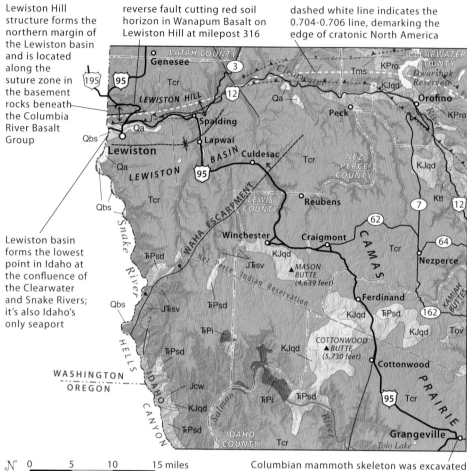

Lewiston Hill structure forms the northern margin of the Lewiston basin and is located along the suture zone in the basement rocks beneath the Columbia River Basalt Group

reverse fault cutting red soil horizon in Wanapum Basalt on Lewiston Hill at milepost 316

dashed white line indicates the 0.704–0.706 line, demarking the edge of cratonic North America

Lewiston basin forms the lowest point in Idaho at the confluence of the Clearwater and Snake Rivers; it's also Idaho's only seaport

Columbian mammoth skeleton was excavated from Pleistocene loess deposits at nearby Tolo Lake in 1994; on display at Eimers Park at the junction of Pine Street and US 95

QUATERNARY-TERTIARY
- Qa — alluvial deposits; includes landslides
- Qbs — Bonneville Flood deposits (Pleistocene)
- Tms — sedimentary rocks associated with flood basalts (Miocene)

VOLCANIC ROCKS
- Tcr — Columbia River Basalt Group (Miocene)
- Tov — volcanic rocks (Oligocene)

INTRUSIVE ROCKS OF THE BLUE MOUNTAINS PROVINCE
- Ktt — tonalite and trondhjemite (Cretaceous)
- KJqd — quartz diorite (Cretaceous and Jurassic)
- TRPi — intrusive rocks (Triassic and Permian)

- ---- fault; dashed where concealed
- normal fault
- thrust fault
- syncline
- anticline

BLUE MOUNTAINS PROVINCE WALLOWA TERRANE
- KPro — Riggins Group, Orofino Metamorphic Suite, and related rocks (Permian to Cretaceous)
- Jcw — Coon Hollow and Weatherby Formations (Jurassic)
- JTRsv — sedimentary and volcanic rocks (Jurassic and Triassic)
- TRPsd — Seven Devils Group (Triassic and Permian)

Geology along US 95 between Grangeville and Lewiston Hill. See the cross section on page 200.

highway as it descends into Lawyer Creek. The tropical climate in the Pacific Northwest during the Miocene Epoch produced red soils when there was a long enough hiatus between basalt eruptions. Mason Butte, a steptoe that exposes Jurassic to Cretaceous quartz diorite intruded into Wallowa terrane rocks, is visible to the south at milepost 277.

Between mileposts 279 and 290, the highway descends Winchester Grade along Lapwai Creek, dropping down the Waha escarpment, a monocline fold in the basalt. The nearly horizontal basalt of Camas Prairie dips gently to the northwest into the Lewiston basin, a northeast-trending trough along the Lewiston syncline. The basin is bordered to the south by the northeast-trending zone of folds and faults of the Waha

The Pleistocene Tolo Lake Columbian mammoth remains on display at Eimers-Soltman Park in Grangeville.

View to northwest across Camas Prairie.

View south from Cottonwood Butte. The southern edge of the Camas Prairie lies in the foreground with the canyon of the lower Salmon River beyond. Snow-clad peaks of the Seven Devils Mountains lie in the distance to the right.

escarpment and to the north by another structural zone of folds and faults that trend east-west along Lewiston Hill. Farther to the west, the Snake River flows north into the Lewiston basin from Hells Canyon. The Clearwater River drains from the east along the axis of the Lewiston basin to its confluence with the Snake River in Lewiston.

At milepost 288, large, nicely formed columns in roadcuts along the road are in the Imnaha Basalt, which lies below the Grande Ronde Basalt. After crossing the axis of the Lewiston syncline at Lapwai and crossing the Clearwater River between mileposts 304 and 305, US 95 merges with US 12 at the base of Lewiston Hill, which forms the northern side of Clearwater Canyon and the northern margin of the Lewiston basin. Views to the northwest of Lewiston Hill on the approach to the Clearwater crossing are deceptive. Basalt layers on both sides of the Clearwater River here dip gently south. Thus, it appears that the basalt layers once simply connected across Clearwater Canyon before being cut by the river. However, completely hidden by erosion of the canyon here but apparent elsewhere, is a major anticline that trends along this section of the canyon and disrupts the south-dipping panels of rock to either side of the canyon. This anticline, the main expression of the Lewiston Hill structure in this area, is a typical example of how folds in Columbia River basalt are expressed. The brittle basalt in the tight fold axis is crushed, and these zones of broken rock are preferentially eroded by streams to form canyons. The main evidence of the fold is now gone. It is almost as if a devious geology professor hid the structure to challenge the skills of students.

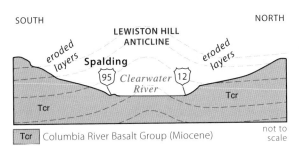

View to the west down the Clearwater River near the junction of US 95 and US 12 shows folded layers of basalt in the anticline that is part of the Lewiston Hill structure. The Clearwater River preferentially eroded the broken rock along the hinge of the fold to form the modern canyon.

At milepost 312, US 95 veers northward from combined US 95/US 12 along the Clearwater River just outside of Lewiston and climbs Lewiston Hill. Most of the Lewiston Hill structure in this area consists of an east-trending asymmetrical anticline with a long southern limb that underlies most of the hill and a short northern limb exposed toward the top of the hill.

Clearwater River stream gravels are preserved on top of the Columbia River Basalt Group up to 300 feet above the modern river (to milepost 313), attesting to the enormous amount of gravel that has accumulated in the Lewiston basin since approximately 12 million years ago. Recall that the entire Snake and Clearwater Rivers and their tributaries—draining most of central and southern Idaho—flow into the basin. Farther uphill from the gravels, the Grande Ronde Basalt dips parallel to the hillslope on the southern limb of the Lewiston Hill anticline. A fairly obvious red paleosol is visible in roadcuts on the uphill side of the road about halfway between mileposts 314 and 315, and its orientation follows the transition from the southern limb through the hinge and into the northern limb of the anticline that cuts across the highway downhill from the major switchback.

Uphill from the big switchback, a large outcrop of the Grande Ronde Basalt north of the highway displays a beautiful breccia at the top of one flow overlain by the colonnade and entablature fracture zones of an overlying basalt flow. At milepost 316, a reverse fault cuts a bright-red soil horizon preserved between two flows of the Wanapum Basalt. The fault, part of the Lewiston Hill structural zone, offsets the rocks by a few feet.

A spectacular overview of the Lewiston basin is available between mileposts 317 and 318, where a turnoff to the south leads a short distance down old US 95 (the Spiral Highway) to parking areas. The Lewiston basin lies in the Clearwater embayment, a lobe of Columbia River basalt less than 3,000 feet thick that covers much of the suture zone between the Wallowa island arc and the North American continent. Although extremely poorly exposed because of basalt cover, the suture zone likely underlies Lewiston Hill and has been reactivated by the Lewiston Hill structure. Basement rocks along the highway north of the boundary are in the old North American continent. The exit of Hells Canyon, where the Snake River cuts through the Waha escarpment, is visible to the south. To the west, the Lewiston basin is bounded by the Blue Mountains, a broad anticline in rocks of the Columbia River Basalt Group. Poking out of the skyline to the southwest, in Washington between the entrance to Hells Canyon and the Blue Mountains, is Puffer Butte, a preserved volcanic vent for some of the Wanapum Basalt flows that erupted about 14.5 million years ago.

Cresting Lewiston Hill, US 95 descends into the upper reach of Hatwai Canyon, which exposes horizontal basalt lava flows to the east, indicating that the highway has completely traversed the Lewiston Hill anticline into the unfolded basalt plateau of the Palouse to the north.

A reverse fault in the roadcut on Lewiston Hill at milepost 316 cuts the Wanapum Basalt and a red paleosol.

Overlook at top of Lewiston Hill of Lewiston basin to the south. (46.4621, -116.9807)

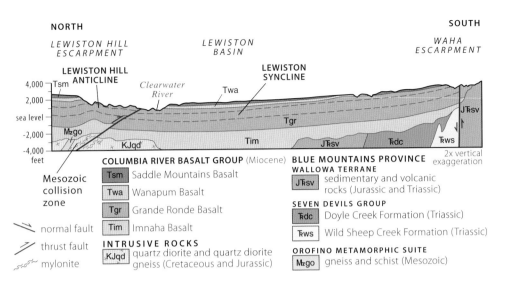

Cross section across the Lewiston basin from the Lewiston Hill to the Waha escarpment.

ID 13
Grangeville—Kooskia
26 miles

ID 13, a busy shortcut, connects US 95 with US 12 by dropping down the Grangeville Grade from Grangeville on the Camas Prairie to the South Fork of the Clearwater River, which ID 13 follows to Kooskia on the Main Fork of the Clearwater River. The highway parallels the Salmon River suture zone in rocks of the eastern edge of the Wallowa terrane. The suture zone lies east of here and is crossed by US 12 and ID 14 east of their junctions with this route.

Northeast of Grangeville, ID 13 traverses Grande Ronde Basalt on the Camas Prairie, a flat plateau on the Columbia River Basalt Group. As the road begins to descend the Grangeville Grade, it crosses a large ridge in rocks of the Saddle Mountains Basalt that overlie the Grande Ronde Basalt. Continuing down, it crosses back into Grande Ronde Basalt between mileposts 6 and 7, where a roadcut exposes a light-tan, mud-rich sedimentary interbed sandwiched between the Saddle Mountains and Grande Ronde Basalts. Several more miles of roadcuts expose Grande Ronde Basalt and more sedimentary interbeds. Just east of milepost 10, the road passes below the base of the Columbia River Basalt Group into greenstone—low-grade metamorphosed volcanic and sedimentary rocks—of the Seven Devils Group of the Wallowa terrane. At milepost 11, ID 14 diverges from ID 13 and continues up the South Fork of the Clearwater River (see the next road guide). ID 13 continues north down the South Fork of the Clearwater River to Kooskia.

Roadcuts for the next 1.5 miles consist of more Seven Devils Group greenstone followed by quartz diorite of the 160-million-year-old Harpster pluton that intruded the Seven Devils Group rocks. Near Harpster, roadcuts in these plutonic rocks are

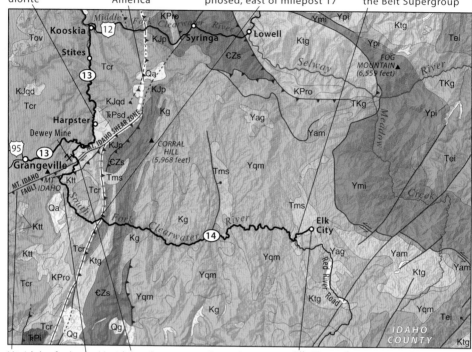

Geology along ID 13 and ID 14.

Diorite and gabbro enclaves (blobs of magma) mixed into the 160-million-year-old magma chamber of the Harpster quartz diorite pluton. (45.9976, -115.9645)

chock-full of dark blobs of fine-grained diorite and gabbro that formed when mafic magma was mixed into the quartz diorite magma chamber. Reaction zones between the two magmas are preserved as hornblende-rich boundaries. A nice quiet place to contemplate these features far away from certain death along busy ID 13 is found in outcrops about a half mile up Sally Ann Creek Road, which heads east to the small community of Clearwater (turn off to east, just north of milepost 15).

ID 14
ID 13 Junction—Elk City
49 miles
See map on facing page.

ID 14 heads up the South Fork of the Clearwater River, crossing the suture zone and all the wonderful geology and structures associated with it, to Elk City deep within the Idaho batholith. Colorful roadcuts in a long curve just south of milepost 2 consist of interbedded volcanic-rich sandstone, shale, siltstone, and chert that are likely the Permian Hunsaker Creek Formation of the Wallowa terrane. They are light green due to the minerals epidote and chlorite that crystallized during low-grade metamorphism.

Across from a concrete bridge spanning the river between mileposts 3 and 4, a very rusty-looking roadcut consists of altered pyrite-bearing oceanic rhyolite with quartz crystals and metamorphosed volcanic and sedimentary rocks. The oceanic rhyolite, shallowly emplaced into the Wallowa island arc complex, is not a true rhyolite because it is low in potassium. Oceanic rhyolites were very common in the Permian Wallowa island arc and are associated with hydrothermal alteration, meaning that as they intruded, hot fluids coursed through and altered the rocks around them, depositing gold and copper-bearing minerals in veins. Copper-gold mineralization across the Wallowa terrane is typically linked with these Permian oceanic rhyolites, including that at the Dewey Mine, located just to the west of here.

At milepost 4, the highway crosses broken and ground-up rocks of the Mt. Idaho fault, the relatively young northeast-trending thrust fault that uplifted Mt. Idaho and cuts 15-million-year-old basalt lava flows near Grangeville (see US 95: New Meadows—Grangeville section). The faulted rocks along this structure were deformed in a brittle fashion because they were near the surface when faulting occurred. The rocks along the highway for the next couple of miles represent a very different type of fault rock that occurs in the Mt. Idaho shear zone, which predates the Mt. Idaho fault. They are still part of the Wallowa terrane but have been metamorphosed to chlorite- and epidote-bearing schist. Minerals in the schist are aligned; these rocks experienced higher-grade metamorphism than the greenstone rocks to the northwest. The metamorphic breakdown and recrystallization of chlorite and epidote accommodated movement in the Mt. Idaho shear zone. Continuing farther southeast into the shear zone, progressively higher-grade metamorphic rocks are encountered, in this case hornblende gneiss and marble, and these rocks continue on the other side of the shear zone.

The Mt. Idaho shear zone cuts northeast across the Wallowa terrane from Pittsburg Landing (where it is called the Klopton Creek thrust), through Mt. Idaho, and

Mylonite developed in 155-million-year-old Permian tonalite of the Wallowa terrane within the Mt. Idaho shear zone near milepost 4. Asymmetric fabrics developed by the drawn-out crystals indicate that the top (upper left) moved up and to the right relative to the bottom (lower right), the same sense of motion in the shear zone. View is to the southwest. (45.9093, -116.0092)

northeast to Syringa on the Clearwater River, thrusting more highly metamorphosed rocks up on the southeast side relative to the lightly metamorphosed rocks on the northwest side. The shear zone cuts both the Salmon River suture zone and the Western Idaho shear zone. In addition to thrust faulting, a large component of right-lateral strike-slip movement displaced the suture zone and the Western Idaho shear zone 5 to 6 miles.

At approximately milepost 6, the highway crosses into the 111-million-year-old Blacktail pluton on the southeast side of the Mt. Idaho shear zone. This pluton consists of tonalite intruded by a younger magma phase of trondhjemite. The tonalite contains very distinctive biotite crystals that are up to an inch in size. The trondhjemite, which contains visible crystals of muscovite and garnet, varies from coarse- to fine-grained and occurs in dikes and blobs. The Blacktail pluton is part of the belt of tonalite and trondhjemite plutons ranging in age from 120 to 110 million years old that solidified more than 18 miles deep in the crust during a time when convergence between the collided Wallowa terrane and North America was thickening the crust. The best place to look at these rocks is in fresh boulders placed around the perimeter of a large pullout on the river side at milepost 6.

East of milepost 13, the road exits the Blacktail pluton and enters hornblende gneiss of the Riggins Group, high-grade metamorphic rocks that originated as mostly sedimentary rocks of the Wallowa terrane. They also include schist, marble, and quartzite and were buried in the thickened crust of the Salmon River suture zone that magmas of the Blacktail and related plutons later intruded. A good pullout to look at these rocks is at the entrance to Jungle Creek Campground between mileposts 13 and 14. Impressive roadcuts between mileposts 15 and 16 display a thick section of

Biotite-bearing tonalite of the Blacktail pluton. Lighter colored dikes are trondhjemite. (45.9089, -116.0086)

the marble. Across the river are piles of gravel that were part of an extensive placer operation on a large gravel bar in the South Fork of the Clearwater River. The river sediment was dug up and washed over grates to separate gold from the sand and gravel.

At milepost 17, the highway crosses the boundary between the Wallowa terrane and North America, the 0.704-0.706 line discussed in the introduction to this chapter. The roadcuts here (pullout on south side of road just east of Meadow Creek Campground) are typical for the boundary and show nearly vertical sheets of metamorphosed plutonic rocks that completely intruded the original boundary between 110 and 100 million years ago. The intensely developed metamorphic banding and lineation from shear along the Western Idaho shear zone overprinted the boundary between 105 and 90 million years ago. Roadcuts for the next mile east of here show 90-million-year-old tonalite containing conspicuously large hornblende crystals that caught the very tail end of deformation on the margin of the Western Idaho shear zone. A good spot to see these rocks is at the turnoff for FS Road 484 between mileposts 17 and 18.

East of milepost 19, Precambrian North American rocks appear in roadcuts. The first few outcrops consist of quartzite with biotite and garnet of the metamorphosed Windermere Supergroup, the 700- to 530-million-year-old sedimentary sequence that formed as the supercontinent Rodinia rifted to open the paleo–Pacific Ocean. The rocks were metamorphosed in the suture zone at about 90 million years ago.

The rocks farther east are quartzites of the metamorphosed Middle Proterozoic sedimentary rocks of the Lemhi subbasin of the Belt Supergroup. As is the case for

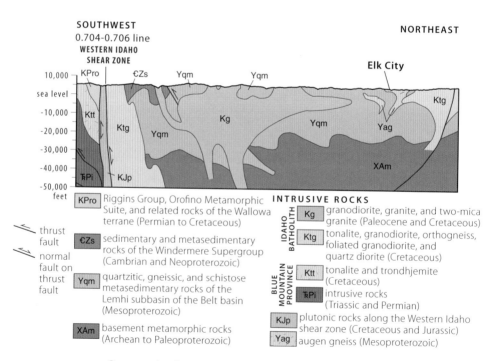

Cross section between the 0.704-0.706 line and Elk City.

most Belt Supergroup quartzites, these examples contain conspicuous amounts of feldspar crystals, indicating that the feldspathic parts of the sandy sediment deposited in the Belt basin didn't travel far from its sedimentary source. Feldspar grains break down quickly during weathering, so they don't survive lengthy travel. A good place to see these rocks is in the outcrop next to the trailhead parking lot for Johns Creek Trail between mileposts 19 and 20.

Much younger granitic rocks of the Idaho batholith appear along the canyon walls east of milepost 20. These rocks are biotite-bearing granodiorite of the 80- to 67-million-year-old Atlanta lobe of the batholith. They form distinctive granite domes produced by exfoliation slabs that peel off the outside surfaces of the exposed granite like the outer layers on an onion. As the granite bodies are brought to the surface during uplift and erosion, the stress from the weight of the overlying rock is unloaded and exfoliation occurs. A nice large pullout to examine Idaho batholith rocks lies on the south side of the road between milepost 20 and 21.

Over the next 15 miles, the highway passes through mostly granodiorite of the Idaho batholith with sporadic sections of quartzite of the Lemhi subbasin of the Belt Supergroup that was caught up in the batholith magma. East of milepost 36, metamorphic rocks derived from Lemhi subbasin sedimentary rocks become more prevalent than the Idaho batholith rocks that intrude them. These metasedimentary rocks consist of biotite gneiss and schist with minor quartzite and calc-silicate rocks intruded by Cretaceous granodiorite and pegmatite of the batholith. Good spots to examine outcrops with these rocks are just west of milepost 41, at the junction with Crooked River Road just west of milepost 43, and at milepost 45.

Exfoliation slabs developed in granodiorite of the Idaho batholith near milepost 22. View is down the valley. (45.8218, -115.8459)

Biotite gneiss and schist near milepost 41 originated as Mesoproterozoic muddy sediments of the Lemhi subbasin of the Belt Supergroup. Light-colored units are dikes and sills of the Cretaceous Idaho batholith. The vertical lines are drill holes from road construction blasting. (45.8321, -115.5574)

The scar in the hillslope is from the 2016 landslide near milepost 39. (45.8297, -115.5822)

Between milepost 38 and 39, the highway passes the scarred slopes left by a major landslide on the north side of the highway. The event occurred in February 2016, closing the highway to all traffic for nearly six months as the slope was stabilized and the debris cleared. The geologic conditions on the slope played a large part in its failure. Foliation and layering in the schist and gneiss parallel the slope, forming weak slip-planes that easily failed under wet conditions.

The canyon opens up dramatically east of milepost 43 where the South Fork of the Clearwater River changes to a low-gradient meandering stream. This transition from the steep gradient stream confined to its canyon to the broad, meandering stream in the high meadows is a knickpoint. Central Idaho to the east is a plateau uplifted in Neogene time. Streams are cutting their channels eastward and deeper into the interior of the plateau, but this takes considerable time in the hard granitic and metamorphic rocks of central Idaho. The knickpoint at this location on the stream will continue to migrate eastward in the geologic future.

Google Earth image from 2016 of the lower Crooked River (junction of Crooked River Road near milepost 43) showing the historical placer mining disruption. The natural stream was transformed into a low-gradient stream with little vegetation to provide shade, an ecological nightmare for fish. Since the image was taken, this section of the Crooked River has been restored to a straighter, more natural stretch to renew the robust fishery it had once hosted. Image is about three-quarters of a mile long, and north is toward the top.
—Google Earth image

Around milepost 44, the river is lined by placer gravel piles where miners washed the modern stream gravels to extract gold. Between milepost 46 and 47, beautiful outcrops of augen gneiss appear in roadcuts. The original rock was a granite that intruded sedimentary rocks of the Belt Supergroup at about 1,370 million years ago. It was metamorphosed to gneiss in Cretaceous time, causing recrystallization and banding of the biotite, quartz, and plagioclase feldspars. The augen (German for "eyes") are large potassium feldspar crystals from the original granite that metamorphism streamlined into the shape of eyes.

The white eye-shaped forms in this 1,370-million-year-old augen gneiss from Red River Road were potassium feldspar crystals in granite prior to metamorphism. (45.8026, -115.4121)

Red River Road turns off on the south side of the highway between mileposts 46 and 47. This road can be followed to its junction with the Magruder Corridor (FS 468), a dirt road that continues 101 tedious miles across the Bitterroot Mountains to Darby, Montana. The route follows an ancient Nez Perce trail that was also used by early settlers. Its namesake, a well-known Elk City merchant, was murdered with four companions on the route in 1863. The road cleaves a corridor between the Selway-Bitterroot Wilderness to the north and the Frank Church–River of No Return Wilderness to the south. This route, with its magnificent views of the central Idaho high country, crosses through mostly Cretaceous and Eocene granite and granodiorite plutons of the Idaho batholith and Challis magmatic events. Belt Supergroup rocks occur as pieces of wall rock in places.

At milepost 48, ID 14 crosses open meadows developed in Tertiary sediment deposits and reaches the small settlement of Elk City at milepost 49. Elk City was a major mining district that experienced its first gold strike in 1861, very early in Idaho's mining history. The region right around Elk City is in the roof zone of the Idaho batholith, the rock that lay above the intruding batholith magma. Gold veins formed along the contacts between the Cretaceous and Eocene granitic intrusions and the metasedimentary roof rocks. Heated water streaming upward from the crystallizing granitic magmas below dissolved and reprecipitated metals in veins in the metamorphic roof rocks. Faults in these rocks provided permeable pathways for these fluids. Tertiary streams in the Elk City basin eroded gold from these deposits and redeposited it in stream sands and gravels. Modern streams reworked the older stream deposits, further concentrating the gold into rich placer deposits. Mining these alluvial deposits involves digging up the gravels and washing them over grates, which trap the small gold flakes. The washed gravels are left behind in piles as the miners move on to fresh gravels.

Eocene granodiorite gneiss near Burnt Knob in Frank Church–River of No Return Wilderness. Roadcut in distance is the Magruder Road. —Photo by Reed Lewis

Looking northwest at the Sawtooth Range from the Galena Summit viewpoint on ID 75. Alturas Lake sits out of view, behind the prominent moraine that stretches horizontally across the far side of the valley. The more rounded mountains to the south (left) are gray Cretaceous granitic rock. To the north the jagged peaks are pink Eocene granitic rocks. (43.87, -114.72)

CENTRAL MOUNTAINS

Main highways circumnavigate central Idaho because at its heart is the immense roadless area of the Frank Church–River of No Return Wilderness. Here, the resistant rocks of the Cretaceous Idaho batholith form rugged mountains and steep canyons. This geographic barrier to railroads and roads is the fundamental cause for the cultural diversity of Idaho since territorial days in the 1870s. The batholith effectively divides the state into two, separating the northern portion, with its connections to eastern Washington, from the Snake River Plain in the south, which provides an easy travel corridor through flat agricultural lands where most of Idaho's population lives.

The Idaho batholith intruded a variety of older rocks, with the oldest exposed rocks in central Idaho being metamorphosed 2.65-billion-year-old Archean granites in the Pioneer Mountains. Also in the Pioneer and Salmon River Mountains and extending northeast to the northern Beaverhead Mountains and into Montana are Paleoproterozoic metavolcanic rocks and sills that cooled about 1,770 million years ago in the Great Falls tectonic zone. This ancient fault zone in the deep North American crust, extending from east-central Idaho through Montana to the Canadian border, is more fertile for gold and base metal mineral deposits than the Archean crust to the south. These Archean and Paleoproterozoic basement rocks were part of the supercontinent Columbia, which included the Laurentian tectonic plate, the cratonic core of North America.

LEMHI SUBBASIN

Overlying the Paleoproterozoic rocks in the Lemhi, Beaverhead, and Salmon River Mountains are rock layers of the Lemhi Group and related units, which were deposited in the Lemhi subbasin of the Belt basin between 1,450 and 1,385 million years ago. This huge body of Mesoproterozoic sedimentary rock correlates with the upper three parts of the Belt Supergroup, the Ravalli, Piegan, and Missoula Groups. (The Belt Supergroup is further discussed in the introduction to the Northern Idaho chapter.)

Much of the rock in the Lemhi Group is quartzite and siltite, a slightly metamorphosed quartz-rich siltstone. Like their counterparts in the Ravalli, Piegan, and Missoula Groups to the north, these rocks were largely deposited on alluvial flats that fed into a huge shallow inland sea or lake. The primary source for sediment of the Lemhi subbasin was a huge magmatic arc, known as the Big White, that was active

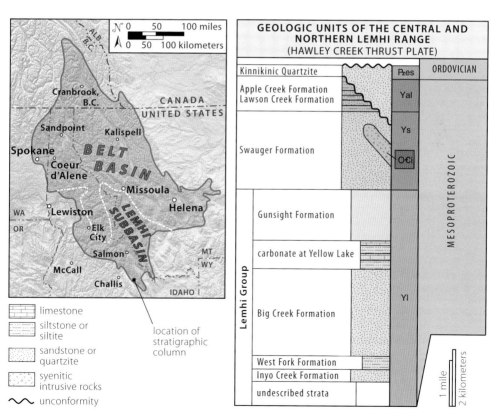

Dashed white line shows the general northern boundary of the Lemhi Subbasin within the Belt basin. Map is based on the present distribution of rocks in the Belt and Purcell Supergroups. The stratigraphic column is for the central and northern Lemhi Range. —From Lewis et al., 2012

about 1,720 million years ago and lay to the south, perhaps in what is now southern Nevada. There is evidence that this source also fed sediment into the main Belt basin that now lies to the north of the Lemhi subbasin.

The upper section of the Lemhi subbasin, which likely correlates with the upper Missoula Group of the main Belt Supergroup, is the most extensive part of the package. The uppermost Lemhi Group and overlying Swauger and higher formations are 35,000 feet thick in the Lemhi Range and Beaverhead Mountains. These cross-bedded quartzites were deposited by north-flowing sheet floods, generated by periodic huge storms. In contrast to modern stream deposits, they do not contain incised channels.

Cross-bedded quartzite of the Swauger Formation in Allan Mountain quadrangle, northwest of Gibbonsville. The sheet floods that deposited these cross beds flowed from left to right. (45.60, -114.04)

The Lemhi subbasin was clearly connected to the main Belt basin—they share the same sediment sources—but their histories are different. Although there is little evidence for normal faults growing during deposition, the formation of the main Belt basin was initiated with stunningly rapid subsidence. The extremely thick Prichard Formation in northern Idaho and northwestern Montana was deposited in the subsiding basin. At the same time, central Idaho remained a high-standing region with no deposition. The Lemhi subbasin began subsiding during deposition of the Ravalli and Piegan Groups of the Belt Supergroup but only collected minor amounts of sediment. Then, during late Missoula Group deposition in the main Belt basin, the bottom dropped out of the Lemhi subbasin to deposit the very thick Swauger Formation and related rocks as deposition in the main Belt basin petered out.

Following deposition of sedimentary rocks of the Lemhi subbasin, central Idaho underwent a fairly extensive period of magmatism that included the intrusion of both granitic (felsic) and mafic magma between 1,380 and 1,370 million years ago. These now-metamorphosed intrusive rocks—augen gneiss for the granite and amphibolite for the mafic igneous rocks—can be found in several locations cutting Lemhi subbasin rocks. The main Belt basin to the north does not share this magmatic event. We normally associate the combination of felsic and mafic magmatism with extensional environments—they allow both mantle (mafic) and continental crust (felsic)

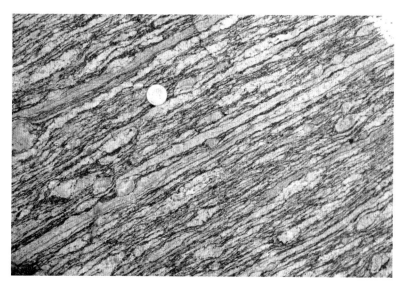

The augen, or "eyes" in this 1,370-million-year-old layered augen gneiss are the white potassium feldspar crystals. The gneiss was metamorphosed from a felsic pluton intruded about 1,370 million years ago into the Belt Supergroup. In places the augen gneiss has been extensively sheared and has a mylonitic texture (exhibited in this photo), demonstrating grain-size reduction of the mica-rich portions. Photo taken west of Shoup along the Salmon River near the mouth of Panther Creek. (45.35, -114.28) —Photo by Terry Maley

melts to rise into the overlying crust—but, again, there is little evidence for extensional normal faults during Mesoproterozoic time. Zircon crystals eroded from the 1,380- to 1,370-million-year-old augen gneiss have turned up in sediments far from Idaho, and we'll come back to that later.

The supercontinent Columbia, including a part called Nuna, broke into fragments after 1,350 million years ago, following deposition of the Belt Supergroup and the intrusion of the mafic and felsic magmas. Blocks of crust were jostled along strike-slip faults and moved laterally. From 1,200 to 1,000 million years ago, a continent-continent collision between Laurentia and Baltica produced the supercontinent Rodinia. The collision, known as the Grenville orogeny, occurred far to the east and north of present-day Idaho and did not directly impact the rocks of Idaho.

RODINIA RIFTING

Rodinia rifted apart in the Neoproterozoic, starting perhaps 700 million years ago. Volcanic rocks erupted as the crust was stretched apart, and eventually a sea, the paleo-Pacific Ocean, filled the rift. The volcanic rocks and rift sediments are called the Windermere Supergroup, which is exposed in several places in Central Mountains, including at Edwardsburg and Bayhorse, but it is much more accessible in southeastern Idaho. See the introduction to the Basin and Range chapter for more discussion of the Windermere rocks and the continental shelf along the western margin of ancient North America in Paleozoic time.

ANTLER OROGENY

At the end of Devonian time, tectonic plates began to converge to the southwest of Idaho in an event known as the Antler orogeny. In Nevada, early Mississippian strata record deposition in what is known as a foreland basin, a trough pushed down by thrusting of rising highlands. Also in Nevada, the Roberts Mountains thrust fault served to uplift the Antler highlands to the west. In Idaho, the relations are less clear, and if there is an equivalent Roberts Mountains thrust fault, it has been overprinted by Cretaceous and Eocene faulting. However, the late Devonian Antler orogeny did produce a drastic change in Mississippian strata between central and eastern Idaho. Across much of the Northern Rockies, Mississippian rocks were part of an extensive carbonate bank on the western margin of ancient North America. However, east of the Pioneer thrust fault in the Pioneer Mountains, early Mississippian rocks consist of shallow marine conglomeratic rocks and turbidite submarine fans deposited in a deep basin. These basin deposits belong to the Copper Basin Group. Geologists think that the Antler thrusting in Nevada transitioned to a left-lateral strike-slip fault to the north through central Idaho. The coarse quartzite conglomerates of the southern Copper Basin Group had sources from ancient North America and were deposited in a pull-apart basin at the south end of this strike-slip fault.

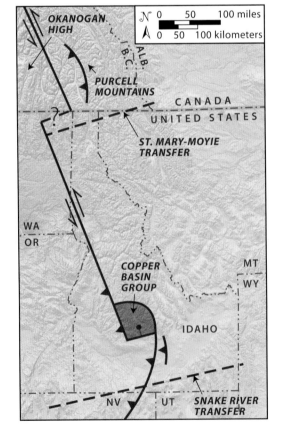

During the Antler orogeny in late Devonian time, the thrusting in Nevada transitioned to left-lateral strike-slip faulting in Idaho. The St. Mary–Moyie and Snake River transfers are strike-slip faults that offset thrust systems.
—From Beranek et al., 2016

BIG COLLISIONS AND THE IDAHO BATHOLITH

Terranes of the Blue Mountains Province collided with the North American margin between 150 and 140 million years ago, initiating a long-lived orogeny across much of Idaho that lasted from Late Jurassic to Paleogene time. The details of these terranes and the suture zone between the terranes and the continent are discussed in the Western Margin chapter. Here, we pick up the geologic story in central Idaho on the North American continent to the east of the suture zone.

In response to terrane accretion and the subduction that followed, the Sevier fold-and-thrust belt developed across central Idaho. The entire sedimentary section of Mesoproterozoic through Late Paleozoic sedimentary rocks that had accumulated on the continental margin was folded and thrust eastward in a series of thrust plates. From west to east in the Pioneer Mountains, the big thrust faults are the Pioneer and Copper Basin faults. The Hawley Creek thrust fault in the northern Beaverhead Mountains is the next major Sevier thrust fault going east. Sevier thrusting was followed by the Laramide orogeny from the Late Cretaceous to the Paleogene in Montana and Wyoming to the east.

Crustal thickening of central Idaho by thrust faulting in the Sevier orogeny led to the formation of a high plateau by Late Cretaceous time that probably looked similar to the high Andean Plateau of South America today. We know this because zircon crystals with approximately 1,375-million-year-old peak ages determined by uranium-lead dating have been found in sedimentary rocks as far away as Vancouver Island, northwestern Washington, and the Central Valley of California. The source of the zircons is the distinctive 1,380- to 1,370-million-year-old augen gneiss of central Idaho, so this area had to be uplifted 100 to 80 million years ago. Major rivers carried grains of sediment from their headwaters in Idaho to depositional basins and deltas along the Pacific Ocean.

Magmatic arc activity, which had been prevalent on crust of the Blue Mountains terranes, migrated eastward onto the continental crust and began forming the Idaho batholith, one of the great Cordilleran batholiths of western North America, from Late Cretaceous to Paleogene time. The batholith intruded in a series of distinct phases. The widespread initial phase of the Idaho batholith was intruded between about 98 and 85 million years ago. Part of this phase is known as the border zone because it occurs in a belt along the eastern side of the suture zone. Border zone magmas produced tonalite and granodiorite plutons, basically compositions that are a mix from melting of both oceanic and continental mantle and crust. These rocks are commonly deformed, demonstrating that plate convergence was ongoing along the suture zone during this time.

The batholith then entered its main phases of magmatism, producing two distinct lobes, a generally older southern one, called the Atlanta lobe, and a generally younger northern one, called the Bitterroot lobe. Both lobes underwent the same evolutionary sequence beginning with intrusion of mostly hornblende- and biotite-bearing granodiorite that was derived from melting of both continental mantle and some crust, followed by intrusion of biotite-bearing granodiorite and true granites that contain both biotite and muscovite (called two-mica granites) and were derived almost entirely by melting the continental crust. The Atlanta lobe developed between 80 and 67 million years ago. The Bitterroot lobe formed between 70 and 53 million years ago, and its continuation to the east is exposed as parts of the Boulder batholith and

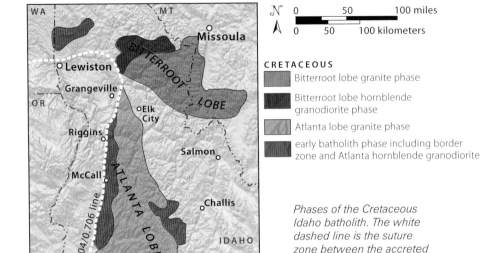

Phases of the Cretaceous Idaho batholith. The white dashed line is the suture zone between the accreted terranes to the west and the North American continent to the east. —From Gaschnig et al., 2011

Two hand samples of Idaho batholith rocks side-by-side: hornblende biotite granodiorite (left) and muscovite biotite (two-mica) granite (right). In both samples the gray mineral is quartz and the white is feldspar. Shiny cleavage faces of muscovite are visible in the lower part of the granite. —Granite photo by Reed Lewis

related plutons in western Montana. In places, the intrusion of the batholith did not completely remove blocks of preexisting sedimentary rock. These are now present as roof pendants, or screens, and in several places contain metallic mineral deposits in locations such as Bellevue, Bayhorse, Stibnite, Edwardsburg, and Elk City.

The huge batholith is not traversed by many roads. ID 21 from Stanley to Idaho City and Boise follows the Middle Fork of the Payette River through spectacular batholith canyon country. The urban area of Boise is hard against Idaho batholith rocks, and Bogus Basin is within it.

EOCENE EXTENSION AND THE CHALLIS MAGMATIC EVENT

A second wave of arc magmatism, known as the Challis magmatic event, swept across Idaho between 51 and 43 million years ago in the Eocene Epoch, following closely on the intrusion of the Idaho batholith. At the same time, tectonic compressive stresses relaxed and much of the Pacific Northwest was extended in a northwest-southeast direction. The Challis magmatic event produced andesite, dacite, and rhyolite volcanic rocks that are extensively exposed in the region around Challis. In many places, such as the southern Wood River Valley, the Idaho batholith had been uplifted and eroded by 50 million years ago, such that Challis volcanic rocks erupted directly onto batholith rocks. Probably the most common preserved expression of Challis magmatism in Idaho are andesite, dacite, and rhyolite dikes. In many areas of central Idaho, it is difficult to find an outcrop that doesn't contain a Challis dike. The dikes are mostly northeast-trending; the rising Challis magmas filled cracks that were oriented in this direction.

Challis dike and volcanic rocks can be differentiated relatively easily by the minerals that form visible crystals in the fine-grained groundmass. Rhyolite dikes and crystal-bearing volcanic rocks typically contain crystals of quartz (dark gray and clear), potassium feldspar (largest tabular crystals that weather to a pink color), and plagioclase feldspar (tabular, white-weathering crystals), with minor amounts of

Eocene Challis rhyolite dike (left) and dacite lava flow (right).

biotite (platy black mineral). Dacite dikes and lavas contain mostly plagioclase feldspar and hornblende (tabular black crystals), with some biotite. Andesite is notable for its dark color and paucity of crystals that may include plagioclase feldspar and hornblende.

The northeast-trending Challis dikes are parallel to northeast-trending Eocene normal faults that are part of a 60-mile-wide regional belt of extension that cuts across central Idaho as the Trans-Challis fault zone. The northeast trend of the Trans-Challis fault zone is parallel to the Paleoproterozoic northeast-trending Great Falls tectonic zone, which marks the northern boundary of Archean crust in central Idaho and extends northeast through Montana. Further work is needed to determine if the similarities in position and trend are related or merely a coincidence.

Unerupted magma cooled underground into intrusive rocks, including diorite, granodiorite, and granite in much of the rest of central Idaho. In many areas the magma cooled very near to the surface (within a half mile). Miarolitic cavities (gas bubbles) preserved in the intrusive rocks attest to low pressures at the time of crystallization because of the lack of much overlying rock. Some regions such as the Sawtooth Range and Boulder Mountains are underlain by extensive shallow-level Challis intrusive complexes. Along US 12 in the northern part of central Idaho, deeper Challis magmatic systems are preserved.

Another expression of Eocene extension was the development of core complexes across Idaho. Core complexes develop when crust is over-thickened by long episodes of orogeny such as that experienced across Idaho during the Sevier orogeny in Cretaceous and early Paleogene time. As the over-thickened crust relaxes after the compressive forces are removed, the crust gravitationally collapses, causing the upper crust to slide off the top of the lower crust on low-angle detachment faults. These faults typically accommodate tens of miles of displacement and bring rocks that were once more than 10 miles deep in the crust to the surface.

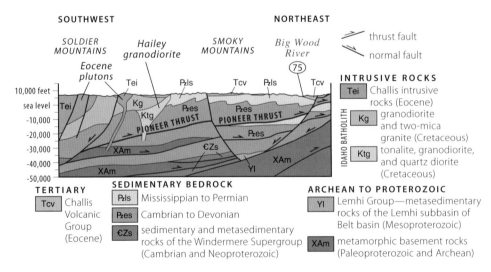

This section across the Soldier and Smoky Mountains west of Hailey shows the complexity of the Central Mountains, with various igneous rocks of Cretaceous to Eocene age intruding the thrust faulted and folded Paleozoic rocks that overlie the Archean basement rocks.

Many of these core complexes, such as the Pioneer Mountains core complex, were intruded by deeper level Challis intrusions that were subsequently exposed by the detachment faulting and then covered by younger Challis volcanic rocks that erupted across the newly exposed surface. These relationships demonstrate that the Challis magmatism and core complex formation occurred at the same time.

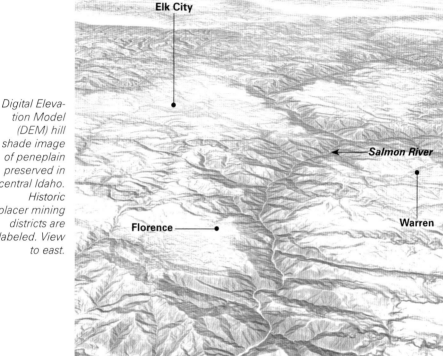

Digital Elevation Model (DEM) hill shade image of peneplain preserved in central Idaho. Historic placer mining districts are labeled. View to east.

Challis Volcanic Group along the Salmon River at Deadman's Hole west of Challis. (44.255, -114.339)

Although the high plateau that had been constructed in Late Cretaceous time had collapsed to a mere shadow of its former greatness, most of central Idaho was still high-standing by Oligocene time. On the top of the elevated region was a broad, low-relief erosional surface called a peneplain, remnants of which can still be viewed in many parts of central Idaho. Rivers flowed across this surface, and basins formed that collected gravel in places such as Elk City, Pierce, and Florence. Many of these gravels are rich in placer gold, and the first wave of gold strikes in Idaho in the 1860s were centered in these older basins. The rugged relief that characterizes much of central Idaho results from later (middle Miocene) faulting and river canyon cutting that disrupted the original peneplain. The several forks of the Salmon River flow through canyons several thousand feet deep.

MIOCENE, PLIOCENE, AND PLEISTOCENE TIME

Basin and Range normal faulting (discussed in depth in the introduction to the Basin and Range chapter) began around 17 million years ago. In the Basin and Range, the faults are mostly north-trending with upthrown sides forming north-trending mountain ranges and downthrown sides forming the intervening valleys. These faults extend north into central Idaho, and many are active, causing earthquakes when movement occurs. Some quakes have been large enough to result in significant property damage and loss of lives. The Sawtooth Range is uplifted on an east-dipping Basin and Range normal fault. Likewise, the flat Long Valley, from McCall south to Cascade, is a Basin and Range valley, with the east-dipping normal fault on the west side.

Superimposed on the Basin and Range faulting is a wave of eastward-progressing extensional faulting that follows the passage of the region over the Yellowstone hot spot during the last 10 million years. As this hot spot and the topographic high it produced migrated eastward, the Snake River Plain subsided in its wake. The effect on central Idaho was rapid mountain uplift. Streams that formerly drained north and east to Montana were captured in Pliocene time by the west-flowing Snake River, reversing their drainage direction.

In central Idaho, a secondary effect of the hot spot uplift was enhanced erosion in the last 5 million years. Elevated regions are always subjected to more erosion. They gather more snow and rain, and gravity increases stream gradients, creating dramatically steep mountain sides and spectacular canyons. As these mountains rose, the increased elevation also brought the mountain tops within reach of alpine glaciers in the Pleistocene Epoch, when the Earth experienced cycles of glaciation and deglaciation.

During the peak of the most recent glacial episode, many of the high mountains of central Idaho were home to alpine glaciers. The elevations that these glaciers resided at varies from west to east across central Idaho due to the rainshadow effect. As storms from the Pacific barreled eastward across the state, they dropped more snow in the western mountains, progressively becoming more depleted in moisture to the east. Less snowfall means glaciers have less mass and are more limited or confined to their higher elevations. As an example, the average elevation of glaciers in the Sawtooth Range was about 7,800 feet whereas glaciers to the east in the Lost River and Lemhi Ranges occurred at an average elevation of about 9,200 feet.

The warmer climate of the last 15,000 years melted the last of Idaho's glaciers. A small glacier that existed on the north face of Borah Peak in 1975 had disappeared by

1990. Many perennial snowfields still exist in the Sawtooth Range and a few other high peaks in central Idaho. Isostatic uplift after the ice melted, combined with many annual cycles of freeze-thaw, contribute to the steep and craggy mountains we have today.

US 12
Kooskia—Lolo Pass on Montana Border
101 miles

This section of US 12, a curvy route across the Clearwater and Bitterroot Mountains, passes right through the heart of one of the great granite batholiths of western North America, the Idaho batholith of Cretaceous to Paleogene age. The road mainly follows the Lochsa River, one of the renowned whitewater rivers of the western United States. The route begins at the suture zone between ancient North America and the accreted Wallowa terrane. See the introduction to the Western Margin chapter for more information about the Wallowa terrane, suture zone, and shear zones along the margin of the Idaho batholith.

Around Kooskia, the Middle Fork of the Clearwater River cut into an ancient predecessor valley that was filled by basalt lava flows of the Columbia River Basalt Group around 16 million years ago in Miocene time. Outcrops of the basalt persist to the east to approximately milepost 81, east of which outcrops of Jurassic to Cretaceous quartz diorite rocks of the Wallowa terrane are prevalent. Between mileposts 82 and 83, the highway crosses 110-million-year-old, biotite-bearing tonalite, which yields strontium isotopic values in the range of 0.704 to 0.706. These rocks lie within the 0.704-0.706 line, or transition zone, which is actually a few miles wide here. The line indicates the transition in the lower crust between rocks of the Wallowa island arc terrane of oceanic origin (isotopic ratios 0.704 and less) to the west and much older rocks of North American continental origin (ratios of 0.706 and greater) to the east.

East of milepost 87, outcrops along US 12 include hornblende-bearing gneiss of the Orofino Metamorphic Suite, part of the highly metamorphosed Permian to Jurassic sedimentary and volcanic rocks of the Wallowa terrane near the suture zone with North America. At about milepost 88, the road crosses over the trace of the Woodrat Mountain suture, the fault that places rocks of the accreted Wallowa terrane against the continental rocks, which has been intruded by granodiorite at this location. Younger plutons commonly overprint the suture, disrupting older rocks. However, in the mountains just north of here, portions of the Woodrat Mountain suture zone are preserved, and these exposures represent one of the very few places where it is possible to place a finger on the suture itself in outcrop.

The Woodrat Mountain suture, likely of Late Jurassic or Early Cretaceous age, juxtaposes Orofino metamorphic rocks of the Wallowa terrane to the west and quartzite of the Windermere Supergroup of North America to the east. The quartzite represents metamorphosed sandstone that was deposited along the nascent shoreline of western North America as other continents were rifting away to open the Pacific Ocean at approximately 700 million years ago in Neoproterozoic time. Beautiful outcrops of these quartzites that have been intensely sheared in the Woodrat Mountain suture are accessible about a quarter mile up Smith Creek Road (turn off between mileposts 88 and 89). Signs of shearing in quartz and uncommon muscovite mica grains indicate that the North American side of the fault zone moved up and to the west, while the

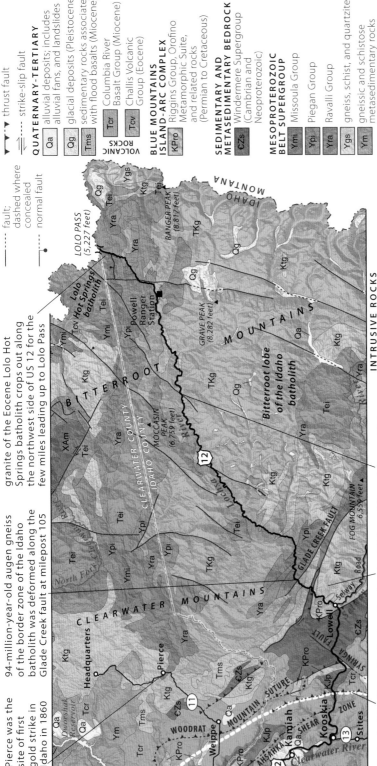

Geology along US 12 between Kooskia and Lolo Pass on the Montana border.

Windermere Supergroup quartzite outcrop in the Woodrat Mountain suture on Smith Creek Road (between mileposts 88 and 89).(46.1316, -115.7602)

Orofino Metamorphic Suite moved east and beneath the North American rocks. At other locations along the Woodrat Mountain suture, the Neoproterozoic quartzite is associated with schist containing kyanite and sillimanite, metamorphic minerals that indicate that the original sedimentary rocks attained high pressure (kyanite) and high temperature (sillimanite) as they were deeply buried and metamorphosed during suturing of the Wallowa terrane to North America.

The highway crosses the northeast-trending Syringa fault at milepost 90. This late-stage fault was active sometime between 90 and 73 million years ago and offsets the Woodrat Mountain suture zone and Ahsahka shear zone 6 to 8 miles in a right-lateral sense. A very straight southwest-trending section of the Clearwater River follows the Syringa fault for several miles here because rock ground up along the fault has eroded much more easily than the surrounding intact rock. Outcrops along the road for several miles consist of well-layered schist and gneiss of the Windermere Supergroup.

At Lowell, between mileposts 96 and 97, the Selway River, flowing from the southeast, joins the Lochsa River, flowing from the northeast, and the two rivers combine to become the Middle Fork of the Clearwater River. The geology in this area is complicated by faulting and folding that postdates deformation on the Woodrat Mountain suture zone and Ahsahka shear zone. A blister of more deeply seated rock, called the Coolwater culmination, has been brought to the surface by doming and faulting. The rock in the culmination consists of gneiss of the Orofino Metamorphic Suite intruded by 86-million-year-old tonalite of the Idaho batholith that was subsequently deformed and metamorphosed into the Coolwater Ridge gneiss between 76 and 64 million years ago. The popular explanation for this mysterious blister of Wallowa rocks within North America proper is that as the Orofino suite rocks were subducted eastward beneath the North American margin, this portion buckled and popped up into the overriding North American crust on a series of faults.

Between Lowell and the Powell Ranger Station, US 12 follows the Lochsa River northeast for about 65 miles. At milepost 101 the highway crosses out of the Coolwater culmination and back into North American geology. The western margin of ancient North America in this part of Idaho consists of metamorphosed rocks of the Mesoproterozoic Belt and Neoproterozoic Windermere Supergroups that have been intruded in turn by granitic rocks of the Late Cretaceous to Paleocene Idaho batholith and Eocene Challis magmatic event. The first outcrops of Idaho batholith rocks

COOLWATER CULMINATION ALONG SELWAY ROAD

Parts of the Coolwater culmination, a blister of deeply seated rock brought to the surface by doming, can be seen along Selway Road, accessed by crossing the bridge over the Lochsa River at the confluence at Lowell. Outcrops of well-banded gneiss of the Orofino Metamorphic Suite at 1.9 miles from US 12 are highly metamorphosed sedimentary and volcanic rocks of the Wallowa terrane that are caught up in the structural culmination. An ultramafic unit occurs in the shear zone that forms the southwestern side of the Coolwater culmination, 2.8 miles up the road (park at Swiftwater Bridge and walk across the bridge to outcrops on the other side of the river). The rocks here consist of the alteration minerals talc, actinolite, and chlorite, and carbonate minerals that have replaced the original peridotite rock that consisted of mostly olivine. This outcrop is part of a belt of ultramafic rocks that occur in the Orofino Metamorphic Suite and Riggins Group within the suture zone. At 5.9 miles from US 12, the strongly deformed biotite tonalite of the 86-million-year-old Coolwater Ridge gneiss is beautifully exposed.

Tonalite of the Idaho batholith was metamorphosed to Coolwater Ridge gneiss, exposed here along Selway Road. (46.0927, -115.5269) —Photo by Karen Lund, US Geological Survey

appear at milepost 103. These are rocks of the border zone, which persists for a few miles, followed by the massive Bitterroot lobe of the batholith that dominates the rest of the route along the Lochsa River.

The 98- to 85-million-year-old rocks of the Idaho batholith border zone for the next few miles are strongly deformed by northwest-trending faults that cut the rocks approximately 73 to 66 million years ago at approximately the same time that the Coolwater culmination formed. At milepost 105, beautiful 94-million-year-old augen gneiss that originated as granodiorite of the Idaho batholith is exposed in outcrops on the north side of the road. These rocks have been sheared and metamorphosed along the Glade Creek fault zone. The large potassium feldspar crystals, the "eyes" in the augen gneiss, are rimmed by sodium-rich plagioclase feldspar, a mineral texture called rapakivi. Rocks in roadcuts from east of here to about milepost 108 are calc-silicate gneiss, a section of metamorphosed, originally limestone-rich rocks of the Belt Supergroup that is completely surrounded by intrusive rocks of the Idaho batholith.

For the 50 miles between milepost 111 and Powell, the highway passes through the 66- to 54-million-year-old Bitterroot lobe, and outcrops along the road are dominated by biotite granodiorite and muscovite-biotite (two-mica) granite. Light-colored pegmatite (very coarse-grained) and aplite (very fine-grained) granite dikes are typically found in these roadcuts. As crystallization of a typical magma chamber was nearly complete, the small amount of residual magma at the very end was concentrated with water and other constituents that don't occur in the usual granitic minerals. This last gasp of crystallizing magma became over pressurized, shooting

A block of migmatite (partially melted metasedimentary rock of the Belt Supergroup) encased in biotite granodiorite (gray) of the Bitterroot lobe of the Idaho batholith in a roadcut near milepost 116. Lighter-colored dikes and sills are pegmatite and aplite associated with the batholith rocks. View across photo is approximately 3 feet.

dikes of pegmatite and aplite magma into the surrounding crystallized granitic rocks. Chunks of metamorphosed Belt Supergroup rocks in some of these outcrops appear to have been very gooey as they were intruded by dikes and stringers of granodiorite and granite.

During Challis magmatism from 51 to 43 million years ago in Eocene time, lava flows and tuffs were likely erupted at the surface here. Subsequent erosion removed all of the volcanics, however, and along US 12 we only see the deeper intrusive levels of the Idaho batholith cut by Challis dikes of mostly andesite and rhyolite, with some dacite. The andesite and dacite dikes are pretty typical for central Idaho, as described in the chapter introduction, but most of the rhyolite dikes are fairly poor in crystals. At several locations along the highway, larger Challis plutons are exposed in roadcuts, including granodiorite and syenite (a potassium feldspar-rich granitic rock) near Fish Creek at milepost 120 and diorite and gabbro near milepost 129.

Two natural hot springs along the Lochsa River, Weir Creek Hot Springs at milepost 142 and Jerry Johnson Hot Springs between mileposts 151 and 152, have marked parking areas along the highway. The hot springs occur along Eocene-age, northeast-trending normal faults that cut across the Bitterroot lobe of the batholith. Faulting has crushed the crystalline igneous rocks along the faults, providing permeable conduits for groundwater to circulate deep into the crust where rocks are much warmer. Once warmed at depth, the hot, buoyant water flows back toward the surface, dissolving ions such as sulfur from rock in the fault zones and discharging at the surface as springs.

East of milepost 161, Belt Supergroup rocks metamorphosed by the Idaho batholith become increasingly common. Many of the outcrops along the road consist of gneiss of the Belt's Piegan Group. Other gneiss, schist, and quartzite originated as sedimentary rocks of the Ravalli Group. In outcrops of metamorphic rocks that are close to the Bitterroot lobe, dikes and sills of granodiorite and granite are still common, but they become increasingly rare farther east away from the magma body. Outcrops of the Belt Supergroup in Montana between 8 and 12 miles east of Lolo Pass are only slightly metamorphosed, as compared to these outcrops of high-grade gneiss and quartzite that are much closer to the batholith.

Between milepost 170 and Lolo Pass on the Montana border, the highway follows a very young north-trending fault that has pulverized the rock in outcrops on the ascent to the pass. To the west of the fault, pink granites of the Lolo Hot Springs batholith of Eocene age intruded the Belt Supergroup and Idaho batholith rocks. These granitic rocks, distinct from the older Idaho batholith rocks, are coarse-grained and contain smoky-colored quartz, commonly pink-weathered potassium feldspar, and well-formed hexagonal biotite. They also commonly contain gas cavities with well-formed smoky quartz crystals, indicating the granite batholith crystallized within 1 mile of the Earth's surface where magma chamber pressures were low enough to permit water vapor and other gases to form bubbles in the magma. The smoky color of the quartz results from radiation damage to the quartz crystal structure; Eocene granites are much more radioactive than their Cretaceous-Paleocene counterparts in the Idaho batholith. Other features that contrast the Eocene Challis granites with Cretaceous Idaho batholith granites include a paucity of pegmatite dikes in the Eocene rocks, and their tendency to weather into very rounded outcrops. Good outcrops of the Lolo Hot Springs batholith occur in Montana between 5 and 8 miles east of Lolo Pass.

US 93
CHALLIS—SALMON—LOST TRAIL PASS AT MONTANA BORDER
104 miles

Between Challis and Salmon, US 93 follows the Salmon River, which eroded its path along northeast-trending normal faults in the Trans-Challis fault zone. These faults formed in Eocene time because of extensional forces. Several volcanic centers of the Challis Volcanic Group erupted along this zone of extension, including the Twin Peaks caldera northwest of Challis. The ash and pyroclastic debris accumulated in basins dropped down along the normal faults. One of the thickest accumulations of Challis Volcanics occurs along the Salmon River between Challis and Ellis within the Trans-Challis fault zone.

The Custer Motorway (beginning on Challis Creek Road north of milepost 246) heads northwest from Challis and goes through the old mining town of Custer before circling back to ID 75 on the Salmon River at another mining area, the Yankee Fork (also see map on page 289). Where Challis Creek Road heads west from town, it climbs through red Eocene ignimbrites (tuff of Challis Creek) that dip south, parallel with the road.

From milepost 248 north of Challis is a fine view to the east of Pennal Gulch, with red Eocene Challis sedimentary and volcanic basin fill at the base of the hill, and Neoproterozoic and Cambrian quartzite of the Wilbert Formation at the top. The red rocks to the south across the Salmon River are also basin-filling sandstone altered by hydrothermal waters of Challis Hot Springs.

Between mileposts 259 and 260, hard layered rocks of the Belt Supergroup rise steeply from the edge of the highway. These are the Lawson Creek and Swauger Formations deposited into the Belt basin 1.4 billion years ago. The Lemhi Range, looming steeply above the Pahsimeroi Valley south and east of the Pahsimeroi junction at Ellis, has been uplifted along the large, active Lemhi normal fault. The northern Lemhi Range contains upper Belt Supergroup strata, stacked by the Iron Lake and Poison

Red flat-lying sedimentary beds of the Challis Volcanic Group along the Salmon River viewed from north of Challis near milepost 248. (44.532, -114.201)

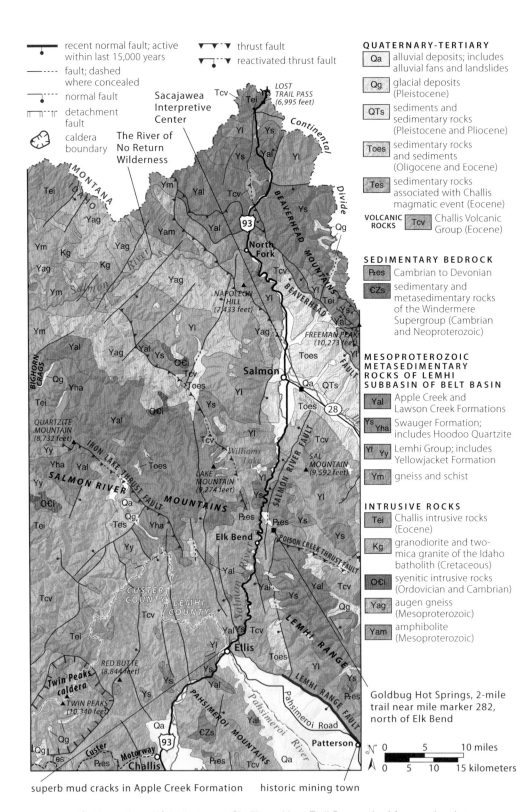

Geology along US 93 between Challis and Lost Trail Pass at the Montana border.

Creek thrust faults. The Iron Lake thrust south of US 93 is hidden at depth beneath the Pahsimeroi Valley.

On the east side of the Pahsimeroi junction, bedding planes of the Apple Creek Formation of the Lemhi Group dip steeply westward toward the road, and you can see ripple marks on the rock surfaces. These sediments were deposited in and along the edge of a shallow inland sea or huge lake in the Belt basin in Mesoproterozoic time. Depositional features such as mud cracks and ripple marks are well preserved in the Belt rocks because land plants did not yet exist. High rates of erosion on the vegetationless land caused high rates of sedimentation in the basins, covering up and preserving depositional features. Plus, there were no worms and other creatures to stir up the sediment. You can see more Belt rock for several miles north of the junction.

Between Ellis and Salmon, the Salmon River eroded its path along the recurrently active Salmon River fault zone, an Eocene northeast-trending normal fault, which may have initiated as part of the rift margin in Neoproterozoic time and was later reactivated in Mesozoic time. Cliffs of white and red rocks on both sides of the Salmon River, particularly dramatic between mileposts 271 and 273, are Challis Volcanic Group. Most of the rocks are lavas and tuffs, east dipping on the down-dropped side of the Salmon River fault. The curving highway passes through volcanic debris flows and welded tuffs.

Elk Bend, an isolated settlement about 35 miles north of Challis, offers access to the Salmon River Mountains to the west via Iron Creek. East of the Iron Creek Road junction, pinnacles in Challis volcanic rocks decorate the reddish hillsides. The access road to the trailhead for Goldbug Hot Springs lies a little farther north near milepost 282, just north of vertical red cliffs of Challis volcanic rocks that rise directly up from the east side of the highway.

Apple Creek Formation with ripple marks (parallel to the hammer handle) just east of Ellis and west of milepost 264. (44.6926, -114.0485)

The Salmon River cuts through the gap in the middle of the photo. The distant northern Lemhi Range, composed of Mesoproterozoic Lemhi Group, peeks through the gap. Much younger Eocene Challis volcanic rocks form the cliffs and roadcuts in the closer ridge. The contact between the young and old rock is the Salmon River fault, a normal fault that forms an abrupt change in slope near the base of the high mountains. View looks northeast from just east of Ellis. (44.6967, -114.0485)

About 5 miles south of Salmon, Williams Creek Road provides access west to Williams Lake, Panther Creek, Bighorn Crags, and the Yellowjacket mining district. The Cobalt mining district on Panther Creek has been explored for much of the later twentieth century, and a new mine was under construction in 2018. The Forest Service road continues west over the summit of Quartzite Mountain and provides access to the north to Eocene granite of the Bighorn Crags. Going west from the summit and down Shovel Creek brings one to Yellowjacket, another old silver mining district in Belt Supergroup rocks. West of here is the deep canyon of the Middle Fork of the Salmon River and the Frank Church–River of No Return Wilderness.

Just north of the Williams Creek Road junction the valley widens, and toothy peaks of the Beaverhead Mountains form the skyline to the north. The prominent spike of Freeman Peak is the middle of the view. The rocks in the Beaverhead Mountains are folded upper Belt Supergroup and total close to 9 miles thick. Paul Link and students have demonstrated that more than 80 percent of the zircon sand grains in these rocks are about 1,720 million years old. The mineral zircon is particularly resistant to weathering, so if you find zircon grains in sedimentary deposits, you can determine what rocks they eroded from based on their age. These grains were eroded from a huge magmatic complex to the south known as the Big White arc and deposited into the Lemhi subbasin about 1,400 million years ago.

In the Salmon Basin, US 93 lies on the west side of the west-dipping Salmon Basin detachment fault that accommodated extension in Eocene time when core complexes were forming in Idaho. Eocene and Oligocene tuffaceous lakebeds on this side of the fault are evidence that down-dropping on the fault formed a low area, the Salmon

View east to northern Beaverhead Mountains from north of Salmon. The white rocks near the Salmon River are Eocene lakebeds of the Salmon Basin.

Basin, that hosted a lake. The steeply west-dipping Beaverhead normal fault, an active Basin and Range structure, has since cut the Salmon Basin detachment, further dropping the Salmon Basin. Belt Supergroup rocks on the uplifted block of these faults make up the steep Beaverhead Mountains to the east.

At North Fork, the Salmon River turns west and flows through the heart of Idaho's central mountains, forming a legendary section of whitewater that is on every rafter's bucket list. US 93 runs north to Lost Trail Pass following the route of Lewis and Clark, who astutely decided to avoid what was called the River of No Return. North of North Fork, the highway is entirely surrounded by the Belt Supergroup and passes large roadcuts in places. Faults control the location of the valley. Placer workings in gravel deposits associated with streams draining gold-bearing Eocene intrusive rocks are prominent in Hughes Creek, west of the highway on a gravel road just south of milepost 332. The old mining settlement of Gibbonsville (mile 337), a gold camp from 1877 to 1899, has a colorful, but not affluent, history.

The highway switchbacks up to Lost Trail Pass, passing roadcuts in rocks of the Swauger and Lawson Creek Formations and Lemhi Group. Northwest-trending faults cut the rocks here. An Eocene pluton lies west of the highway at the pass. You can inspect an outcrop of this granodiorite on the uphill side of the road from a pullout on the downhill side at milepost 347.

ID 21
Boise—Stanley
131 miles

ID 21, also known as the Ponderosa Pine Scenic Byway, cuts through the heart of the Idaho batholith, from the gold-rich deposits of the Boise Basin to the lofty Sawtooth Range. The highway boasts some of the most outstanding mountain scenery in the state and transects parts of three major river systems—the Boise, Payette, and Salmon—all of which have their origins in the rugged Sawtooths.

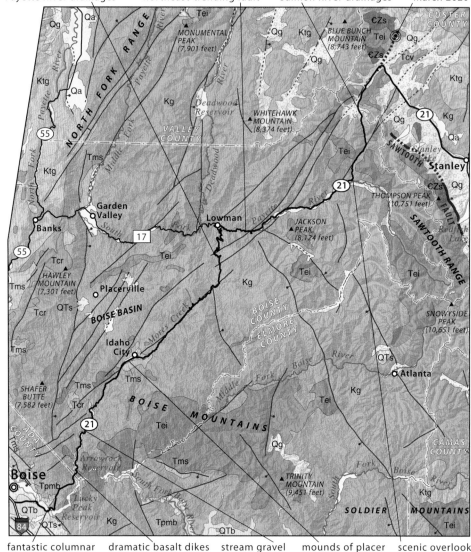

Geology along ID 21 between Boise and Stanley.

The southern terminus of ID 21 begins at I-84 a few miles southeast of Boise on old river terraces from the Boise River. The Boise foothills lie to the north and east, an uplifted region bounded by northwest-trending normal faults that help define the western Snake River Plain. Near milepost 2, the highway descends through dark-gray basalt that erupted about 572,000 years ago from volcanic vents located about 30 miles to the east near Smith Prairie. The vents were associated with the northern edge of the Snake River Plain. The lava flowed west down the ancestral Boise River drainage, temporarily damming and diverting the river before it cut a new channel along the edge of the lava.

The highway crosses the Boise River near milepost 3, and mileposts jump ahead about 5 miles due to a realignment of the highway from Warm Springs Road to Gowen Road. A diversion dam just upstream of the bridge redirects a portion of the Boise River water into canals for irrigation.

The next several miles of highway pass beneath spectacular exposures of columnar jointing north of the highway in the Basalt of Lucky Peak, another canyon-filling lava that erupted about 1.36 million years ago from the Smith Prairie area. The photogenic hexagonal and pentagonal columns form as lava uniformly cools and contracts, creating vertical fractures separated by angles of about 120 degrees. This scenic cliff section, known as Black Cliffs, is popular with local rock climbers.

A tall roadcut of river gravel and sand on the north side of the highway near milepost 10 attests to flood events in Pleistocene time when the Boise River was at a higher level. As the road leaves the river and climbs to the top of Lucky Peak Dam, roadcuts along the highway expose Cretaceous granodiorite of the Atlanta lobe of the Idaho

Columnar jointing in 1.36-million-year-old basalt above the Boise River downstream from Lucky Peak Reservoir. (43.5339, -116.0825)

Black basalt dikes cut through white granodiorite of the Idaho batholith near milepost 18. (43.6000, -115.9895)

batholith. Past the dam, the highway heads north over a small pass before dropping down to the north arm of Lucky Peak Reservoir along Mores Creek. At milepost 17, ID 21 crosses the Mores Creek arm, where roadcuts of white, 85-million-year-old granodiorite stand in contrast to the black, canyon-filling basalts surrounding the reservoir. The basalts are about 107,000 years old. Near milepost 18, snaky basaltic dikes slice through the granitic rocks in roadcuts on the east side of the road. The highway descends to creek level and back into the canyon-filling basalts near milepost 21.

An instructive roadcut between mileposts 24 and 25 and near the bridge over Mores Creek captures the geologic essence of this region. Here, Cretaceous granodiorite is overlain by up to 4 feet of stream gravel. This, in turn, is covered by pillow basalt, indicating the basalt flowed into stream and interacted with water. Orange palagonite, an alteration product, coats many of the pillows and accentuates their rounded shape. About 8 feet above the gravel, the pillow basalt grades into typical basalt that caps the roadcut. Collectively, this roadcut (and others nearby) illustrate the following sequential geologic events: erosion of the Idaho batholith by streams, deposition of gravel along the streambed, and lava filling the streambed.

As the highway approaches Idaho City, it enters the heart of the Boise Basin, a region centered on the richest gold strike in Idaho history. The discovery of gold here in 1862 spawned a large gold rush. Towns like Idaho City, Placerville, Centerville, and others popped up quickly, making the area the most populated region in the Pacific Northwest by the mid 1860s. The Boise Basin area yielded over 2 million ounces of gold from placer deposits from 1863 to 1896. Unnatural mounds of river gravel scattered throughout the basin and along ID 21 near Idaho City are a testament

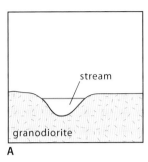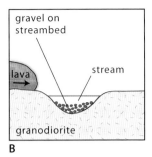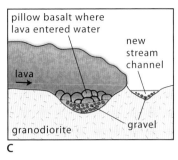

Diagram depicting geologic events near Mores Creek drainage and milepost 25.

The roadcut near Mores Creek bridge and milepost 25 shows Cretaceous granodiorite (base) overlain by stream gravels, pillow lava, and massive basalt (top). (43.6772, -115.9717)

to the glory days of placer gold mining. The extensive mining activities significantly disrupted the local topography and drainage systems, rerouting rivers and streams, and creating standing water bodies where none previously existed.

Near milepost 49, the highway begins its ascent out of the Boise River drainage basin. Rocks exposed in roadcuts are mostly biotite-bearing granodiorite of the Idaho batholith but vary somewhat in color from white to light gray to orange brown, depending on the exact type of rock, the degree of fracturing, and how much iron oxidation occurred as groundwater infiltrated the rock. Brown basalt dikes cut through white granitic rock between mileposts 52 and 53 and again near milepost 54. In some roadcuts, large piles of angular, sand-sized material are found below the

As granitic rocks weather, the individual crystals separate into a coarse-grained material called grus. (43.9321, -115.6682)

white granitic outcrops. Called grus, this is the product of granitic rock disaggregating due to chemical and physical weathering processes.

Near milepost 63, Beaver Creek Summit divides the Boise River drainage basin from that of the Payette River. A dark-gray Eocene andesite dike with large crystals of hornblende is exposed in a roadcut between mileposts 66 and 67. The highway crosses one of the most scenic rivers in Idaho, the South Fork of the Payette River, between mileposts 72 and 73 and then follows the South Fork eastward toward its headwater region. Cretaceous granodiorite dominates this section of roadway. Kirkham Hot Springs lies near the river between mileposts 76 and 77 and is one of many public-access hot springs found along the South Fork. Northeast-trending faults and fractures create pathways for groundwater heated at depth to rise to the surface. In general, hot springs are common within the Idaho batholith.

Near milepost 94, the highway abandons the canyon of the South Fork of the Payette River and follows a smaller tributary, Canyon Creek, as it heads northward. A scenic overlook for the Sawtooth Range lies between mileposts 95 and 96 on the east side of the road. More rugged and craggy outcrops near milepost 99 are part of the Sawtooth pluton, an intrusion of magma that cooled and crystallized in this region about 45 million years ago in Eocene time during the Challis magmatic event. From a distance, the Sawtooth intrusive rocks look weathered and gray, but up close, they are distinguished by large pink crystals of potassium feldspar that contrast with the salt-and-pepper appearance of the much older Idaho batholith.

Banner Summit, between mileposts 105 and 106, marks the drainage divide between the Payette and Salmon Rivers. Roadcuts near the 7,037-foot summit reveal a mixture of boulders, gravel, and sand from till deposited by alpine glaciers in Pleistocene time. Between Banner Summit and ID 21's big southward turn, the highway follows a northeast-trending fault, part of the Eocene Trans-Challis fault zone, eroded by Cape Horn Creek. The highway crosses the creek between mileposts 112 and 113 near the northern end of the Sawtooth Range and the most northerly point of ID 21.

The 6.5 magnitude Stanley earthquake on March 31, 2020 was located on a northwest-trending Basin and Range fault a few miles north of this point and felt throughout much of Idaho and western Montana. The strong shaking triggered avalanches and rockfall in the surrounding mountains but caused no notable property damage in the sparsely populated area.

The highway swings sharply southeast and follows Marsh Creek into a wide glacial outwash valley now occupied by meadows. Near milepost 117, the highway offers on and off glimpses southward into the airy Sawtooth Range rising more than 10,000 feet above sea level. A large active Basin and Range normal fault, the Sawtooth fault, cuts the dramatic eastern side of the range, uplifting the Sawtooths while dropping the adjacent valley.

The serrated profile of these mountains is the product of glacial erosion over the past few hundred thousand years. As the glaciers grew and moved downslope, the ice gouged and scoured the underlying granitic bedrock, creating sharp ridgelines and pointy peaks. Larger canyons in the Sawtooths are characteristically U-shaped from glaciers advancing down former stream valleys, straightening and widening them. The farthest downslope progression of each glacier onto the valley floor is marked by a steep ridge of sediment, deposited by melting ice at the toe of the advancing glacier. Called moraines, these hills contain a hodgepodge assortment of rocks carried by the ice. As the glaciers receded and melted, the moraines dammed streams draining the mountains, forming many of the iconic lakes in the area such as nearby Stanley and Redfish Lakes.

The small mountain town of Stanley is situated at the northern end of the Sawtooth Valley. Low hills north of town are intrusive rocks of the Idaho batholith. ID 21 terminates with ID 75, adjacent to the upper reaches of the Salmon River.

ID 55
Boise—New Meadows
111 miles

ID 55 runs north from Boise to New Meadows, generally following the western edge of the Idaho batholith, parallel to, and inboard of, the collisional boundary between ancient North America and the accreted Blue Mountains terranes. It crosses this boundary and the Western Idaho shear zone in the final stretch between McCall and New Meadows.

North of Eagle, a northern suburb of Boise, ID 55 climbs to Spring Valley Summit of the Boise Front at milepost 57. Roadcuts along the highway are in mostly Miocene sedimentary deposits of gravel, sand, and mud of the Payette Formation, a unit that includes all sedimentary deposits associated with the Columbia River Basalt Group and Weiser volcanics. Between mileposts 50 and 51, and then again across Spring Valley Summit, granodiorite of the Cretaceous Idaho batholith, which lies below these sedimentary deposits, pokes out at the surface to form very weathered outcrops. Descending to Horseshoe Bend, roadcuts along the highway consist of the Idaho batholith rocks. The drainage to the west of the highway contains more Payette Formation sediments. Obvious white tuffs within these sediments are deposits of ash erupted from volcanoes associated with passage of the Yellowstone hot spot to the south of here. Most of the slope has been disrupted by landslides because the ash,

wind gap in West Mountains visible between mileposts 133 and 134 preserves an ancient west-flowing stream that existed before the Payette River captured the Long Valley drainage basin

mylonite rocks of the 105-million-year-old Little Goose Creek granite were deformed by the Western Idaho shear zone

Payette Lakes are ponded behind end moraines of the once-glaciated upper North Fork of the Payette River

migmatite rocks of the Wallowa terrane near Snowbank Mountain were deeply buried, metamorphosed, and partially melted along the Western Idaho shear zone

QUATERNARY-TERTIARY

Qa	alluvial deposits; includes alluvial fans and landslides
Qg	glacial deposits (Pleistocene)
QTs	sediments and sedimentary rocks (Pleistocene and Pliocene)
QTpms	sedimentary rocks associated with Basin and Range extension (Quaternary, Pliocene, and Miocene)
Tms	sedimentary rocks associated with flood basalts; includes Payette Formation (Miocene)

VOLCANIC ROCKS

Tpmb	basalt (Pliocene and Miocene)
Tmr	rhyolite (Miocene)
Tmfo	older rhyolite, latite, and andesite (Miocene)
Tcr	Columbia River Basalt Group (Miocene)

INTRUSIVE ROCKS

| Tei | Challis intrusive rocks (Eocene) |

IDAHO BATHOLITH

| Kg | granodiorite and granite (Cretaceous) |
| Ktg | tonalite, granodiorite, and quartz diorite; includes deformed and metamorphosed units (Cretaceous) |

BLUE MOUNTAINS PROVINCE

KJqd	quartz diorite (Cretaceous and Jurassic)
KJp	plutonic rocks along the Western Idaho shear zone; includes Little Goose Creek Complex (Cretaceous and Jurassic)
Kis	syenite and related rocks (Cretaceous)

WALLOWA TERRANE

| KPro | Riggins Group, Orofino Metamorphic Suite, and related rocks (Permian to Cretaceous) |

PALEOZOIC TO MESOPROTEROZOIC

| PzYs | metasedimentary rocks of uncertain age |

The North Fork of the Payette River drops 100 feet/mile through a renown Class V whitewater section

dashed white line indicates the 0.704-0.706 line demarking the edge of cratonic North America

— recent normal fault; active within last 15,000 years
---- fault; dashed where concealed
— normal fault
▼▼ reactivated thrust fault

Geology along ID 55 between Boise and New Meadows.

View to the north of the hummocky topography developed in the Miocene-age Payette Formation sediments on the slope leading from Spring Valley Summit to Horseshoe Bend (middle distance). ID 55 lies mostly out of sight to the left. —Photo by John Burch

now altered to clay, is extremely weak. Instability in hillslopes like this one is relatively straightforward to identify by the lumpy forms at the base of the slumps, a feature that geologists call hummocky topography. The original highway was located on the weak sedimentary deposits and was plagued by problems with slumping on the slope. In the early 1990s the highway was moved to its present location on the much more stable Idaho batholith.

The town of Horseshoe Bend is built on the inside curve of a large meander, or horseshoe bend, in the Payette River. This relatively wide stretch of the valley that continues for a few miles upstream from Horseshoe Bend results from the easily eroded Miocene sediments. Downstream from Horseshoe Bend, the river flows through Cretaceous granodiorite of the Idaho batholith and Eocene dikes that are difficult to erode, resulting in a narrow valley and relatively straight river.

North of Horseshoe Bend, the underlying granodiorite of the Idaho batholith pokes through the Miocene sediments just north of milepost 67, and a small section of Columbia River basalts that also lie on top of the granodiorite appear in a roadcut at milepost 70. North of milepost 71, the Payette River canyon is cut into granodiorite

BLACK CANYON

ID 52 follows the Payette River west from Horseshoe Bend and downstream through a narrow canyon to Emmett and skirts Black Canyon Reservoir west of Montour. The reservoir drowned the Black Canyon of the Payette River, which was cut into dark basalts of the Miocene Columbia River Basalt Group. Visible to the north of Black Canyon Reservoir is Squaw Butte, a long north-oriented ridge of the basalt that was uplifted along Basin and Range normal faults. Farther to the west, a thick section of soft, easily eroded Payette Formation sediments surround Emmett and form badlands to the south of town.

and tonalite in the border zone of the Idaho batholith. Both rock types contain biotite. If you see the black, needlelike hornblende mineral, you are looking at the granodiorite. The batholith rocks here are around 84 million years old and are intruded by Eocene dikes of the Challis magmatic event.

Banks, between mileposts 78 and 79, lies at the confluence of the North and South Forks of the Payette River. County Route 17 follows the South Fork of the Payette to Lowman, crossing Idaho batholith rocks that have been extensively intruded by northeast-trending rhyolite, dacite, and andesite dikes of the Eocene Challis magmatic event. They are well exposed in the impressive gorge of the South Fork.

North of Banks, ID 55 follows the North Fork of the Payette River. The steep river gradient is immediately apparent as the road winds around tight corners through a dramatically narrow canyon. A nearly continuous ribbon of roaring and frothing Class V whitewater has replaced the relatively placid lower Payette River with its only occasional stretches of whitewater downstream from here. For comparison, the stream gradient on the placid Main Fork averages close to 15 feet of elevation drop per river mile; the 17-mile-long stretch along the North Fork of the Payette River north of Banks averages nearly 100 feet of drop per mile.

Why the sudden change on this section of the Payette River? Basin and Range normal faulting in the last 15 million years has disrupted the Long Valley region to the north of Banks between Cascade and McCall. Before this disruption, the Long Valley area drained to rivers that flowed to the west. As the north-trending normal faults formed a set of north-trending ranges that blocked westward drainage, the Payette River to the south was able to pirate or capture the Long Valley drainage, effectively stealing the upper part of the drainage basin from its now dismembered basin to the west. Once captured, it takes time for rivers to adjust their gradients. The Payette River is in the process of reworking and flattening out its gradient on its newly claimed over-steepened North Fork by advancing the upper knickpoint farther north toward McCall. Thus, the stretch downstream from Banks represents the older, more mature gradient of the Payette River.

Note that the gradient on the North Fork of the Payette River is quite variable. Long stretches of dramatic whitewater are punctuated from time to time by quiet pools. This variability indicates several factors are at work on this disequilibrium stream. In the big picture the river drops 1,675 feet in 17 miles, from an elevation of

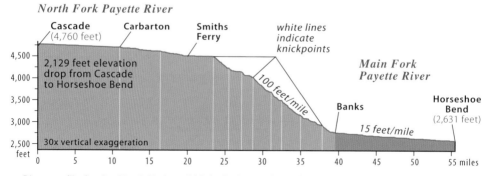

River profile for the North Fork and Main Fork stretches of the Payette River along ID 55. The disequilibrium gradient of the North Fork between Smiths Ferry and Banks is readily apparent.

4,555 feet at Smiths Ferry to 2,880 feet at Banks. However, at a finer scale, the oversteepened canyon sides are prone to landslides that send debris down into the canyon, commonly damming the river. As the river cuts down through these deposits, the smaller material is moved downstream, but the larger boulders are left in the river channel, creating a pool upstream of the landslide deposit and a steep, rapid downstream. A nice example of an old landslide deposit that is now largely grown over on the banks occurs at milepost 89. Needless to say, frequent landslides in the river canyon and rockfall from high roadcuts in the canyon sides provide a nearly constant source of frustration for the Idaho Transportation Department, which is charged with keeping this highway open year-round.

Another factor that influences the finer-scale river dynamics is the presence of normal faults within the canyon of the North Fork. A dramatic example of this occurs at Smiths Ferry. The highway crosses a major normal fault at milepost 95. North of the fault the steep narrow canyon opens, and the river slows to nearly a standstill as it flows through meanders in this sediment-filled, down-dropped fault block, a graben in the Idaho batholith. The steep stretch of river downstream of the fault flows through the uplifted fault block. The river was forced to cut down dramatically as faulting blocked its path.

North of milepost 98, the river gradient picks up, and rapids are common once again. After another mile, the highway crosses the North Fork and climbs up to Round Valley north of milepost 101. Round Valley is ringed by normal faults that raise the surrounding rim in uplifted blocks and lower the valley in the intervening graben. North of milepost 108, Snowbank Mountain comes into view to the west of the highway. At its top are migmatite rocks of the Wallowa terrane, discussed in the Snowbank Mountain sidebar on page 276.

Between mileposts 113 and 114, the highway crosses the North Fork of the Payette River as it meanders its way through Cascade Valley. The quiet water in the very low gradient stream at this locale seems oblivious to the oversteepened gradient it will encounter once it crosses the knickpoint and cascades down the canyon downstream from here. At Cascade, the North Fork of the Payette River flows out of Lake Cascade, which is partially impounded by a ridge of granite that has been upthrown on the west of Cascade Valley along a north-trending normal fault.

North of Cascade, the highway climbs to a summit at milepost 120 with deeply weathered muscovite- and biotite-bearing granitic rocks of the Cretaceous Idaho batholith in roadcuts on both sides of the road. The highway then descends into Long Valley, another north-oriented, fault-bound graben, through which the North Fork of the Payette River flows into Cascade Reservoir, which is visible to the south. The basin contains more than 3,000 feet of sedimentary fill. The West Mountains on the west side of the graben are granite of the 105-million-year-old Little Goose Creek Complex that intruded into the Salmon River suture zone and later was sheared along the Western Idaho shear zone. It was then covered much later by Miocene basalt lavas of the Columbia River Basalt Group before being uplifted by normal faulting. To the east, the graben is bounded by the Salmon River Mountains that contain younger, undeformed granitic rocks of the Idaho batholith.

North of Donnelly, between mileposts 133 and 134, a distinctive notch is visible in the West Mountains to the west. This feature, known as a wind gap, represents a now uplifted ancient river valley that pre-dates the creation of the graben that forms

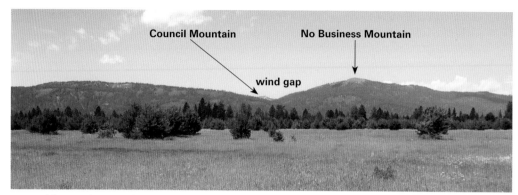

View to the west between mileposts 133 and 134 of a wind gap in the West Mountains.

SNOWBANK MOUNTAIN

Carbarton Road, a gravel route that takes off from ID 55 between mileposts 107 and 108, provides access to Snowbank Mountain, a high point in the West Mountains. Snowbank Mountain lies along the Western Idaho shear zone in rocks of the Wallowa terrane that have been partially melted to migmatite during burial. In other words, these rocks attained the highest metamorphic grade possible for a metamorphic rock and then started to melt. What's really interesting about this part of the collisional boundary along the western side of Idaho is that these rocks were deeply buried, metamorphosed, and partially melted twice in their history, first during the initial collision of the Wallowa terrane with the North American continent along the Salmon River suture zone from 145 to 128 million years ago, and then again from 104 to 90 million years ago as they were caught up in shearing along the Western Idaho shear zone. To see these rocks, turn west onto Carbarton Road and drive 2.5 miles to the junction with FS 446. Turn left on FS 446, which climbs up Snowbank Mountain, ending at a communications facility after 12.5 miles.

Migmatite in the Western Idaho shear zone on Snowbank Mountain. (44.4101, -116.1354)

modern Long Valley. Within the wind gap are river gravels with rock compositions that match other ancient gravels and their source rocks that lie to the east of Long Valley. The wind gap and its gravels give us a glimpse into a lost world in this region, in which streams flowed west from highlands to the east, cutting valleys into the early Columbia River basalts that overlay the older granites. This ancient world, which existed prior to 17 million years ago, was disrupted by north-trending normal faults that formed north-oriented valleys, causing drainage patterns to realign such that they now flow south. This turn of events permitted the Payette River drainage basin to capture the Long Valley drainage basin.

The mountain resort town of McCall is largely built on end moraine deposits from a Pleistocene valley glacier that occupied the North Fork of the Payette River Valley upstream and to the north of here. Payette Lakes are dammed by these moraine deposits.

North of McCall, the highway crosses a low summit by the town ski hill (capped by basalt), then follows Little Goose Creek west down a steep grade through a narrow, winding canyon into Meadows Valley. Roadcuts along the highway show outcrops of the 105-million-year-old Little Goose Creek granite that has been walloped by the Western Idaho shear zone as the already docked Wallowa terrane continued to converge with, and slide northward along, the North American margin. Rocks in this shear zone have been intensely stretched, as recorded by minerals in the rock that are

Little Goose Creek granite mylonite outcrop in the Western Idaho shear zone. Top photo looking down on the north-trending mylonite foliation. Bottom photo looking along mylonite foliation from the side; view to north. (45.1981, -116.1430)

flattened into a nearly vertical north-oriented foliation and drawn out into steeply oriented rods that align in a linear fashion.

At the bottom of the grade, just west of milepost 152, is the boundary between the Little Goose Creek Complex to the east and the Hazard Creek Complex of the Wallowa terrane to the west. A pullout on the south gives access to a nondescript overgrown outcrop that consists of tonalite and quartz diorite gneiss of the Hazard Creek Complex that was metamorphosed 120 to 110 million years ago along the western side of the Salmon River suture zone.

Meadows Valley, another north-oriented graben, is traversed by the northward-flowing Little Salmon River, which has faced the same recent disruption experienced by the North Fork of the Payette River farther south.

ID 75
Timmerman Junction—Ketchum—Stanley
88 miles

As ID 75 passes through Idaho's resort country, where outdoor recreational opportunities abound, it also crosses some of the most scenic and complicated geology in central Idaho. Big mountains have been uplifted along faults, many still active, that expose old Paleozoic rock, younger granites, and loads of Eocene Challis Volcanics. The first non-native settlements in this remote region were mining camps filled with men seeking metals introduced by hot waters associated with the magmatism.

North of Timmerman Junction, ID 75 follows the Big Wood River upstream through irrigated farms. The rugged hills on the east side of the valley are Challis volcanic lavas. Spring-fed Silver Creek, which flows southeast at the base of those hills, offers world-class trout fishing, first documented by Ernest Hemingway.

South of Bellevue near milepost 109, the hills east of the highway become smoother because they are composed of fine-grained Paleozoic rocks—the Devonian Milligen Formation and Pennsylvanian to Permian Wood River and Dollarhide Formations of the Sun Valley Group. Black shales of the Milligen Formation were deposited in a deepwater, low-oxygen basin during the Antler orogeny and host the historic Wood River mining district. In the late nineteenth century, mining districts dictated the settlement of central Idaho. The Wood River Branch of the Oregon Short Line railroad was completed to Hailey in 1883 and extended to Ketchum the next year.

The circulation of hot fluids during the intrusion of granitic rocks in Cretaceous time mobilized silver, lead, and zinc and reacted with carbonaceous shale of the Milligen Formation. Metal-rich veins, mainly composed of siderite (an iron carbonate), were mined at the Minnie Moore Mine west of Bellevue. The mine is tucked away in the hills to the southwest from the Wood River Mines historical sign between Bellevue and Hailey (between mileposts 112 and 113).

The 95- to 80-million-year-old Hailey and Croesus granodiorite stocks, which form rugged peaks to the west, intruded the Milligen Formation. The Milligen forms the smoother slopes closer to the road. At Hailey the Big Wood River undercuts the base of the hills, causing small slides in the weak shale of the Milligen Formation. West of Hailey are tall cliffs of the Pennsylvanian Hailey Conglomerate Member of the Wood River Formation next to the Big Wood River. These rocks are folded and faulted such that they now stand on-end. The coarse gravel, sand, and thin limestone

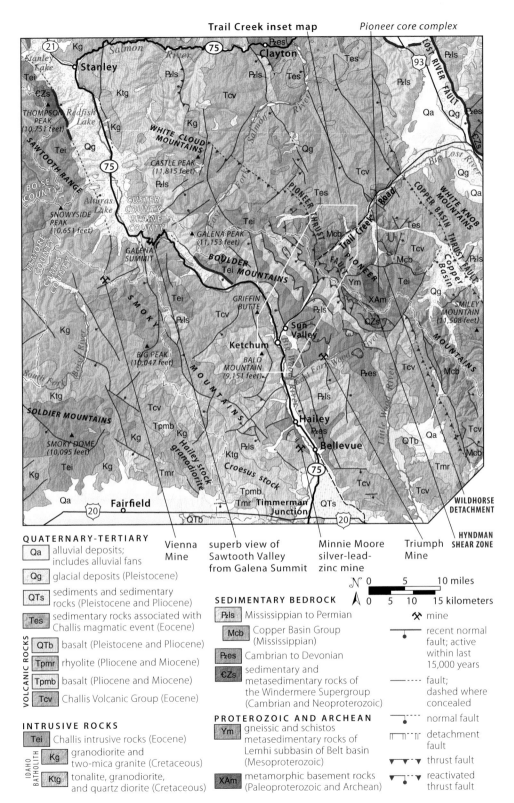

Geology along ID 75 between Timmerman Junction and Stanley.

layers in this unit were deposited on the angular unconformity that developed as the shales of the Milligen Formation were uplifted and eroded. The Carbonate Mountain Trailhead just west of Hailey on Croy Creek Road affords easy access to Hailey Conglomerate outcrops and fantastic views of the Wood River Valley.

Pennsylvanian Hailey Conglomerate of the Sun Valley Group on Carbonate Mountain, west of Hailey. Bedding is vertical. View is looking northeast across the Wood River Valley with the Pioneer Mountains in the distancem. (43.5128, -114.3250).

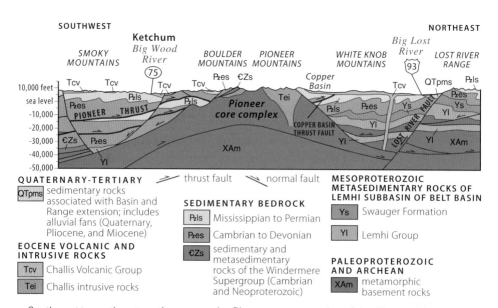

Southwest-to-northeast section across the Pioneer core complex of the Pioneer Mountains.

Up the East Fork of the Wood River, north of Hailey, is the underground Triumph Mine, the largest mine in the Wood River Valley area. It was active until after World War II. Black shale of the Milligen Formation was the primary host for silver and lead minerals, including galena and argentite. Normal faults cut the previously formed veins, challenging the miners who followed veins underground.

The hard, crystalline rocks of the Pioneer core complex form the higher peaks in the Pioneer Mountains, a remote region mostly inaccessible by vehicle. The metamorphic rocks in the core of the Pioneer core complex were originally Archean to Paleoproterozoic basement rocks and overlying Mesoproterozoic Belt Supergroup. Metamorphosed 2.65-billion-year-old Archean granites in the Pioneer Mountains are part of the Grouse Creek block, probably a western exposure of the Archean Wyoming craton. Neoproterozoic and Paleozoic sedimentary rocks deposited on the western passive margin of ancient North America are also present, folded by the Sevier orogeny in late Cretaceous time. A Neoproterozoic pluton, with an age of 695 million years, is exposed in the Pioneer core complex and is probably related to similarly aged rift volcanic sequences of the Pocatello and Edwardsburg Formations of the Windermere Supergroup.

As with most exposures of Precambrian basement in Idaho, the crystalline rocks in the Pioneer Mountains are exposed because they have been exhumed by an Eocene core complex. The late Paleozoic sedimentary rocks slid to the northwest along the Wildhorse detachment fault, off the top of metamorphic rocks in the Pioneer core complex. Motion indicated by slickenlines on the detachment fault surface trend northwest. Exhumation of the metamorphic rocks began at the same time as Challis magmatism and continued after the volcanism had ended. A large 47-million-year-old, Challis-age granitic pluton (the Summit Creek stock of the Pioneer Intrusive Suite) intruded into the core complex prior to motion on the detachment fault. It was split in two by the faulting, and the two halves are now separated by at least 10 miles.

You can get within a couple miles of the metamorphic rocks in your car by heading west from the Triumph Mine on the East Fork Wood River Road and then turning left to the Hyndman trailhead. The road to the trailhead crosses moraines left by the glaciers that carved the Pioneer Mountains. For the first 2 miles from the trailhead, the trail crosses the Wood River and Milligen Formations of the northwest-sliding upper plate before crossing through the Wildhorse detachment fault and into crystalline metamorphic rocks of the middle and lower plates of the Pioneer core complex.

Bald Mountain, the Sun Valley ski hill, looms to the west of Ketchum. The mountain is composed of folded Pennsylvanian and Permian Wood River Formation, mainly brown-weathering silty limestones. At the main intersection in Ketchum (north of milepost 128), the road heading east to Sun Valley follows the canyon of Trail Creek into the Pioneer Mountains and eventually over a steep summit to Summit Creek.

Griffin Butte, a symmetrical peak northwest of the highway north of Sun Valley, is a Challis volcanic center. The historical site for the Ski Lifts near milepost 132 is a good place to stop and see Griffin Butte and Bald Mountain ski hill to the south. Roadcuts here are all in Challis dacite.

The Smoky Mountains, visible to the west from north of the Sawtooth National Recreation Area headquarters north of milepost 136, are composed of Challis volcanic rocks overlying the Pennsylvanian to Permian Dollarhide Formation. Hoodoos of Challis volcanic rocks lie south of the river near milepost 139.

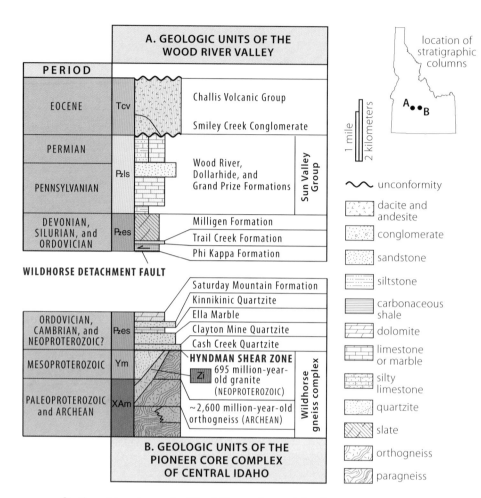

Stratigraphic columns for Wood River valley and the Pioneer core complex.
—From Lewis et al., 2012

North of Baker Creek near milepost 144, the highway offers a spectacular view to the east of the uplifted front of the Boulder Mountains. The lowest exposed rock is 45-million-year-old pink granite. It intrudes 48-million-year-old intrusive dacite of the Challis magmatic episode, which itself intrudes isolated fragments of the Wood River and Milligen Formations that cap the mountains. The west-dipping normal fault forming the western front of the Boulder Mountains is active, cutting Pleistocene glacial moraines and younger sediments.

North of Galena Lodge, the highway heads toward Galena Summit through Challis volcanic rocks, mainly dacite. Several pullouts (near milepost 154, uphill from milepost 155, and just downhill from milepost 157) provide fine views to the west face of the Boulder Mountains and the Wood River Valley to the south. The curved joint, or fracture, surface in Challis volcanic rocks at milepost 155 formed as the volcanic rock cooled.

Southwest front of the Boulder Mountains was uplifted along an active normal fault along its base. Uplifted Challis plutonic rocks include the pink granite. Darker Challis volcanic rocks and Paleozoic strata crop out on the high peaks. (43.7918, -114.5736)

Curved joint surface in Challis volcanic rocks at milepost 155. The joints, regular fractures in the rock, formed as the volcanic rock cooled. (43.8871, -114.6954)

TRAIL CREEK ROAD

This mostly unpaved but graded road heads northeast from Ketchum over the Pioneer Mountains at Trail Creek Summit and traverses a microcosm of Idaho geology. The lower reaches of the road east of Ketchum are paved, but the steep, precipitous upper part is gravel and closed in winter. Start your odometer at the stoplight at Main Street (ID 75) and Sun Valley Road in Ketchum.

Challis volcanic rocks, mainly dacite, surround the resort of Sun Valley. Underneath the volcanics is black, fine-grained siltstone of the Devonian Milligen Formation, which hosts the silver-lead ore that was mined here in the late nineteenth century. The Milligen Formation is exposed on the northwest side of the road where it crosses Trail Creek (mile 3.9). Northeast of Corral Creek, the Pennsylvanian-to-Permian Wood River Formation forms talus-strewn cliffs on both sides of the canyon. Pleistocene glaciers filled the big valleys of Corral and Wilson Creeks on the southeast and deposited moraines visible upstream of Corral Creek (mile 5.0).

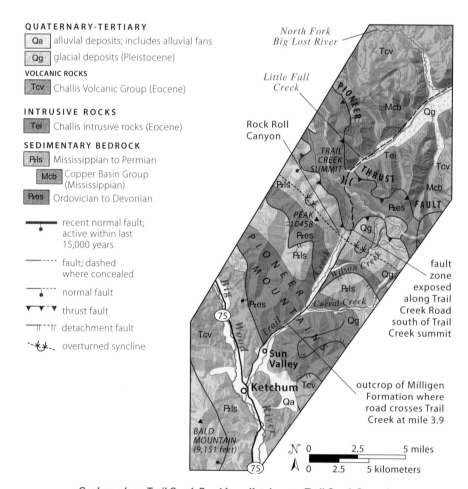

Geology along Trail Creek Road from Ketchum to Trail Creek Summit.

Rock Roll Canyon, viewed from the mile 9.1 pullout, forms the 2,000-foot cliff on the west side of Trail Creek. The southwest-dipping sedimentary beds in the lower part of the cliff are structurally overturned limestones of the Eagle Creek Member of the Pennsylvanian and Permian Wood River Formation. The rocks get younger to the east. (43.7896, -114.2601)

The fault zone at mile 10.3 separates the Wood River Formation on the right (south) from Devonian and older rocks on the left. This fault zone has had a complex history and may have first been active in Mississippian time during the Antler orogeny. Fault-bounded blocks, known to geologists as tectonic fish, are outlined with dashed black lines. (43.8027, -114.2545)

The steep canyon of Rock Roll Creek is across from the wide part of the road at mile 9.1. From the top downward, the highest rocks are the Wilson Creek Member of the Wood River Formation, exposed above a low-angle normal fault below Peak 10458, named for its elevation. Below that fault are layers of the middle Eagle Creek Member of the Wood River Formation. The rocks are folded into a northeastward-overturned syncline, so the west-dipping rocks are overturned. The folding occurred during the Sevier orogeny in Cretaceous time.

A complex fault zone is crossed at mile 10.3. On the upper side of the fault is the Wood River Formation and below the fault are dark-colored siltstones and shales of Devonian, Silurian (Trail Creek Formation), and Ordovician (Phi Kappa Formation) age. You call pull off uphill from the fault and walk back to examine the slickensides (fault grooves) and the dismembered chunks of layered rock (tectonic fish) in the 30-foot-wide fault zone. This fault can be traced across the canyon to the west side, where it forms a discontinuity marked by a white stain from calcite veins high on the canyon wall that can be seen about 0.7 mile up the road, near mile 11.0.

Trail Creek Road continues to Trail Creek Summit (mile 12.3), with shaly dark talus of Ordovician sedimentary rock to the east of the road. Across the canyon is the mouth of Cold Canyon, which forms the upper part of the Trail Creek drainage. If you continue over the summit to the Lost River Valley, you'll cross the Pioneer thrust fault, the beheaded northern flank of the Pioneer core complex, and the Copper Basin thrust fault.

View south toward Sun Valley from mile 11.0, showing the U-shaped glacial valley of lower Trail Creek. (43.8202, -114.2606)

A northwest-dipping fault cuts the Eocene Challis volcanic rocks at Galena Summit near milepost 158. The 2-foot-wide fault zone is soft and usually wet, composed of gooey gouge from ground-up, clay-bearing rock. (43.8705, -114.7127)

On the west side of Galena Summit, stop at the overlook just uphill from milepost 159 to take in the spectacular view of the Sawtooth Valley to the west. Clockwise from the left, or south, is the headwaters of the Salmon River in fine-grained calcareous sedimentary rocks of the Grand Prize Formation, which was deposited in a basin in Pennsylvanian and Permian time along with the Wood River and Dollarhide Formations. To the north and west, the southern Sawtooth Range is composed of light-gray Cretaceous granitic rocks of the Atlanta lobe of the Idaho batholith. The more jagged northern part of the Sawtooths is composed of the pink Sawtooth granite of Eocene age. The contact between the Cretaceous on the south and the Eocene on the north is just north of Alturas Lake. Glaciated horns and arêtes of these rocks justify the name Sawtooths. The roadcut across the highway from the overlook is composed of broken up Challis volcanic rocks. Silty limestone of the Grand Prize Formation forms the roadcuts just north (downhill) of the overlook.

As the highway descends into the Sawtooth Valley and heads north, it crosses moraines between milepost 161 and 162 that were deposited by glaciers that flowed down Pole Creek of the White Cloud Mountains to the east and from the headwaters of the Salmon River to the south. A historical marker near milepost 162 describes Vienna, an 1880s silver-lead district with mineralization associated with the Idaho batholith.

In the Sawtooth Valley, ID 75 follows the Salmon River as it skirts past the edge of moraines deposited from alpine glaciers flowing from the Sawtooth Range on the west. These moraines impounded several big lakes including, from south to north, Alturas, Petit, Redfish, and Stanley. Higher in the Sawtooths are bedrock-floored cirque lakes, carved by the same glaciers that dumped the uneven moraines.

Sawtooth Range viewed to the southwest from the Salmon River near Stanley. (44.2392, -114.9076)

Near the turnoff to Redfish Lake at milepost 185, the highway crosses the moraine that dams the lake. The Salmon River, east of the highway, cascades over moraine boulders. Stanley, which lies along ID 21 west of ID 75, is known for its very cold nightly temperatures, which occur as mountain air sinks into this high-elevation basin that tops out at more than 6,000 feet.

ID 75
Stanley—Challis
55 miles

Between Stanley and Challis, ID 75 follows the Salmon River as it cuts through the Salmon River Mountains of central Idaho. The curvy highway hugs the river, and in summer you'll see rafts and kayaks. Downstream from milepost 191, the Salmon River flows in a narrow canyon through the eastern part of the Atlanta lobe of the Idaho batholith. Most of the granitic rock here is Cretaceous, but there are also intrusions of Eocene granitic rock. A distinctive Cretaceous granodiorite with 1-inch-long crystals of potassium feldspar crops out on the north side of the road across from the Mormon Bend Campground at milepost 196.

At Sunbeam Hot Springs, halfway between mileposts 201 and 202, hot water empties into the Salmon River. When the river level is low enough, bathers customize the water temperature by constructing small rock dams to mix the very hot spring water with the very cold river water.

On the south side of the road just west of Sunbeam village is a pulloff and historical marker for the Sunbeam dam, built in 1909 to provide power to a mineral processing mill up the Yankee Fork. The dam was sabotaged in 1934 as part of the ongoing strife between those who wished to use the Salmon River water for mining or electric generation and advocates of rivers where salmon run free. The dam was never rebuilt.

CENTRAL MOUNTAINS

The historic Yankee Fork gold dredge is 8.4 miles up Yankee Fork Road (part of the Custer Motorway) from Sunbeam. The Yankee Fork of the Salmon River was extensively disturbed by placer mining. East of Sunbeam the highway passes through a narrow canyon where one has to be careful where to pull off to see the batholith. Look for a safe parking area just east of milepost 204.

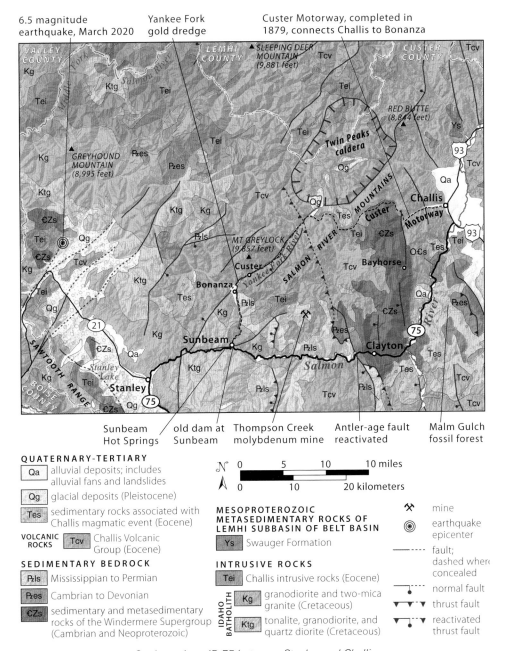

Geology along ID 75 between Stanley and Challis.

View from the overlook near milepost 205 of a pile of gravel, a remnant of gold placer mining, composed mainly of clasts from the Idaho batholith. An outcrop of granodiorite lies behind the pile of gravel. (44.2587, -114.6931)

Halfway between mileposts 211 and 212, the highway emerges from the canyon through the Idaho batholith and passes eastward into Paleozoic sedimentary rock overlain by Eocene Challis volcanic rocks. The fine-grained, dark-colored Paleozoic rocks (Salmon River assemblage of shale, siltstone, and limestone of probably Mississippian age) were deposited in deep water west of ancient North America. The road to the Thompson Creek molybdenum mine, an open pit in the Salmon River assemblage, is at milepost 215.

The highway crosses a major west-dipping fault, mapped as part of the Copper Basin thrust fault, at milepost 220. Light-colored, Mississippian-age shales were originally thrust over the dark, cliff-forming Ordovician Kinnikinic Quartzite. Initial motion on this fault may have been part of the Antler orogeny in Late Devonian time. The fault motion has since reversed, with reactivation as a normal fault.

To the north of Clayton lies a silver-lead mine up Kinnikinic Creek (east of milepost 222). The quartzites and limestones of the Clayton area are Neoproterozoic and Paleozoic, deposited long before the Idaho batholith intruded the sedimentary rocks and mobilized metals in Late Cretaceous time.

Four miles east of Clayton, the East Fork of the Salmon enters from the south and 8 miles farther east, near milepost 234, is the mouth of Malm Gulch, where the fossilized stumps of Eocene trees are preserved. The trees were silicified by hydrothermal groundwater during eruption of the Challis volcanic rocks. In this part of the canyon, Paleozoic sedimentary rocks form the cliffs along the river, and the overlying Challis volcanic rocks reach for the sky in the big mountains above. As the highway pushes eastward, it approaches one of the main volcanic centers of Challis magmatism.

CENTRAL MOUNTAINS 291

The Copper Basin thrust fault dipping to the west, viewed north from milepost 219. The Mississippian Salmon River assemblage (light-colored loose shales at left) lies above older Ordovician Kinnikinic Quartzite (cliff-forming bedding planes on right). The fault has a complex history with at least two periods of activity. (44.2530, -114.4470)

An anticline in the Neoproterozoic-to-Cambrian Clayton Mine Quartzite at milepost 221, west of Clayton. (44.2612, -114.4118)

A side road heading northwest from milepost 238 leads to the old mining district of Bayhorse. Mine buildings at Bayhorse have been restored as part of the Land of the Yankee Fork State Park. The headquarters of this park is north of the highway at the intersection with US 93 south of Challis. The museum at the headquarters contains a variety of old mine equipment and displays about the never-easy life in the Custer mining district. The partially restored 1880s mining town of Custer is accessible by the gravel Custer Motorway loop, which connects Challis to Sunbeam.

Harrison Lake, high in the Selkirk Mountains, lies within a cirque, a glacial basin carved into Cretaceous granitic rocks of the Kaniksu batholith. (48.6777, -116.6550)

NORTHERN IDAHO

Northern Idaho, extending from the Clearwater River north through the Idaho Panhandle to the Canadian border, is a land of heavily timbered broad valleys, majestic snow-capped mountains, magnificent lakes, and clear rivers. Entirely within the northern Rocky Mountains, Northern Idaho transitions from the northern Bitterroot Mountains in the south to the Selkirk and Cabinet Mountains in the north, which bookend the Purcell Trench, a large valley in the Panhandle. Northern Idaho's valleys and rivers loop around in strange drainage patterns created by advances and retreats of the Cordilleran ice sheet from Canada in Pleistocene time. Many of these rivers were repeatedly blocked by ice lobe dams, and some produced catastrophic floods when they failed.

Matching Northern Idaho's spectacular scenery is its geologic history. Archean and Paleoproterozoic gneisses form Northern Idaho's basement rock. These severely deformed and metamorphosed rocks were part of the ancient continent that hosted a sedimentary basin in which the Belt Supergroup accumulated in Mesoproterozoic time. Collectively, these old rocks were thickened by folding and thrusting during the Sevier orogeny in the Jurassic through Paleogene Periods. Parts of the thickened crust melted, and the magma formed granitic batholiths. Beginning about 50 million years ago, the overthickened crust collapsed and extended to produce a series of core complexes, each intruded by even more magma. This extensive geologic history culminated in a world-class mining district that developed in the South Fork of the Coeur d'Alene River drainage basin. These silver, zinc, lead, and gold deposits have been mined for almost 150 years, setting a number of world records for mineral production and leaving a legacy that includes an ongoing Superfund cleanup operation.

ANCIENT CONTINENTAL BASEMENT ROCKS

Archean and Paleoproterozoic basement rocks in Northern Idaho are exposed in the Priest River metamorphic core complex in the Panhandle and the Boehls Butte–Clearwater core complex farther southeast. These rocks were deformed deep in Earth's crust and are exposed now because overlying rocks slid off them along low-angle faults. The metamorphic rocks include 2.6- to 2.5-billion-year-old granite gneiss and amphibolite of the Pend Oreille gneiss in the Priest River core complex. These Archean rocks were initially granite, gabbro, and basalt, igneous rocks from an episode of magmatism into older crustal rocks that are not exposed. The granite was metamorphosed to gneiss, and the basalt and gabbro were metamorphosed to amphibolite. The Archean magmatism was followed by another episode of granite, gabbro, and basalt intrusion at around 1.86 billion years ago to form Paleoproterozoic gneiss and amphibolite. An enigmatic granitic intrusion at about 1.58 billion years ago formed the Laclede gneiss. Many geologists point to this early continent-forming geologic activity as evidence for the construction of a very ancient supercontinent called Columbia.

The first sedimentary rocks preserved in the geologic record are sandstones that were deposited directly on older, eroded plutonic rocks between 1.7 and 1.5 billion years ago in Mesoproterozoic time. These thin sandstones (now metamorphosed to quartzite) include the Gold Cup Quartzite near Priest River and the Razorback Quartzite south of Coeur d'Alene.

BELT SUPERGROUP OF A LONG-LIVED CONTINENTAL BASIN

By 1.5 billion years ago, continued continental construction had enlarged the old supercontinent Columbia, and the western side of the present North American continent lay somewhere near its interior on a portion of Columbia that some geologists refer to as Nuna. For reasons not fully understood, the interior of Nuna downwarped to create a subsiding basin that stretched across 77,000 square miles of what is now northern and central Idaho, southern British Columbia and Alberta, eastern Washington, and western Montana. Over the course of the next 70 million years, this basin accumulated a whopping total thickness of almost 60,000 feet of mostly mud and silt and some sand and limestone. The sediment was lithified into the sedimentary rocks of the Belt Supergroup (and the Purcell Supergroup in Canada) and later slightly metamorphosed, although we use sedimentary rock names in this chapter to avoid confusion.

The rocks of the Belt basin give us an amazing snapshot of climate, terrestrial environments, and life on our planet in the Mesoproterozoic Era. The Belt basin contained a large inland sea or lake, and the sediments that accumulated in underwater environments are preferentially preserved in the sedimentary sequence. Ripples preserved

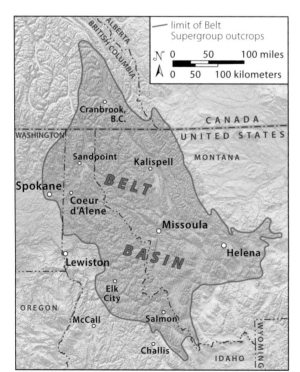

The outline of the Belt basin based on the present distribution of rocks in the Belt and Purcell Supergroups.

in the rocks were left by water currents and the back-and-forth motion of waves in shallow water.

Carbonate sedimentary rocks such as limestone were deposited in shallow water conditions that were oxidizing, saturated in calcium carbonate, and full of Proterozoic life. Animals and plants had yet to develop on our planet in the Mesoproterozoic, so microbes were the dominate life form. In shallow water, photosynthesizing bacteria constructed muddy carbonate mounds called stromatolites. In the muddy substrate, bacteria produced methane and other gases as they scavenged, and the gas periodically escaped, disturbing the sediment and producing underwater cracks in the mud that later filled with calcite precipitated from the surrounding water. These calcite-filled features are known as molar tooth structures.

Flanking the inland sea was a barren land bereft of plants and thus the root structures that hold modern soil together. Following rainstorms, massive sheet floods swept across the land surface, washing sediment into the water and depositing it as flat layers of silt and sand. Mudflats developed in quiescent times, and these environments periodically dried to produce mud cracks. When sheet flooding resumed, many mudflats were stripped of their cracked tops, and chips of the dried mud were incorporated into the newly deposited sediment to be preserved as mud rip-ups, or mud chips.

Missing in the Belt sedimentary rocks are marine deposits containing bidirectional tidal channels and erosional surfaces that mark the diurnal rise and fall of sea level with tides. This suggests that the Belt basin contained a lake or inland sea with no connection to a larger ocean. The cyclic pulses of sediment that we see in the Belt rocks were most likely river and sheet-flow floods into a basin that was subsiding just as fast as it was being filled by sediment.

Only the eastern side of the Belt basin is preserved on North America. The western side lies on another continent, or continents, that faulted away in Neoproterozoic time or earlier. Present ideas about which continents might contain the western side of the Belt include Siberia, Australia, and Antarctica. By studying the sediment grains contained in the Belt Supergroup, geologists have determined the sediment was eroded from two different areas: a North American source that lay to the south and southeast and one or more sources on Siberia, Australia, or Antarctica to the west.

The Belt Supergroup in Northern Idaho consists of the Prichard Formation, Ravalli Group, Piegan Group, and Missoula Group. These units correspond to two major periods of rapid subsidence of the basin, each followed by a period in which the rate of sedimentation outpaced the rate of subsidence and the basin began to fill. Initial rapid subsidence of the basin corresponded with melting of upper mantle rocks below the basin to produce mafic magmas. The magma intruded between layers, forming nearly horizontal sheets of gabbro, now known as the Purcell sills.

Sedimentation in the Belt basin began 1.47 billion years ago with rapid subsidence and deposition of sand and silt of the Prichard Formation in deep water. Turbidite sequences are common in these rocks, representing sediment-laden currents that cascaded off shallower shelves on the margins into deeper water farther out in the basin. These beds consist of characteristic fining-upward sequences of sand, silt, and mud that were deposited as each current event slowed and ever-finer sediment settled out of the water. The Prichard Formation sediments are extensively intruded by mafic Purcell sills. The addition of these gabbro sheets to the basin added considerable weight to the upper crust, causing it to subside even further and maintaining

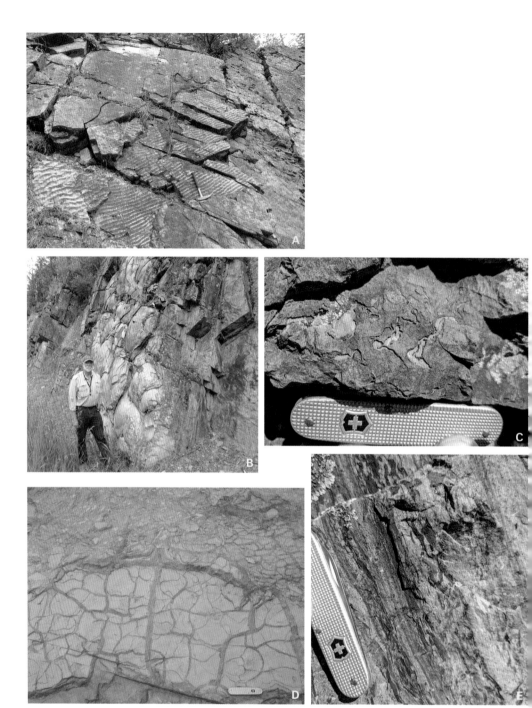

A. Ripples formed when water currents worked the upper surface of Belt sediments. **B.** Domal stromatolites, originally bacterial mats that formed mounds on the nearly horizontal limy sediment of the Belt basin, are well displayed in a steeply dipping layer of limestone along US 2. Geologist Mark McFaddan provides scale. **C.** Molar tooth structures, the wormlike features perpendicular to the bedding, developed in Belt sediments as microbes living in the muddy substrate released methane gas as they broke down organic matter in the sediment. **D.** Mud cracks developed in muddy sediments when the sediment dried. More sediment added on top of the mud-cracked surface filled the cracks. **E.** Mud chips or rip-up clasts formed as sheet flows washed across cracked, dried mudflats, ripping up mud chips and incorporating them in sediment transported in sheet flows.

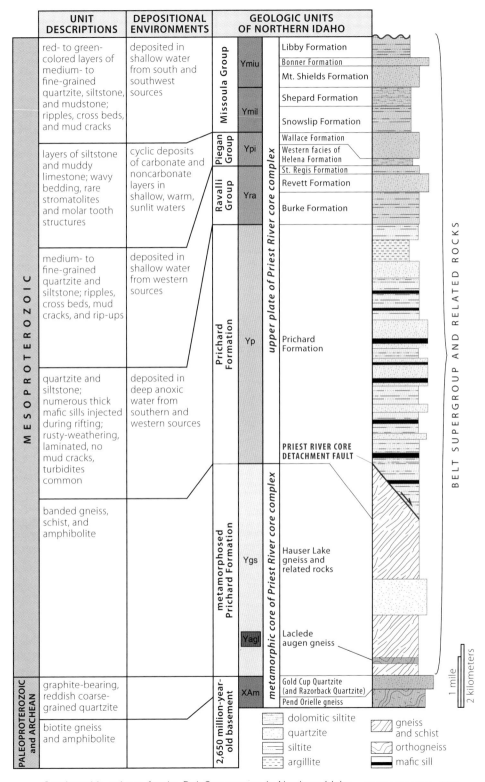

Stratigraphic column for the Belt Supergroup in Northern Idaho. —From Lewis et al., 2012

deepwater conditions over a long period of time. We know that conditions were oxygen-poor in these deep bottom waters because trace amounts of reduced iron sulfide minerals such as pyrite and pyrrhotite were deposited with the sediment. Outcrops of the Prichard Formation have a characteristic rusty coloring because the sulfide minerals oxidize in our modern oxygenated atmosphere and alter to iron oxide and hydroxide minerals.

Following deposition of the Prichard Formation, subsidence of the Belt basin and intrusion of Purcell sills waned, and filling of the basin with sediment outpaced subsidence. The red beds of the overlying Ravalli Group were deposited under more oxidizing conditions. Sedimentary rocks of the Ravalli Group consist of thinly bedded siltstone and mudstone that were deposited as vast sheets of sediment, delivered from a source to the west, in shallow water and in mudflats. Sedimentary structures such as ripples and mud cracks are common.

Following deposition of the Ravalli Group, the Belt basin deepened once again, and the next sedimentary package in the Belt Supergroup, the Piegan Group, was deposited. This limestone-bearing, muddy to silty sedimentary package accumulated in fairly shallow water that was hospitable to life. We know this because these rocks contain stromatolites, carbonate mounds of algae that require sunlight for photosynthesis and water chemistry that is similar to modern oceans and lakes. No fossil shells or bones are preserved because animals with shells and other hard structures didn't evolve for another billion years. Molar tooth structures are also common in Piegan Group rocks.

In Idaho, the Piegan Group consists mostly of the Wallace Formation, which includes distinctive sedimentary sequences called "black and tan couplets" in the vernacular of Belt geologists. These sequences consist of thinly layered, alternating dark-colored shale (the black) and light-colored limy siltstone and sandstone (the tan), both deposited underwater on the floor of the basin. These deposits look like turbidite beds, but the couplets likely result from storms that churned up and redeposited sediment in shallow water. In each of these events, the coarser-grained tan sands fell back out of suspension first, followed by the finer-grained dark muds. Each storm repeated the pattern, leading to thick sequences of black and tan couplets. The

Black and tan couplets of the Wallace Formation with their characteristic fining-upward sand-mud sequences formed as underwater sediment was churned up by storms and then redeposited. Arrow delineates a single couplet and its original up-direction; note additional beds both above and below this one.

original "up" direction for the rocks is always given by the direction in which sand layers grade into mud layers in the couplets.

The uppermost sedimentary package in the Belt Supergroup is the Missoula Group, which marks a return to basin filling with intervals of subsidence. This package consists of mostly fine-grained sandstone, siltstone, and shale with some carbonate beds that represent mostly shallow water deposits. The units of this group in Northern Idaho are the Snowslip, Shepard, Mt. Shields, Bonner, and Libby Formations. Ripples, cross beds, and mud cracks are common in these rocks, and you can even find impressions of raindrops. Also common are salt casts, small cubic cavities formed when cubic salt crystals precipitated from the evaporating briny water and then the salt later dissolved away in groundwater.

Based on ages of intrusive rocks into the Belt Supergroup in central Idaho, Belt basin sedimentation ended by 1.385 billion years ago. Strata of the Deer Trail Group, just younger than the Belt, were deposited in northeastern Washington and the northwestern Idaho Panhandle.

Collisions on the east coast of what is now North America around 1.1 billion years ago resulted in the construction of the next supercontinent, Rodinia. By 700 million years ago, Rodinia began to rift to form the western margin of Laurentia (ancient North America) and the nascent Pacific Ocean. The rift was located in what is now eastern Washington, and the sedimentary sequence associated with rifting, the Windermere Supergroup, mainly occurs in the Okanogan region of Washington. These Neoproterozoic and Cambrian rocks are overlain by limestone, mudstone, and sandstone that were deposited on a gently sloping continental shelf in Cambrian and Ordovician time.

SEVIER FOLD-AND-THRUST BELT AND THE KANIKSU AND IDAHO BATHOLITHS

Pacific Ocean tectonic plates began subducting along the western margin of North America by late Paleozoic time. The west coast at that time lay just west of the western edge of Idaho. In Northern Idaho, the major impact of this long-lived process was crustal deformation of the Sevier orogeny and intrusion of granitic magma. From Jurassic to Cretaceous time, the Belt Supergroup rocks were folded and cut by east-directed thrust faults as the collision of volcanic island arcs and other landmasses in the Pacific Ocean ensued.

Extensive subduction-related magmatism occurred along the entire west coast in the Cretaceous Period, and two distinct intrusive events impacted Northern Idaho. Between 120 and 90 million years ago, the Kaniksu batholith was intruded into southern British Columbia and the Panhandle region of Idaho, and it now forms the backbone of the Selkirk Mountains. The second event was the intrusion of the Idaho batholith between 78 and 65 million years ago in the Coeur d'Alene and St. Joe Mountains of the southern part of Northern Idaho. Some of these granite and granodiorite rocks are associated with gold mineralization in the Coeur d'Alene mining district.

The Kaniksu batholith contains two distinct groups of igneous rocks: (1) granite, granodiorite, and tonalite containing the black, needle-like mineral hornblende; and (2) granite containing muscovite mica (a shiny, platy mineral) and occasionally garnet. The presence of muscovite in a granitic rock tells us that the magma likely melted from sedimentary rocks of the continental crust.

EOCENE MAGMATISM AND THE PRIEST RIVER CORE COMPLEX

Subduction-related magmatism was rejuvenated about 50 million years ago in the Eocene Epoch across the region from Washington to Montana. Farther south in Idaho this episode led to the Challis magmatic event that includes the intrusion of plutons and dikes, and the eruption of volcanic sequences. In Northern Idaho, the rocks that are preserved from this episode consist of granite, granodiorite, and diorite plutons and rhyolite, dacite, and andesite dikes.

The long period of Jurassic to Paleogene crustal shortening thickened the crust to the point that it began to collapse under its own weight in Eocene time to produce a series of metamorphic core complexes. These complexes are linked together with strike-slip faults along a zigzagging line that trends northward from southern and central Idaho to the western border of the Idaho Panhandle and into British Columbia. As the crust was pulled apart, the upper crust slid off the lower crust on low-angle detachment faults. The unloaded lower crust rose up to form dome-shaped structures that expose the highly metamorphosed cores of the complexes.

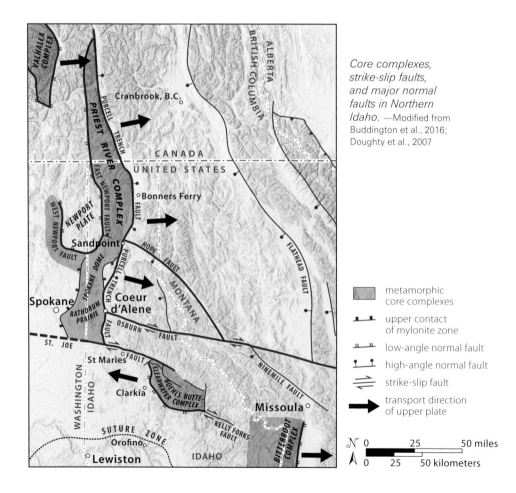

Core complexes, strike-slip faults, and major normal faults in Northern Idaho. —Modified from Buddington et al., 2016; Doughty et al., 2007

Core complexes are particularly well developed in Northern Idaho. The Boehls Butte–Clearwater core complex in the southern part of the region links to the south with the Bitterroot front along the Idaho-Montana border on the St. Joe fault, a large northwest-trending, right-lateral strike-slip fault that cuts almost completely across Idaho. The St. Joe fault links with the Priest River core complex farther north in the Panhandle. The Priest River core complex underlies the entire western half of the Panhandle and hosts highly metamorphosed Archean, Paleoproterozoic, and Mesoproterozoic rocks intruded by Cretaceous and Eocene granitic plutons.

Many rocks now exposed at the surface of the Priest River core complex are in the Spokane dome mylonite zone, a 2.5-mile-thick zone, now bowed into a dome, that was sheared as the upper crustal section slid to the east along the detachment fault. Once the weight of 10 miles of upper crust was removed from above the metamorphic core, it uplifted, warping the originally more planar, shallowly east-dipping detachment zone into the shape of a dome. In places, such as Round Mountain, the detachment fault at the top of the Spokane dome mylonite zone is exposed. A steeply dipping normal fault, the Purcell Trench fault, runs north-south right down the middle of the Purcell Trench and cuts the detachment fault of the Priest River core complex, dropping the upper plate rocks down on its eastern side. The Purcell Trench fault was active for a relatively short interval following shearing on the detachment fault around 48 million years ago.

Belt Supergroup rocks in the upper plate of the Priest River core complex are presently exposed across the eastern half of the Panhandle and include relatively unmetamorphosed Prichard Formation that contrasts with its highly metamorphosed counterpart in the metamorphic core to the west. The metamorphosed Prichard Formation rocks are called the Hauser Lake gneiss. We know the sedimentary and metamorphic rocks share the same origin because the mineral zircon in the Hauser Lake gneiss gives radiometric uranium-lead ages that are identical to the Prichard Formation. Even after extensive metamorphism, outcrops of the gneiss still weather to

Both rock samples are from Purcell sills that intruded the Prichard Formation of the Belt Supergroup in Mesoproterozoic time. The sample on the right, from the Cabinet Mountains on the eastern side of the Panhandle, is relatively unmetamorphosed gabbro. The sample on the left, from the core complex in the Selkirk Mountains, is metamorphosed, a garnet amphibolite. The circular reddish areas are garnets.

a characteristic rusty-brown color, just like the unmetamorphosed Prichard Formation rocks. The Purcell mafic sills have been metamorphosed to garnet amphibolite.

The timing of crustal thickening, detachment faulting, and uplifting of the core complex can be bracketed by several events. The peak metamorphism that corresponds with maximum crustal thickening at the end of the Sevier orogeny occurred 64 million years ago and pre-dated development of the core complex. Eocene igneous rocks in the core complex bracket cessation of detachment faulting and uplift. Older Eocene intrusions such as the 50-million-year-old Silver Point pluton along US 2 near Newport, Washington, were emplaced deep in the crust and metamorphosed and sheared during the tail end of the development of the complex. Younger Eocene intrusions such as the 48-million-year-old Wrencoe pluton along US 2 near Sandpoint were shallowly emplaced and are undeformed, indicating that they intruded after the complex had fully developed.

MIOCENE FLOOD BASALTS ON THE EDGE OF THE COLUMBIA PLATEAU

Northern Idaho lies on the fringe of the Columbia Plateau, and only a few of the massive basalt lava flows of the Columbia River Basalt Group made it this far northeast. Most of the lava flows were erupted from fissures well to the south and west. The larger lava flows were able to travel some distance up drainages toward the north and east, but the ones that did are thin. Flows of the Imnaha, Grande Ronde, and Wanapum Basalts occur in the southern part of the region. The only units that made it all the way to the Rathdrum Prairie are flows of the Wanapum Basalt. See the Western Margin chapter for the main discussion of the Columbia River Basalt Group.

Sedimentary basins were very common along the margin of the plateau, typically forming when basalt lava flows dammed major drainages, creating wetlands and lakes upstream from the lava dams. Sedimentary deposits preserve fossils of the wetland and lakeside forests, and the species of fossils indicate a very warm, humid, subtropical climate in the Pacific Northwest during the Miocene Epoch. The Coeur d'Alene and Clarkia areas hosted two of the larger basins. The Clarkia area preserves such an extraordinary fossil assemblage of middle Miocene plants, insects, and fish fossils that it has been designated a lagerstätte, an exceptional fossil assemblage that defines the life and environments of a particular time in Earth's history. There are fewer than one hundred lagerstätte around the world.

PURCELL ICE LOBE AND THE RATHDRUM PRAIRIE

The final chapter in the extensive geologic history of Northern Idaho is a doozy. For the past few million years, ice sheets have been present in the arctic region, and the Pleistocene Epoch in North America was marked by alternating southward advances of these ice sheets during ice ages and northward retreats during interglacial periods. In the largest advances, such as the last one 33,000 to 14,000 years ago, the ice sheet crossed the 49th parallel into the region that is now the northern United States. Along the southern edge of this ice sheet, a series of ice lobes extended down north-south-trending valleys and trenches, dramatically changing the landscape. Ice lobes in the Pacific Northwest included the Puget Sound, Okanogan, and Columbia lobes. In the Idaho Panhandle, the Purcell lobe extended south to fill the Purcell Trench and the basin presently occupied by Lake Pend Oreille.

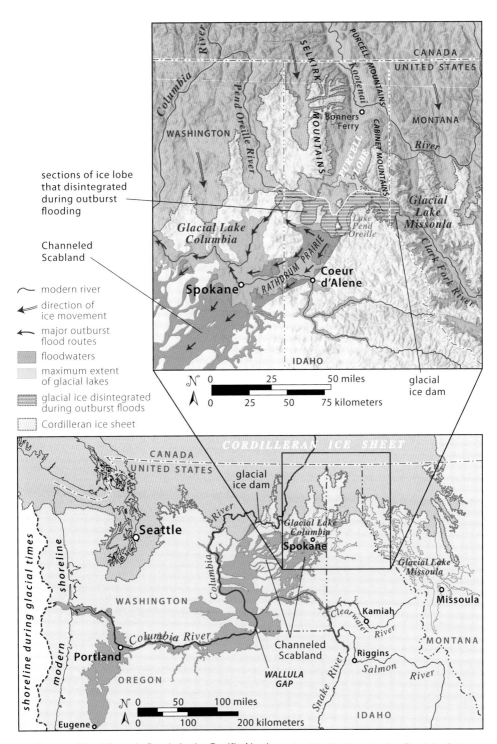

Extent of the Missoula floods in the Pacific Northwest. —Modified from Ice Age Floods Institute

As the ice lobes filled major valleys, they dammed rivers, creating temporary glacial lakes in the upper valleys. One of the largest glacial lakes that formed during glacial advances was in the Clark Fork drainage basin behind the Purcell lobe. Known as Glacial Lake Missoula after the extensive sequence of shorelines preserved in Montana's Missoula Valley, the temporary lake covered 2,900 square miles. Between about 19,000 and 14,000 years ago, the lake repeatedly formed and broke-out of its temporary ice dam, sending torrents of water and icebergs streaming down the Rathdrum Prairie, through the Spokane Valley, and across the scabland of eastern Washington. The volume of water discharged during each of these events was on the order of 500 cubic miles. Similar dams and outburst floods were created by other glacial lobes that blocked rivers to the west in northern Washington. At times multiple glacial lakes were backed up behind ice dams, and as upstream dams such as the Purcell lobe burst in Northern Idaho, the excessive load of floodwater entered downstream glacial lakes, such as Glacial Lake Columbia in the Spokane Valley, causing their ice dams to burst as well.

The Rathdrum Prairie collected colossal volumes of coarse-grained flood sediment every time the Purcell ice lobe failed. Almost all of the normal glacial features such as moraines, kettles, eskers, and outwash gravels are missing from the Rathdrum Prairie because the outburst floods repeatedly removed any preexisting features in their path, replacing them with flood channels, giant current ripples, and oversized gravel bars. Coarse flood gravels with boulder-size material reach thicknesses of almost 1,000 feet

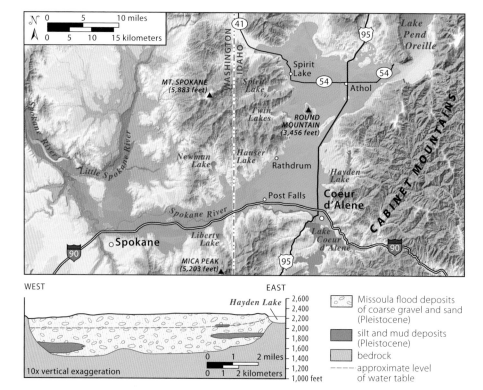

Spokane Valley–Rathdrum Prairie with cross section. Green indicates extent of the underlying aquifer. —Modified from Spokane Valley–Rathdrum Prairie Aquifer Atlas, 2009 update

beneath the Rathdrum Prairie. This massive blanket of material dammed most side valleys that feed into the Rathdrum Prairie, creating flood gravel dams that impound Lake Coeur d'Alene, Hayden Lake, Spirit Lake, and many other lakes. The legacy of this history of floods and dams is a huge aquifer system beneath the Rathdrum Prairie that practically flows like an underground river through the coarse flood gravel. Creeks that you would expect to flow out onto the prairie disappear almost immediately, pouring into the porous subsurface. The aquifer serves as a tremendous groundwater resource for the cities of Coeur d'Alene and Spokane, and for the region's agriculture, but its high permeability makes it susceptible to infiltrating pollutants. Any surface contamination is, for all practical purposes, instantly transmitted directly to the aquifer, and then throughout the entire downstream aquifer and the rivers with which it interacts.

I-90
Post Falls—Lookout Pass
73 miles

I-90, the major east-west route across Northern Idaho, follows river valleys where possible but also crosses two summits, Fourth of July Pass over the Coeur d'Alene Mountains and Lookout Pass over the Bitterroot Mountains at the Idaho-Montana border. Most of the route east of Coeur d'Alene parallels a broad zone of faults that occurs along the northwest-trending Lewis and Clark zone from eastern Washington through western Montana.

I-90 also cuts right through the heart of the truly remarkable Coeur d'Alene mining district. Operating for more than 150 years, from the late 1860s to the present, this district broke all kinds of world records including most total mineral production from the smallest area, largest silver ore vein, and most silver from a single mine. It also broke numerous national records including the largest underground mine and deepest mine. The main metals produced in the mining district include silver, lead, and zinc, with minor amounts of gold. A nexus of geologic events contributed to the ore mineralization in a relatively small area. The extensional setting of the Belt basin in the Mesoproterozoic Eon provided faults and fractures through which hot water flowed and dissolved the dispersed metal minerals from the sedimentary rocks. The dissolved metals were then precipitated in metal sulfide layers on the floor of the basin. In the Jurassic, Cretaceous, and Paleogene Periods, faulting and folding of the Belt Supergroup rocks along the Lewis and Clark zone further fractured the rocks, and two major intrusions of magma, in the Cretaceous Period and Eocene Epoch, produced heat and circulated fluids through the fractures in the Belt Supergroup rocks. The original ores that were stratified within the sedimentary layers were redissolved and concentrated into veins that filled the fractures.

At the Washington-Idaho border, I-90 passes across Missoula flood gravel deposited when Purcell lobe ice dams burst repeatedly during the Pleistocene ice ages. The gravel is thick here, more than 700 feet in places, and hosts a very unusual aquifer that operates more like an underground river than a traditional aquifer. No surface water exists on the prairie because it rapidly seeps into the subsurface through the gravels. The Spokane River is the only exception, and its very existence hinges on the water table; it must be close enough to the surface to prevent the river water from draining completely into the subsurface.

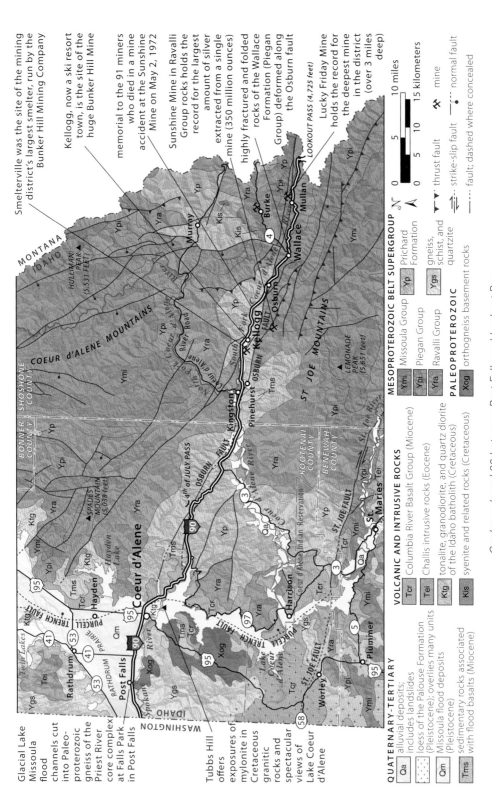

Geology along I-90 between Post Falls and Lookout Pass.

View west down the Spokane River in one of the Missoula flood–carved channels cut into Paleoproterozoic gneiss of the Priest River metamorphic core complex at Falls Park near Post Falls. (47.7096, -116.9543)

At Post Falls (milepost 5), a dam on the falls of the Spokane River controls the level of Lake Coeur d'Alene, which lies just east of here. The falls were originally harnessed to run a saw and grist mill. The land around the falls was purchased from the Coeur d'Alene tribe in 1891 by Frederick Post. By 1906, the Washington Water & Power Company had built a dam and turbines at the site and was producing much of the power for the Coeur d'Alene mining district. The falls expose banded schist and gneiss of the Priest River metamorphic core complex. At Falls Park and Q'emiln State Park (exit 5, take North Spokane Street south and turn right on West 4th Avenue), hiking trails and climbing walls provide nice exposures of these metamorphic rocks in bedrock channels cut by the Missoula floods.

I-90 continues along the south side of the Rathdrum Prairie on Missoula flood deposits. Low hills to the south are high-grade metamorphic rocks of the Priest River metamorphic core complex. The city of Coeur d'Alene (milepost 11) is built on Missoula flood deposits that dammed the combined flow of the St. Joe and Coeur d'Alene Rivers at Rathdrum Prairie, creating Lake Coeur d'Alene.

On the east side of Coeur d'Alene, the highway winds through roadcuts in Wanapum Basalt of the Columbia River Basalt Group. These flows, relatively thin this far north, erupted in Miocene time and filled an ancestral valley. Beneath the basalt lies the Eocene north-trending Purcell Trench fault, which cuts the Spokane dome mylonite zone and drops rocks on the east side of the fault. Because of this fault, the highway now traverses through the unmetamorphosed rocks in the upper plate that

TUBBS HILL

Tubbs Hill, a scenic park overlooking Lake Coeur d'Alene on the south side of the city, has hiking trails and beautiful exposures of the Spokane dome mylonite zone that was involved with unroofing the Priest River core complex during the Eocene Epoch. Tubbs Hill is a body of Cretaceous granodiorite cut by Eocene granite dikes. Mylonite texture in the granodiorite, developed by the slipping and shearing of hot rocks along the deeply buried fault, consists of well-developed foliation and lineation. The foliation dips consistently east-northeast and represents the slip surfaces along the detachment

View to north of Lake Coeur d'Alene and the Rathdrum Prairie. Rocks to the west (left) of the Purcell Trench fault, including those at Tubbs Hill, are high-grade metamorphic rocks of the Priest River core complex, whereas rocks to the east (right) are relatively unmetamorphosed rocks of the Belt Supergroup in the upper plate, above the detachment fault. —Photo by Andy Buddington

slid off the top of the Priest River core complex on the Spokane dome mylonite zone during Eocene time. Over the course of 56 miles, between the Purcell Trench fault and Lookout Pass, I-90 generally traverses through the Mesoproterozoic Belt Supergroup sequence from the oldest rocks, the Prichard Formation, through progressively younger rocks of the Ravalli Group and Piegan Group.

For the next 20 miles east of milepost 16, roadcuts display typical rusty-weathering outcrops of siltstone with minor shale and sandstone of the lower Prichard Formation at the bottom of the Belt Supergroup. The rusty weathering in these rocks results from oxidation of the originally reduced iron minerals that were deposited in the deep, oxygen-poor water of the developing Belt basin in Mesoproterozoic time.

Between Coeur d'Alene and Lookout Pass, I-90 more or less follows the Osburn fault, a major west-northwest-trending structure that pre-dates the development of the Priest River metamorphic core complex. The Osburn fault and related faults and folds form what's known as the Lewis and Clark zone, an ancient Mesoproterozoic crustal weakness that trends west-northwest to east-southeast across northeastern Washington, the Idaho Panhandle, and well into western Montana. Since its initiation

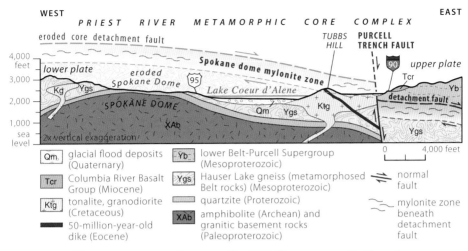

West-to-east cross section across the Priest River core complex through Tubbs Hill in Coeur d'Alene. —Modified from Wintzer and Lewis, 2016

fault. Lineation plunges shallowly east-northeast and tracks the transport direction of the upper plate of rock that slid off the top of the Priest River core complex. The mountains visible to the east of Coeur d'Alene are Belt Supergroup rocks in the upper plate. The Spokane dome mylonite zone was offset later in the Eocene Epoch along the north-trending Purcell Trench fault, located just east of Tubbs Hill and beneath Lake Coeur d'Alene. The lake fills a north-trending valley eroded along the fault. Rocks west of the fault have been uplifted relative to rocks east of the fault. To reach the park from I-90, take exit 15 and head west on East Mullan Avenue, then continue south to a trailhead on South 10th or head for Tubbs Hill Drive at the end of South 8th Street.

during formation of the Belt basin, it has been reactivated numerous times to deform younger rocks; one important episode in its history includes right-lateral strike-slip faulting that has offset mappable features such as mineralized zones by an estimated 8 to 16 miles. The Osburn fault and related structures along it have also served as conduits for fluids, heated by generations of magma intruded at depth, that deposited and repeatedly reconcentrated ore in the Coeur d'Alene mining district. At Fourth of July Pass (exit 28), rocks of the Prichard Formation have been broken up along the Osburn fault, and fluid flow within these fractured rocks altered them to the yellow, orange, and red colors visible from the highway.

East of Fourth of July Pass, I-90 drops into the Coeur d'Alene River valley, which has been intensely impacted by more than one hundred years of mining. Repeated seasonal floods carried large sediment loads downstream from mine tailings dumped directly on the valley floor farther upstream. Wetlands along the river are contaminated with heavy metals, mostly lead, arsenic, cadmium, and zinc, a consequence of the easily dissolved sulfide minerals in these tailings piles. Heavy metal contamination persists farther downstream in Lake Coeur d'Alene and poses a threat to the

Spokane River and the Rathdrum Prairie aquifer. It wasn't until the 1950s that mining operations were required to contain their tailings piles to protect the river and downstream communities.

In the 1980s, the area around Kellogg, which has the greatest concentration of tailings and slag refuse in the mining district, was listed as a Superfund site. Other parts of the district benefit from a variety of governmental and privately funded organizations that operate to clean up adversely impacted waterways and soils today. One aspect of these environmental cleanup efforts was the construction of a nearly 70-mile-long Rails-to-Trails bicycle path, the Trail of the Coeur d'Alenes, which stretches from Plummer to Mullan along the Coeur d'Alene River. This project paved the original railroad bed that had been constructed using the readily available mine tailings. The paved bike path now forms a barrier to water infiltration by rain and snow, limiting the contribution of heavy metals that can be leached out of the old tailings.

Kingston (exit 43) sits at the confluence of the South and North Forks of the Coeur d'Alene River. The South Fork drains most of the mining district and was once heavily contaminated. Boulders and cobbles along the stream are commonly iron-stained. In contrast, the North Fork has mostly run clear and clean. The mining legacy along this branch has been mainly gold placer mining, starting with the first ore discovery in the region near Murray in 1881. The resulting gold rush didn't last long, and disappointed prospectors soon fanned out to explore other areas. In 1884, a showy galena (lead sulfide) deposit was found near the site of the future Bunker Hill Mine at Kellogg on the South Fork of the Coeur d'Alene River. This discovery initiated the mining activity in the Coeur d'Alene district.

The Osburn fault zone passes right through Kingston and juxtaposes rocks of the Prichard Formation to the south with rocks of the Burke Formation in the Ravalli Group, the next unit up in the sedimentary sequence, to the north. A nice outcrop of finely laminated quartzite and less common siltstone of the Burke Formation can be seen by driving north 2.7 miles on Coeur d'Alene River Road from exit 43.

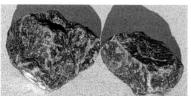

Major ore minerals of the Coeur d'Alene mining district. Clockwise from bottom left: sphalerite (zinc sulfide); galena (lead sulfide); tetrahedrite (silver-copper-antimony sulfide).

East of Kingston, I-90 is on the north side of the Osburn fault. At milepost 45, the highway passes Pinehurst, the gateway to the Silver Valley and the Coeur d'Alene mining district, in which there have been more than ninety producing mines from 1884 to the present. Covering an area of only about 300 square miles, this district held the record for the largest documented cumulative silver production in the world in the 1980s, when it largely shut down. The Coeur d'Alene district has produced over 1 billion ounces of silver, 8 million tons of lead, 3 million tons of zinc, and considerable amounts of antimony, cadmium, copper, and gold. It rivals, in production, the much larger silver mining districts of the world such as Potosi, Bolivia, and Durango, Mexico, that have operated for more than half a millennium. Ore veins that fill faults and fractures in sedimentary rocks of the Belt Supergroup contain the main ore minerals galena (lead sulfide), sphalerite (zinc sulfide), and tetrahedrite (silver-copper-antimony sulfide). Other common vein minerals include quartz, siderite (iron carbonate), pyrite and pyrrhotite (iron sulfides), and chalcopyrite (iron-copper sulfide).

Just east of Pinehurst, the valley of the South Fork of the Coeur d'Alene River opens up. The low gradient and relatively flat valley floor here led to chronic flooding when tailings from the surrounding mines were dumped into the valley. Unable to transport the excess material supplied to the stream, the South Fork of the Coeur d'Alene River repeatedly spilled over its banks at high water, turning the valley into a temporary lake.

The unnatural-looking berm that continues for more than a half mile along the south side of the highway east of milepost 45 is the first of many tailings, slag, and reclamation piles that line the sides of the valley here. This particular one is a tailings

Morning Mill in Mullan discharged directly into the South Fork of the Coeur d'Alene River circa 1909. —Photo from Barnard-Stockbridge Collection, University of Idaho Library

Aerial view of the Bunker Hill smelter, circa 1984. Note denuded hillslopes due to both timber production and nearly a century of acid rain produced by smelters in the mining district. —Photo by E. H. Bennett, Digital Geology of Idaho

containment and stabilization project that is part of the Bunker Hill Superfund Site. If you had been traveling this route before 1996, the view before you would have been dominated by two massive smokestacks that were part of the Bunker Hill smelter. Sulfide ore from the Bunker Hill Mine, located on the slopes to the south, and other mines in the district was shipped to the smelter to be refined into lead, zinc, and silver metals. At its peak in 1980 the smelter and associated metallurgical plant were producing about 20 percent of the nation's lead and zinc and 25 percent of its silver. The operation was shuttered in 1981. The gigantic smokestacks were toppled dramatically and buried in 1996.

East of Smelterville (milepost 48), the highway passes the main slag pile for the Bunker Hill smelter on the south for more than 1 mile. At Kellogg (exit 49), the mine site for the Bunker Hill Mine is visible to the south. The Bunker Hill deposit, one of the first discovery sites (1884) in the Coeur d'Alene district, was discovered by Noah Kellogg. Much of Kellogg Mountain, south of town, is honeycombed with more than 150 miles of tunnels in rocks of the Prichard Formation. The main workings are several miles inside the mountain from the mine entrance that is visible from town. A 2-mile-long tunnel was constructed in 1903 to connect the mine entrance to these workings and to transport ore to the smelter. A worthwhile stop in Kellogg is the Shoshone County Mining and Smelting Museum in the historic Staff House on McKinley Ave off exit 50.

It's hard to imagine now, looking at the fresh new face of Kellogg with its upscale ski resort economy, but not long ago in the 1990s when most of the mines and mills were only recently closed, the hillslopes around the town were nearly completely denuded of trees, largely killed off by acid rain from the smelter. The hillsides have since been reclaimed by terracing and replanting to limit erosion. In 1974, Kellogg children had four-and-a-half times the average blood lead levels of other children in the country. Even after the smelter closed in 1981, they remained at more than double the national average. To remedy this issue, one project of the Bunker Hill Superfund reclamation effort involved removing lead-tainted soil from many of the residential yards of Kellogg and replacing it with clean soil. The 150 miles of tunnels beneath Kellogg Mountain constantly drain acid mine water containing heavy metals. The large treatment ponds for this waste are visible from the highway on the west side of Kellogg.

At Big Creek (exit 54) is a memorial on the north side of the highway to the ninety-one miners who died in the tragic 1972 Sunshine Mine disaster, the worst fire-caused hard rock mining disaster in US history. A fire within the mine workings generated carbon monoxide that poisoned the victims. The silver lining of this disaster was that in 1977 the US Congress updated underground mining safety regulations to protect miners, including those who still work in the Coeur d'Alene district.

The Sunshine Mine, located 2.1 miles up Big Creek Road to the south of exit 54, lies on the south side of the Osburn fault. This mine set the world record for the largest silver production from a single mine with more than 350 million ounces of silver produced before it closed in the late 1990s. The mine workings below the massive headframe are in the St. Regis and Revett Formations of the Ravalli Group. Thinly bedded shale and sandstone (metamorphosed to phyllite and quartzite) of the St. Regis Formation are exposed in the roadcut right across Big Creek Road from the Sunshine headframe; white quartzite of the Revett Formation is well-exposed another 1.6 miles up the road.

At milepost 58, the highway recrosses the Osburn fault. East of milepost 58, I-90 passes through rocks of the Wallace Formation (Piegan Group) on the south side of the fault. The Wallace Formation was deposited in warm, shallow water that chemically favored deposition of limestone and dolostone. Good exposures of the middle part of the Wallace Formation are accessible by taking exit 60 for Silverton, turning east on Silver Valley Road (on the south side of the interstate) for a quarter mile,

Water treatment pond for the Bunker Hill Mine sits in front of a huge reclaimed tailings pile. The Trail of the Coeur d'Alenes bike path in the foreground is also part of the reclamation effort. (47.5389, -116.1391)

The Sunshine Miners Memorial recognizes the ninety-one workers tragically killed in a mine accident on May 2, 1972. Rocks of the Prichard Formation in the background. (47.5281, -116.0480)

then right on Lake Gulch Road. Classic black and tan couplets are exposed in roadcuts of the Wallace Formation, starting about a half mile up Lake Gulch Road. These sequences consist of thinly bedded dark-gray mudstone and light-gray fine sandstone and dolostone.

The town of Wallace (exits 61 and 62) is surrounded by rocks of the Wallace Formation that take its name. The Wallace District Mining Museum on the I-90 business route at 509 Bank Street is a worthwhile stop. East of exit 62, rocks of the Wallace Formation, strongly deformed by the Osburn fault, are visible in roadcuts on the north side of the highway. Exit 62 links to ID 4, which follows Canyon Creek 6 miles north to the historic mining town of Burke. Canyon Creek was severely impacted by mining, so much so that it earned the nickname Lil' Stinker for the foul water it carried during the mining heydays. A high concentration of mines and mills in the deep, narrow confines of the canyon forced the miners and their families to build their houses directly over the creek. Much of the stench in the creek was from sewage discharged from the houses directly into the creek, but tailings piles along the sides of the canyon also discharged a heavy load of metals and acidic water to the creek. An active restoration project underway along the creek is funded through a consortium of private mining interests and the state of Idaho.

East of Wallace, the highway crosses back into rocks of the Ravalli Group. At exit 69, the headframe for the still-operating Lucky Friday Mine is visible to the north. At almost 2 miles deep, tunnels of the Lucky Friday are the deepest in the district. For westbound travelers, a new tailings containment pond for the Lucky Friday Mine is visible in the valley to the north near milepost 73. At Lookout Pass at the Idaho-Montana border, roadcuts consist of siltstone and shale of the Ravalli Group.

Deformed rocks of the Wallace Formation along the Osburn fault east of exit 62. (47.4739, -115.9104)

Burke, Idaho, along Canyon Creek, circa 1888. —Photo from Barnard-Stockbridge Collection, University of Idaho Library

US 2 AND ID 200
Newport—Sandpoint—Montana Border
56 miles

If there was a list of top ten geologic driving routes in the United States, this route across the Idaho Panhandle would surely be on it. Not only is it wonderfully scenic as it follows the north bank of the Pend Oreille River and then the northern shore of Lake Pend Oreille, it also slices through deep time, going all the way back to the Archean, more than half the age of the Earth. In only 56 miles, it crosses the very heart of the Priest River metamorphic core complex, a tiny sliver of Laclede gneiss that possibly belongs to another continent, a nearly complete section of the Belt Supergroup, and the site at which the Clark Fork River was repeatedly dammed by the Purcell Trench ice lobe, leading to catastrophic releases of the Missoula floods. Hold on tight and watch for safe pullouts; you'll want to see these rocks up close.

East of Newport, Washington, US 2 enters Idaho, crosses the Pend Oreille River, and follows it upstream, deep within the Priest River metamorphic core complex. Recall from the chapter introduction that this complex contains extremely metamorphosed renditions of the Belt Supergroup and older rocks. The geology on both sides of the highway consists of Hauser Lake gneiss—the metamorphosed equivalent of the Prichard Formation rocks that form the lower unit of the Belt Supergroup—intruded by Eocene plutons. Some of the Eocene plutons are quite large, including the first one encountered along this route, the 50-million-year-old Silver Point pluton of granodiorite that originally crystallized at mid-crustal depths in the core complex. Roadcuts across from Albeni Falls Dam east of milepost 2 are in the Silver Point pluton, which here is extensively fractured. The fractures are filled by chlorite, a dark-green mica mineral that forms under relatively low temperature (shallow crustal level) conditions, indicating that this part of the Silver Point pluton was mostly exhumed by the nearby Newport detachment fault, described in the Priest Lake sidebar, when the fractures formed.

At milepost 6, the highway passes through the town of Priest River. Some of the most spectacular geology in Idaho can be seen across the Pend Oreille River to the south of Priest River by turning south on Wisconsin Street, crossing the bridge, and then turning left on Dufort Road. The outcrops within the first 2 miles of the bridge show some of the deepest levels of the Priest River core complex. This metamorphosed section of crust, exposed within an anticline in the Priest River complex, includes the following in the order that they are encountered while driving east on Dufort Road: (1) garnet amphibolite, the lowest part of the Hauser Lake gneiss, represents one of the mafic sills within the Prichard Formation that attained granulite grade metamorphism, the highest grade attainable before melting; (2) Archean Pend Oreille gneiss that originated as a 2.65-billion-year-old granitic pluton and forms the basement to the Proterozoic metasedimentary rocks; and (3) Gold Cup Quartzite, the 1.7- to 1.5-billion-year-old metamorphosed quartz sandstone that lies below the Belt Supergroup. The Gold Cup Quartzite is composed of nearly entirely quartz sand. Outcrops along the road in either direction beyond these three units consist of rusty-weathering Hauser Lake gneiss.

East of Priest River, roadcuts of thinly layered, fine-grained, rusty-weathering schist and quartzite of the Hauser Lake gneiss appear along US 2 starting just west

NORTHERN IDAHO 317

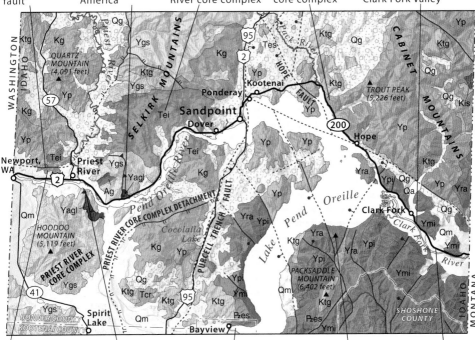

- southern segment of East Newport detachment fault
- 1,576-million-year-old Laclede augen gneiss, a faulted piece of basement rock that may not belong to North America
- 48-million-year-old undeformed granodiorite of the Wrencoe pluton intruded after deformation ended in the Priest River core complex
- Prichard Formation is the parent rock to the Hauser Lake gneiss of the Priest River metamorphic core complex
- outcrops scoured by the Purcell lobe ice dam along the highway east of milepost 56; water in Glacial Lake Missoula ponded 2,000 feet deep behind the dam in the Clark Fork Valley
- 50-million-year-old deformed granodiorite of the Silver Point pluton
- some of the oldest rocks in Idaho are exposed along Dufort Road
- southernmost extent of the Purcell ice lobe at Bayview and Farragut State Park
- Green Monarch Mountain scoured by the Purcell ice lobe
- from Hope to the Montana border are spectacular roadcuts that display a nearly complete section of the Ravalli, Piegan, and Missoula Groups of the Belt Supergroup

QUATERNARY-TERTIARY
- Qa alluvial deposits
- Qm Missoula flood deposits (Pleistocene)
- Qg glacial deposits (Pleistocene)
- Tes sedimentary rocks associated with Challis magmatic event (Eocene)

SEDIMENTARY BEDROCK
- Pzes Cambrian to Ordovician

MESOPROTEROZOIC BELT SUPERGROUP
- Ymi Missoula Group
- Ypi Piegan Group
- Yra Ravalli Group
- Yp Prichard Formation
- Ygs gneiss, schist, and quartzite

VOLCANIC AND INTRUSIVE ROCKS
- Tcr Columbia River Basalt Group (Miocene)
- Tei Challis intrusive rocks (Eocene)
- Kg granodiorite and two-mica granite (Cretaceous) [KANIKSU BATHOLITH]
- Ktg tonalite, granodiorite, orthogneiss, foliated granodiorite, and quartz diorite (Cretaceous) [KANIKSU BATHOLITH]
- Kis syenite and related rocks (Cretaceous)

BASEMENT ROCKS
- Yagl Laclede augen gneiss (Mesoproterozoic)
- Ag Pend Oreille gneiss (Archean)

- - - - fault; dashed where concealed
- normal fault
- detachment fault
- thrust fault

Geology along US 2 and ID 200 across the Idaho Panhandle.

PRIEST LAKE

Priest Lake, the archetypal Northern Idaho lake, can be reached by driving north 22 miles from Priest River on ID 57. The waterbody sits in a glacially carved valley with the spectacular Selkirk Mountains rising to its east. Priest Lake lies within the Newport plate, part of the unmetamorphosed rocks that slid off the top of the Priest River core complex. The Newport detachment system, surrounding the Newport plate on three sides, is in the shape of a horseshoe, open to the north. The detachment fault dips inward on all sides below the Newport plate to form a north-south elongated bowl in which the plate rests. The Newport plate remained fairly intact as it foundered in this bowl, while the lower crust in the core complex below the bowl was lifted up and out from beneath it in three directions. The eastern half of the Newport plate lies in Idaho; the western half in Washington. Coarse gravel fan deposits of the Eocene-age Tiger Formation record the erosion of uplifting rocks to the west of the western detachment fault. The gravel was transported east and deposited on top of the Newport plate in Washington.

Rocks in the Newport plate consist of relatively unmetamorphosed sedimentary rocks of the Prichard Formation that were intruded by Purcell sills shortly after they were deposited. Outcrops of Prichard Formation rocks are easily spotted by their rusty weathering. Much later in Cretaceous and Eocene time, they were intruded by granitic magma prior to activation of the detachment fault. The first few miles of ID 57 north of the town of Priest River pass through Eocene Silver Point granodiorite that has been mylonitized along the East Newport detachment fault. Granitic roadcuts farther north belong to the Cretaceous Kaniksu batholith.

During the Pleistocene Epoch, an ice lobe repeatedly flowed south into the Priest River valley. It carved the depression in which Priest Lake now resides and left glacial till and outwash throughout the valley. Contemporaneous alpine glaciers sculpted the craggy peaks of the Selkirk Mountains to the east, leaving glacial horns, arêtes, cirques, and U-shaped valleys through which tongues of ice fed out on top of the glacial lobe in the Priest River valley. During times of glacial retreat, as the southern edge of the ice lobe receded north into Canada, meltwater accumulated in glacial lakes, depositing very light-colored mud. Glacial till, deposited directly from the melting ice, is notable for its variety of sediment sizes—from silt- to boulder-size. Thick soils and forest cover most of the rocks of the Newport plate.

Priest Lake with Kaniksu batholith granitic rocks in the Selkirk Mountains in background. (48.5482, -116.9230)

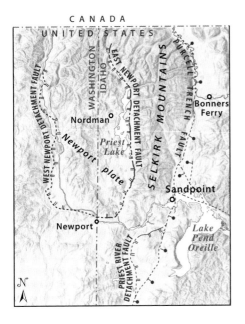
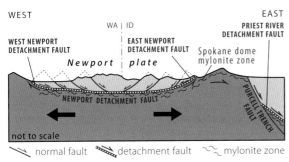

Geology along ID 57 between Priest River and Nordman. The inset map shows the entire Newport plate, the unmetamorphosed rocks that slid off the top of the Priest River core complex. —Inset from Doughty and Chamberlain, 2008; Stevens et al., 2016; cross section from first edition of Roadside Geology of Idaho

of milepost 11. East of milepost 12 in a long northward bend in the highway are beautiful roadcuts in the 1.58-billion-year-old Laclede augen gneiss. The name *augen* comes from the German word for "eye," and this texture is apparent in the large potassium feldspar crystals that have been stretched by deformation to form lozenge shapes that look like eyes. The gneiss began its life as a granite pluton that pre-dates deposition in the Belt basin. This particular lens-shaped body of augen gneiss has been faulted into the surrounding Hauser Lake gneiss as a tectonic sliver. Its age is

Rocks in the Priest River core complex exposed along Dufort Road, south of Priest River. **A.** *Garnet amphibolite, a metamorphosed gabbro Purcell sill in the lower part of the Mesoproterozoic Belt Supergroup. Dark selvages around red garnets result from reaction of the garnet during uplift and cooling to hornblende (black mineral) and feldspar (white mineral). (48.1693, -116.8882)* **B.** *Pend Oreille gneiss, a metamorphosed Archean granitic pluton (light-colored), intruded by dark-colored dikes metamorphosed to schist and amphibolite. (48.1677, -116.8777)* **C.** *Gold Cup Quartzite, the metamorphosed 1.7- to 1.5-billion-year-old sandstone that pre-dates the Belt Supergroup. (48.1657, -116.8746)*

problematic for North American rocks because we don't have granite rocks of that age in the Pacific Northwest. It's possible that the Laclede gneiss is a crustal fragment from another continent left behind as the supercontinent Rodinia rifted to form the Pacific Ocean basin.

Dark-colored amphibolite lenses in the augen gneiss represent mafic intrusions into the original granite that have also been deformed. These units have been squashed in the same direction as the augen: flattened in the vertical direction and pulled apart in the horizontal direction. The amphibolite behaved more brittlely than the augen gneiss during deformation, causing it to break into a series of tablets. The tablets have rotated clockwise as viewed looking north at the outcrop, and this rotation is consistent with top-to-the-east sense of shear that accommodated unroofing of the Priest River core complex.

To visualize how this works, place your right hand on top of your left hand (both palms-down) with a pencil between them pointing away from you as you look to the north at the roadcut with the mafic tablets. Now pull your right hand to the right (east) off the top of your left hand. Note that the pencil between your hands rotates clockwise as it spins in the shear plane formed between your hands. The spaces left behind as tablets pulled apart filled with melted rock, now a light-colored igneous rock that has been dated at 48 million years old. It was derived from partial melting of the surrounding gneiss to form what geologists call *migmatite*, literally "mixed rock" in Greek, referring to the mix of metamorphic and igneous rock that results from this process.

In the Laclede gneiss, we see the highest grades of metamorphism in this part of the Priest River core complex; the rocks were so hot that they were beginning to melt

View to the north of amphibolite tablets in the Laclede gneiss east of milepost 12. The dashed white line shows the tablet hidden behind the tree. (48.1557, -116.7981)

in places. Peak metamorphic conditions were attained at about 64 million years ago, when the rocks had been buried to their deepest levels during crustal shortening and thickening in the Sevier orogeny. Gravitational collapse of the overthickened crust began after 60 million years ago, initiating development of the Priest River core complex. As these deeply buried rocks were brought toward the surface while the upper plate above the complex was sliding off to the east, pressure was decreasing, but the rocks retained most of their heat and were still hot. Minerals in the gneiss with the lowest melting points, primarily quartz and feldspar, partially melted, and this magma was injected as light-colored dikes and sills into the surrounding gneiss. At 48 million years ago, unroofing of the Priest River core complex was complete.

Shortly east of the Laclede augen gneiss outcrop, the highway traverses back into spectacular roadcuts of migmatitic Hauser Lake gneiss. Amphibolite lenses within the Hauser lake gneiss represent Purcell sills that intruded into the original Prichard Formation. The amphibolite contains garnets that grew during metamorphism of the sills. The gneiss is folded, pulled apart, and injected with light-colored igneous intrusions that are likely within the same age range as 60- to 48-million-year-old melts found in deep levels of the core complex. A particularly impressive roadcut in these rocks occurs just west of milepost 17 across from a nice big pullout on the south side of the highway.

East of milepost 19, undeformed and unmetamorphosed granodiorite of the 48-million-year-old Wrencoe pluton appears in roadcuts. Because the Wrencoe is

Migmatitic Hauser Lake gneiss near milepost 17. The darker rock on the right side is amphibolite. The rusty weathering gneiss on the left side is Hauser Lake gneiss. The light-colored rock represents granitic melt that formed and injected these rocks. (48.1885, -116.7242)

undeformed, the deformation in this part of the Priest River core complex can be no younger than 48 million years old. The dark minerals in the granodiorite are biotite and hornblende, and the large pink minerals are potassium feldspar. A nice outcrop of the Wrencoe pluton occurs just east of milepost 25, with a pullout on the south side of the highway.

Sandpoint lies on glacial river and lake deposits along the north shore of Lake Pend Oreille at the southern end of the Purcell Trench. US 2 connects to ID 200 via US 95 in Sandpoint, crossing the north-trending Purcell Trench fault in the process. This fault cuts the main detachment fault for the Priest River core complex, dropping the relatively unmetamorphosed rocks of the upper plate down farther on its eastern side.

East of Sandpoint, ID 200 crosses the Purcell Trench north of Lake Pend Oreille in mostly glacial outwash gravels with some outcrops of Prichard Formation rocks cut by Tertiary and Cretaceous granitic dikes. East of milepost 38, the highway crosses over delta sediment built out into the lake by the Pack River, one of the major streams draining the east side of the Selkirk Mountains. Southeast of the Pack River, ID 200 hugs the northeastern shoreline of the lake, initially passing roadcuts of Cretaceous granodiorite. Southeast of milepost 40, the roadcuts contain rusty-weathering, thinly bedded sandstone and mudstone of the lower Prichard Formation that is intruded by the granodiorite. These Mesoproterozoic rocks of the Belt Supergroup were intruded by gabbro Purcell sills of nearly the same age. If you started this route by crossing the Idaho Panhandle on US 2, these outcrops allow you to compare these relatively unmetamorphosed rocks with their high-grade metamorphosed equivalents, the Hauser Lake gneiss and garnet amphibolite, respectively, in the Priest River metamorphic core complex between Priest River and Sandpoint.

Turbidites in this part of the Prichard Formation settled out from sediment-laden currents that cascaded off a shallow shelf into the deeper water of the Belt basin. Many outcrops show small folds and slumps in the original sediment, formed as the newly deposited, water-saturated sediment slid down the gradual slope of the basin. The source for sediment in this part of the Prichard Formation was from the west, on continents that have long since separated from the North American continent. A good roadcut, with a large lakeside pullout for examining these rocks, appears just northwest of milepost 42 near Trestle Creek.

ID 200 is one of the most scenic routes in Idaho with its fantastic views of Lake Pend Oreille, the largest lake in Idaho and one of the largest in the Pacific Northwest. Its glacially scoured basin is large enough and deep enough (1,150 feet) to host a US Navy submarine research station. The footprint of the lake basin probably owes its curiously contorted shape to the ancestral meandering river that flowed south from Canada through the Purcell Trench in Miocene time.

A great overview stop along the lake southeast of milepost 44 has a very well-informed geologic road sign detailing Glacial Lake Missoula and its ice dams at Lake Pend Oreille. During the late Pleistocene ice ages, between 19,000 and 14,000 years ago, the Purcell ice lobe repeatedly advanced down the Purcell Trench from the north. During each advance, all of the lake before you was filled with ice up to the highest ridges that surround it. The ice lobe continued south from the mouth of the Clark Fork River, scouring the curved mountainous wall of Green Monarch Mountain visible on the eastern side of the lake. The ice reached as far south as Bayview and Farragut State Park at the southernmost shoreline of present Lake Pend Oreille.

To the east, the valley of the Clark Fork River, which flows into Lake Pend Oreille from Montana, was dammed repeatedly by the ice lobe. Water backed up for hundreds of miles across a broad swath of northwestern Montana to form huge temporary lakes known as Glacial Lake Missoula. When an ice dam collapsed, all of the ice filling the present lake basin rapidly disintegrated, unleashing a colossal torrent of water from Glacial Lake Missoula through the Clark Fork Valley, then southward and westward through the Lake Pend Oreille basin, entraining nearly all the ice as icebergs into the floodwaters as they poured southwestward across the Rathdrum Prairie and across eastern Washington, northern Oregon, and out to the Pacific.

Clark Fork delta into Lake Pend Oreille. Ice scoured the flank of Green Monarch Mountain in the distance along the southeast side of Lake Pend Oreille.

Southeast of milepost 45, the highway crosses from the north side to the south side of the Hope fault, a major northwest-trending fault system that forms the Clark Fork Valley and the relatively straight northeast shoreline of Lake Pend Oreille. The Hope fault is one of the northwest-trending structures that make up the Lewis and Clark zone, a system of faults that crosses the Idaho Panhandle between Lake Coeur d'Alene and Lake Pend Oreille. Most of its activity appears to be Cretaceous in age, and it may have served as a tear fault that connected Sevier thrust faults in this region. Rocks on the south side of the Hope fault are significantly offset from those on the north side, and crossing to the south side of the fault corresponds with a major jump in the stratigraphic section from near the middle to the very top of the nearly 5-mile-thick Prichard Formation.

East along the highway to the Montana border, the rest of the Belt Supergroup section is laid out in succession starting with rocks of the Ravalli Group. The best roadcuts in these rocks are found along the old highway that runs parallel to the modern highway between mileposts 48 and 52. To access this section of the old highway, southeast of milepost 48 turn north onto Hope Peninsula Road (signed Old

Sam Owens Road), then right onto Denton Road (unsigned). The old highway reconnects with the modern highway near milepost 52. The Ravalli section represents the transition from deepwater environments of the Prichard Formation to emergent environments at the margin of the Belt basin that were receiving sediment delivered from the west. Rocks of the Ravalli Group exposed along the old highway consist of light-gray sandstone of the Revett Formation that was deposited by sheet-flood events that coursed across gigantic sedimentary fans, leaving ripple marks in the sand. Above the Revett are colorful red and green siltstone and shale of the St. Regis Formation that contain alternating mud-cracked layers with sand layers that incorporated ripped-up mud chips as sheet-flood events poured across the desiccated mudflats. The outcrops at the eastern end of the old highway show green to gray shale and thin white sandstone of the Helena Formation, followed by outcrops of the Wallace Formation, both of which are part of the Piegan Group.

Near milepost 52 on ID 200 (just east of the eastern connection with the old highway), a large roadcut on the north exposes rocks of the Wallace Formation. These limy, alternating dark-gray shale and tan sandstone and siltstone, the classic Wallace black and tan couplets, form wavy beds, affectionately referred to as "pinch and swell" bedding in Belt vernacular. Rare stromatolites and molar tooth structures are also present in the outcrop. The east end of this long outcrop displays thinly laminated siltstone and shale of the Snowslip Formation at the bottom of the Missoula Group, and outcrops of the lower Shepard Formation beyond that.

East of the town of Clark Fork, ID 200 follows the Clark Fork River upstream into a narrow valley. Each time the ice dam at Lake Pend Oreille broke, all the water from Glacial Lake Missoula poured through this valley. At peak flows, the discharge from Lake Missoula is estimated to have topped more than 740 million cubic feet per second at Clark Fork, a whopping number that rivals the total discharge for all major rivers on Earth combined at the same time.

East of milepost 56, rocks of the upper Shepard Formation appear in a long glacially polished roadcut on the north (eastbound travelers should use the large pullout on the south that appears halfway down the hill after the crest in the road). The pullout is safe, but the roadcut is close enough to the highway that it warrants extra caution. The upper Shepard Formation consists of thinly laminated dark-gray, green, and white siltstone and black shale containing stromatolites. Glacial striations and ice-plucked surfaces show eastward (up-river) ice-flow directions for the Purcell lobe ice sheet that flowed several miles *up* the Clark Fork to form the face of the ice dam that temporarily held back the waters of Glacial Lake Missoula. Lake levels in the valley above here were more than 2,000 feet above the present Clark Fork River.

The highway continues east past roadcuts on the north of the lower Mt. Shields Formation of the Missoula Group. This unit consists of pink feldspar-rich sandstone interbedded with occasional red to green shale containing mud cracks and salt casts. Near milepost 58, cliffs to the north consist of red and green sandstone, siltstone, and shale of the Bonner Formation. East of milepost 60, outcrops consist of laminated, multicolored siltstone and shale of the Libby Formation at the top of the Missoula Group. Mud cracks, rip-ups, and ripples are common. Limestone layers contain stromatolites and oolites—tiny calcite mud balls that form as the calcite precipitates on particles as they roll around in the waves on shallow substrate. The entrance road to Cabinet Gorge Dam turns to the south between mileposts 62 and 63, and outcrops

Looking east along ID 200 at a glacially polished outcrop in rocks of the upper Shepard Formation east of milepost 56. Inset shows dome-shaped tan stromatolites in ice-plucked, dark shale. Hammer pick points to plucked surface and ice-flow direction (east). (48.1309, -116.1632)

of the Libby Formation at the dam overlook show some of these features. The prominent terrace visible downstream and across from the dam overlook consists of thick Missoula flood deposits.

US 95
Lewiston—Moscow—Coeur d'Alene
110 miles

See pages 231–232 for a discussion of the Lewiston basin and the climb up Lewiston Hill. This road guide begins at the junction of US 195 on top of Lewiston Hill on the Palouse, the eastern margin of the Columbia Plateau. Most of the rocks visible along the route belong to the Miocene-age Columbia River Basalt Group. The basalt thins northward and overlies basement rocks of the ancient North American continent.

North of its junction with US 195, US 95 passes through the Palouse Hills, a unique landscape clearly visible from the highway. This distinctive topography is developed in silty deposits of loess, wind-blown deposits from the Pleistocene Epoch. Wind that scoured glacial outwash deposits from the Canadian ice sheet to the north picked

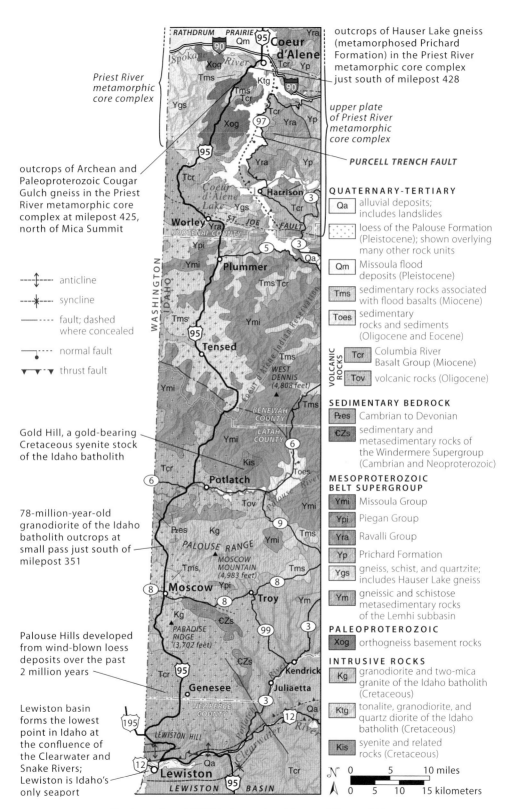

Geology along US 95 between Lewiston and Coeur d'Alene.

View to the north of the Palouse Hills along the Idaho-Washington border from Kamiak Butte, Washington. —Photo by Jane S. Finan

up silt and deposited it across the Columbia Plateau. In many places on the plateau, the loess was heavily eroded by the Missoula floods between 19,000 and 14,000 years ago, but this area on the flood margins preserves some of the thickest deposits. These sediments were deposited on top of weathered basalts of the Columbia River Basalt Group and form a thick, rich soil that is dry-farmed. The Palouse region is renowned for its production of wheat, peas, lentils, and garbonzo beans.

North of milepost 334, the highway crosses Paradise Ridge, a steptoe butte—originally a basement hill that was later surrounded by lava flows. Paradise Ridge is composed of granite and granodiorite of the Idaho batholith that intruded quartzite and schist of the Neoproterozoic Windermere Supergroup. These sediments were deposited in the rift that formed as the supercontinent Rodinia broke apart, opening the ancestral Pacific Ocean beginning about 700 million years ago.

Moscow is nestled in a small basin among steptoe buttes of the Palouse Range. These buttes are mostly granite and granodiorite of the Atlanta lobe of the Idaho batholith. The granitic rocks sometimes contain biotite (a dark brown to black mica) and sometimes both biotite and muscovite (a silver to clear mica). South of Moscow, the granitic rocks intrude metamorphosed rocks of the Windermere Supergroup. North of Moscow, they intrude quartzite and schist of the Mesoproterozoic Belt Supergroup. Roadcuts in the small pass 5 miles north of Moscow (just south of milepost 351) consist of 78-million-year-old granodiorite and granite of the Idaho batholith.

North of milepost 352, the tree-covered butte in Washington to the northwest is Kamiak Butte. Columbia River basalts surround this steptoe butte composed of coarse-grained quartzite with cross bedding. The rock type and ages of the sand grains in the quartzite indicate it is likely Cambrian in age, deposited along the nascent western margin of ancient North America following the rifting of Rodinia.

North of milepost 359, the fractured top of a lava flow, known as entablature, appears in a small quarry on the west and in many roadcuts, a reminder that almost all of the low country on the Palouse is underlain by basalt, here the Wanapum Basalt of the Columbia River Basalt Group. The basement rocks only poke up through the basalt in the buttes. Gold Hill, one of these large buttes to the northeast of Potlatch, consists of the Cretaceous Gold Hill stock of the Idaho batholith. The most common rock type in the stock is syenite that contains mostly potassium feldspar with some plagioclase feldspar, hornblende, and pyroxene. Gold veins are associated with the stock, and several small mines and prospects are evident on the butte.

More entablature basalt appears in a roadcut on the west just south of milepost 367. This basalt lava flow is part of the Onaway Volcanics, a unit of limited aerial extent on the Palouse that pre-dates eruptions of the Columbia River Basalt Group. Ages from the Onaway basalt are around 26 million years old, placing it in the Oligocene Epoch.

As US 95 climbs over another summit, large roadcuts on the west just south of milepost 372 consist of interbedded white quartzite and green shale of the Libby Formation, in the upper part of the Missoula Group of the Mesoproterozoic Belt Supergroup. The valley visible to the north of the pass is filled with silt and clay with some sand and gravel of mainly Miocene age that lie on top of deeply weathered sedimentary rocks of the Belt Supergroup. The climate during the middle of the Miocene Epoch was very warm and wet, and exposed rocks of the Belt Supergoup, Windermere Supergroup, and Idaho batholith were intensely chemically weathered. Deep red, clay-rich soils associated with tropical climates are preserved in the Miocene sediments. Many of the Miocene deposits have produced pure white kaolinite clay used in the ceramic industry. As you cross through the valley for the next dozen miles, note that most roadcuts have been intentionally armored with cobbles and boulders during road construction to prevent the weak, clay-rich sediments from slumping onto the highway.

North of Tensed, the highway climbs over another small range composed of mud chip–bearing sandstone and siltstone of the Striped Peak Formation, equivalent to the Mt. Shields and Bonner Formations of the Missoula Group. Between the towns of Plummer and Worley, US 95 threads between small buttes on both sides of the road that consist of relatively unmetamorphosed sedimentary rocks of the lower Missoula, Piegan, and Ravalli Groups.

Salt and/or barite casts in mudstone of the Missoula Group near Plummer. (47.3129, -116.9412)
—Photo by Jim Cash, Digital Geology of Idaho

For the 15 miles north of Worley, the highway traverses Wanapum Basalt of the Columbia River Basalt Group, which covers the trace of a major fault zone at depth, the west-northwest-trending St. Joe fault. This right-lateral strike-slip fault of Eocene age forms the southern margin of the huge Priest River metamorphic core complex that underlies the western side of the Idaho Panhandle. The upper crustal rocks that slid eastward off the core complex along the Spokane dome mylonite zone consist of relatively unmetamorphosed Belt Supergroup rocks similar to the ones between Plummer and Worley.

The southern edge of the core complex is hidden beneath basalt. For example, a large roadcut through the Wanapum Basalt occurs south of the high bridge over Lake Creek, south of milepost 412, but north of Lake Creek, the Priest River core complex is at the surface. The first outcrops appear by milepost 421 and continue for the next 8 miles to Coeur d'Alene, although parts of the complex remain covered by the younger basalt. These core complex rocks are strongly metamorphosed and consist of the lowermost Belt Supergroup, a distinctive sedimentary package below the Belt rocks, and underlying Paleoproterozoic and Archean igneous rocks that once formed the basement on which Belt Supergroup sediments were deposited. A spectacular set of roadcuts at milepost 425 on the downgrade north of Mica Summit exposes several metamorphosed granite gneisses, including a distinctly light-colored unit at the base of the outcrop, and amphibolite. The granite gneisses yield Paleoproterozoic ages of around 1.865 billion years and probably intrude the amphibolite, which just north of here gives Archean ages of 2.650 billion years.

Lake Coeur d'Alene comes into view on the east, north of milepost 426. Underlying this elongated lake is the Purcell Trench fault, which cuts the Spokane dome mylonite zone. Both of these structures served to slide the relatively unmetamorphosed rocks of the Belt Supergroup that were once above the high-grade gneiss in the Priest River core complex down and to the east to their present location in the hills across the lake. Most of the visible shoreline across the lake is composed of the Prichard Formation of the Belt Supergroup.

The next section of highway traverses upward through the metamorphic sequence in the Priest River core complex, into younger rocks that contain the same heavy metamorphic overprint as the older gneisses. Although not well exposed along the highway here, the next unit above the Archean and Paleoproterozoic gneisses is called the Razorback quartzite, a thin layer of metamorphosed, relatively pure quartz sandstone that was deposited on the older gneiss. It is equivalent to the Gold Cup Quartzite, exposed farther north along US 2, and pre-dates deposition of the lowermost Belt Supergroup sediments, which contain much more feldspar.

Just south of milepost 428, more outcrops appear on the west at a large pullout. These consist of Hauser Lake gneiss, which was the Prichard Formation at the bottom of the Belt Supergroup prior to metamorphism. The feldspar-rich sandstone, siltstone, and shale was metamorphosed to a banded gneiss with alternating light-colored quartz- and feldspar-rich bands and dark-colored biotite-rich bands. The approximately 1.450-billion-year-old mafic igneous sills that intrude the Prichard Formation are now thick amphibolite layers. The lower Prichard Formation was deposited in deep water that lacked oxygen, leading to the formation of dispersed pyrite and pyrrhotite (iron sulfides) in the sediment. Now that the rocks are at the surface, the sulfides in the rock have oxidized to hematite and other iron oxides that

Paleoproterozoic gneiss of the Priest River core complex at milepost 425. (47.6430, -116.8622)

Hauser Lake gneiss at milepost 428 consists of metamorphosed sedimentary rocks of the Prichard Formation and amphibolite (dark gray rock), the metamorphosed Purcell mafic sills that intrude the Prichard Formation. (47.6741, -116.8248)

give outcrops of the Hauser Lake gneiss a distinctly rusty appearance. Granite dikes intrude rocks in the outcrop. Outcrops of Hauser Lake gneiss continue from here to Coeur d'Alene. For geologic information about the Coeur d'Alene area, see the next road guide as well as the sidebar on Tubbs Hill in I-90: Post Falls—Montana border.

US 95
Coeur d'Alene—Sandpoint—Canadian Border
107 miles

US 95 between Coeur d'Alene and the Canadian border traverses north along the Purcell Trench, a wide valley that was repeatedly occupied by the Purcell lobe of the Canadian ice sheet in the Pleistocene Epoch. At full southward extension, the lobe filled the Lake Pend Oreille basin, forming an ice dam that impounded glacial meltwater in the Clark Fork Valley to the east to form Glacial Lake Missoula in Montana. As the dam repeatedly failed, outburst floods raged across and down the Rathdrum Prairie, a relatively long, flat plain that runs from the northeast at Lake Pend Oreille southwest to beyond Spokane, Washington. Hidden beneath the Rathdrum Prairie is a deep trough cut into basalt and underlying Precambrian basement rocks. Filling the trough to a maximum thickness of 1,000 feet is boulder- to cobble-sized gravel, with smaller gravel and coarse sand and silt along the sides. This tongue of unusually coarse sediment was deposited by repeated glacial outburst floods from Glacial Lake Missoula between 19,000 and 14,000 years ago. The tremendous load of coarse gravel piled up at the entrances of the adjoining river valleys.

Coeur d'Alene, on the southeast side of Rathdrum Prairie, is built on the gravel deposits that dam Lake Coeur d'Alene. This dam, of course, leaks like a sieve, because the gravel is highly permeable. The lake water flows directly into the massive Rathdrum aquifer that behaves more like an underground river than an aquifer. The aquifer has one of the fastest groundwater flows and highest discharges known for a basin of its size. Because the substrate beneath the Rathdrum Prairie is so permeable, surface water is very rare on the prairie. For more information about Coeur d'Alene, see the sidebar on Tubbs Hill in I-90: Post Falls—Montana border.

North of Coeur d'Alene, US 95 crosses the Rathdrum Prairie following the buried trace of the Purcell Trench fault. West Canfield Butte, the high point to the east, consists of the Ravalli Group of the Belt Supergroup. At milepost 439, lava flows of Wanapum Basalt of the Columbia River Basalt Group appear in a small butte near the highway on the east. Hayden Lake, another flood gravel–dammed lake along the margins of the Rathdrum Prairie, lies just north of the basalt butte. A few miles farther north, outcrops of Wanapum Basalt continue along the east side of the highway to about milepost 445. The basalt lava flows originally extended north of here and dammed a stream that flowed south from Canada through the Purcell Trench in Miocene time. The basalt north of here was completely removed by the repeated Pleistocene outburst glacial floods. Although very hard, basalt has a natural weakness—its tendency to fracture into vertical columns—that was easily exploited by the fast-moving floods; the raging current simply undercut the columns and toppled them over, one after another. At milepost 447, roadcuts on both sides of the highway expose Cretaceous granodiorite, part of the upper plate that slid east off the Priest River core complex, which lies west of here.

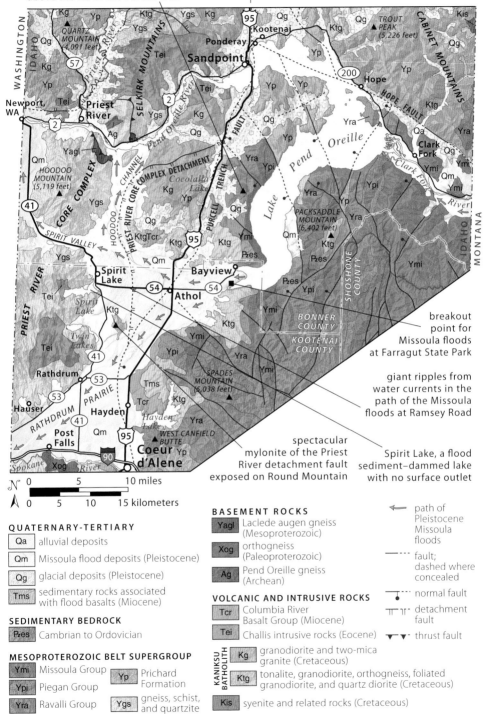

Geology along US 95 between Coeur d'Alene and Sandpoint.

North of milepost 452, roadcuts on the east side of US 95 consist of light-colored Cretaceous biotite granodiorite of the Kelso Lake pluton in the upper plate that also slid east off the Priest River core complex. The pluton contains abundant small, yellowish, stubby crystals of titanite, a calcium-titanium silicate mineral. At the bottom of the hill north of this outcrop, the highway crosses Hoodoo Channel, a dry valley that served as a major floodwater channel during outburst flood events, funneling water northwestward to the Pend Oreille River.

Near Cocolalla Lake (milepost 462), the highway passes through an area of complex drainage divides that was, at times, at the southern end of the Purcell lobe of the Cordilleran ice sheet. Visible from the highway to the east of here, Little Blacktail Mountain was surrounded by ice that was up to 1,800 feet higher than the elevation of the road, leaving only about 450 feet of the mountaintop protruding above the ice. Many of the granitic knobs on the surrounding hillsides have been carved and smoothed by ice and bear the scars of glacial scour.

FARRAGUT STATE PARK

Farragut State Park lies a little over 4 miles to the east of Athol on ID 54 at the southwest end of Lake Pend Oreille. This end of the lake was where the ice dam repeatedly collapsed, disintegrating the 20-mile-long ice tongue that sat in the lake's basin and sending torrents of water from Glacial Lake Missoula and icebergs from the ice tongue raging across the Rathdrum Prairie. Floodwaters flowed 2,000 feet deep across Farragut State Park during these events.

Lake Pend Oreille is the largest lake in Idaho and, at a maximum depth of 1,150 feet, the deepest lake in the inland northwest. Ice of the Purcell lobe repeatedly scoured the basin as it flowed south and then partially filled it with glacial outwash and flood sediment as it melted back north. The size and depth of the lake permitted the operation of a US Navy submarine training station at Farragut State Park during World War II, and the US Navy still maintains a research station on the lake. Flood gravels beneath Farragut State Park now dam the lake at its southern end, and its present outlet, the Pend Oreille River, is north of here through the western arm of the lake at Sandpoint.

Panorama view to northeast of the southern end of Lake Pend Oreille from the breakout point for Glacial Lake Missoula outburst floods at Farragut State Park.

ROUND MOUNTAIN AND THE PRIEST RIVER CORE COMPLEX

Although difficult to see from US 95, the hill rising out of the Rathdrum Prairie about 4 miles west of US 95 from exit 446 is Round Mountain. Spectacular exposures of the detachment fault near the top of the Spokane dome mylonite zone can be found in Cretaceous granite on the east side of Round Mountain. Along this detachment fault, the upper crustal rocks that now lie in the Cabinet Mountains to the east slid eastward off the top of the Priest River metamorphic core complex. The granitic outcrops along the mountain's east side appear to have layers sloping down to the east. This very shallowly east-dipping mylonite fabric consists of recrystallized quartz, feldspar, and mica minerals that have been stretched into ribbon-like shapes during shear. In addition, a well-developed linear fabric trends consistently east-northeast. The planar fabric is the shear-plane on which the upper plate of the core complex slid, and the linear fabric is the direction that the upper plate rocks moved. One final interesting detail observable in these rocks is that they have a slight green coloration. This results from growth of the mica mineral chlorite, a low-temperature mineral that formed very late in the development of the detachment fault. As the once deep-seated rocks of the core complex neared Earth's surface, they reached lower temperatures where chlorite is stable. You can reach Round Mountain by heading west 4 miles on East Brunner Road (exit 446) and then turning south on Ramsey Road. Good outcrops that show these features are located 1.5 miles from this intersection. Ramsey Road becomes Bench Road a little farther south.

View to north-northwest of mylonitized Cretaceous granite in the Priest River detachment fault at Round Mountain. The two geologists are walking on the mylonite shear surface that dips to the east-northeast. Orientation of mylonite lineation is marked by arrow. Inset at bottom (2 inches across view) shows spectacular mylonite fabrics that clearly indicate top-down-to-the-right sense of shear on the detachment fault. Inset at top shows view to the east-northeast, looking down on the shear surface and along the mylonite lineation and transport direction of the upper plate, now exposed in the Cabinet Mountains, visible in the distance. (47.8847, -116.8049)

FLOOD MEGARIPPLES AND SPIRIT LAKE

Missoula floodwater poured nearly due west each time an ice dam ruptured at the southwestern end of Lake Pend Oreille. Spirit Lake, 14 miles away, was in the direct path of the torrents. ID 54 follows this route across the relatively flat Rathdrum Prairie. West of Athol (exit 449 from US 95), between mileposts 5 and 1 (mileposts count down this direction), ID 54 passes through one of the most spectacular giant current ripple fields left by the Missoula floods. In fact, they are so large that you may not appreciate how awe-inspiring they actually are. The highway passes low cuts in rounded gravels at regular intervals spaced about 500 feet apart. Each roadcut is in a separate ripple crest. Although not visible from the highway, megaripple amplitudes reach 45 feet in height and their crests are spaced up to 600 feet apart. Ripple size, spacing, and gravel size all correlate with stream power; increases in each of these three parameters correspond with increasing floodwater velocity and volume. Although difficult to see, the western sides of the ripples have steeper slopes than their eastern counterparts, consistent with a current moving east to west during floods.

Spirit Lake, dammed by outburst flood sediment, can be viewed by turning west on Maine Street in downtown Spirit Lake. Roadcuts just east of the lake expose the coarse cobble- to sand-size sediment that forms the dam. As is typical for these sediment-dammed lakes, Spirit Lake has no surface outlet. All discharge from the lake percolates through the ground, directly into the huge Rathdrum Prairie aquifer.

North of Spirit Lake, ID 41 leaves the Rathdrum Prairie and travels northwest into Spirit Valley, following the route of a significant secondary flow path for Glacial Lake Missoula floodwater that shot northward to the Pend Oreille River valley. This north-trending valley in Washington was invariably blocked by another ice lobe that diverted the floodwater into the Spokane Valley to the south, to join the main flood path. The flat-bottomed Spirit Valley contains gravel deposited by the floods.

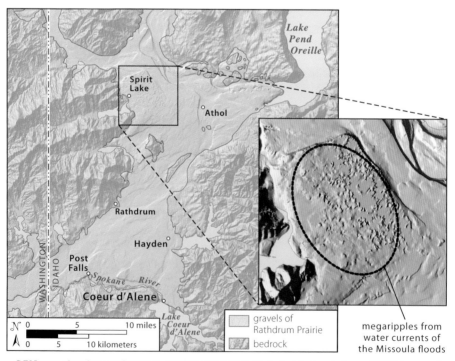

DEM map showing northeastern section of the Rathdrum Prairie. Inset shows LIDAR image of giant current ripples in front of Spirit Lake. —Inset image courtesy of Dean Garwood

These large boulders—some more than 6 feet in diameter—were removed from a recent railroad cut in Missoula Flood gravel bar deposits visible in background. Their position within the gravel bar indicates that they were entrained in the floods rather than simply dropped by melting icebergs floating on the floodwaters. This site lies less than one-tenth mile west of US 95 on Homestead Road. The turnoff is near milepost 452. (47.9869, -116.6995)

This area lacks a certain feature, however, that should be present where continental ice sheets have passed. Normally, the terminus of a continental ice sheet is marked by large terminal moraines, where the melting ice at the snout of the ice sheet leaves huge piles of debris varying from boulder- to silt-sized material. In this case, though, the moraines that developed in front of the advancing Purcell ice lobe were largely washed away by the raging Missoula floods. What's left are relatively thin remnants of moraines and glacial river and lake deposits among the granitic hills that have been deeply eroded into a series of glacial outburst flood channels. Cocolalla Lake is dammed by one of these remnant moraines.

The causeway over the western arm of Lake Pend Oreille to Sandpoint runs right along the Purcell Trench fault below the lake. Outcrops along the eastern shore to the east consist of relatively unmetamorphosed quartzite and siltstone of the lower Prichard Formation of the Belt Supergroup intruded by mafic sills. These lie in the upper plate that slid east off the Priest River core complex. The mountains in the distance to the west consist of highly metamorphosed Prichard Formation rocks and older Proterozoic and Archean gneiss all intruded by Cretaceous and Eocene plutons in the metamorphic core of the Priest River complex. Sandpoint, on the northern shore of the lake, lies on Quaternary glacial outwash and lake deposits.

North of Sandpoint, the highway continues to follow the Purcell Trench fault on the western side of the Purcell Trench, next to the Selkirk Mountains (to the west),

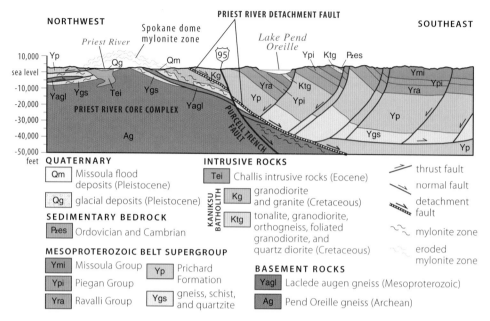

Cross section of the Priest River core complex, showing the Purcell Trench fault offsetting the Priest River detachment fault.

which reach elevations over 7,600 feet. The Selkirks, which lie within the metamorphic interior of the Priest River core complex, are composed of Cretaceous granitic plutons of the Kaniksu batholith, one of the great Cordilleran batholiths of western North America. The Kaniksu batholith intruded between 120 and 90 million years ago into gneiss, schist, and quartzite of mostly metamorphosed Belt Supergroup rocks. Although these rocks were never buried as deeply as those that lie below the Spokane dome mylonite zone in the metamorphic core of the Priest River complex to the south, the rocks in the Selkirk Mountains have experienced relatively deep burial and high-grade metamorphism.

The relatively flat-bottomed Purcell Trench contains mostly glacial lake and river deposits that were deposited by meltwater flowing from and ponding south of the Purcell ice lobe. When the ice lobe was at full extent, this section of the trench was nearly completely filled with ice, and the floor of the trench was mostly eroded by the moving ice. Only the highest elevation ridges and peaks in the Selkirk Mountains were exposed above the great lobe of ice filling the trench, but smaller alpine glaciers carved cirques, arêtes, and horns in these higher regions. During periods of ice lobe retreat, the upper elevations of the Selkirks remained largely encased in alpine ice, sending valley glaciers spilling down to the Purcell Trench far below. In the lower-elevation Cabinet Mountains to the east, only a few of the ridges and peaks are high enough to have been left uncovered by the great ice lobe during maximum ice advance.

Low hills in the Purcell Trench to the east between mileposts 478 and 480 consist of relatively unmetamorphosed sedimentary rocks of the Prichard Formation and mafic sills in the lowest part of the Belt Supergroup. North of milepost 485, the highway crosses the Pack River, a major drainage into Lake Pend Oreille from the Selkirk Mountains to the west.

At about milepost 490, US 95 crosses the present drainage divide for the Purcell Trench. Modern streams to the south of here drain south to Lake Pend Oreille, while those to the north flow to the Kootenai River and into Canada. This divide has shifted dramatically in the past in response to advances and retreats of the Purcell ice lobe.

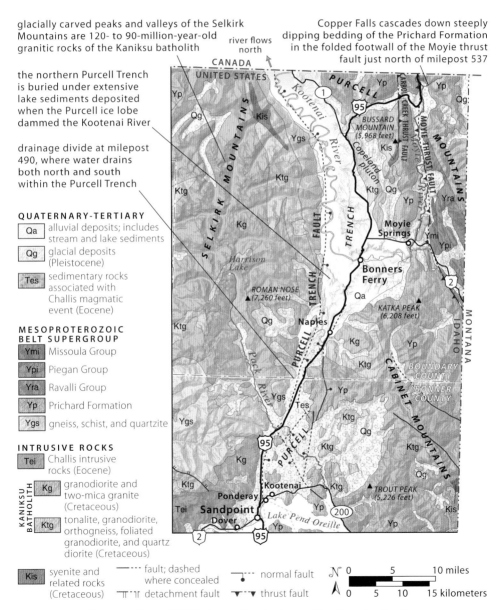

Geology along US 95 between Sandpoint and the Canadian border.

The weight of the ice depressed the crust north of here, forcing a drainage system that would normally drain south to divide, such that drainage in the northern part of the trench has reversed and now drains north into the depression. It takes time for the depressed crust to rebound, or return, to its pre-depressed elevation, so more than 10,000 years after the last ice age, the northern part of the Purcell Trench still drains north even though the ice is long gone. However, the crust is still slowly rebounding, so the drainage divide migrates slowly northward.

The Quaternary sediment fill in the north-draining portion of the Purcell Trench differs significantly from its counterpart in the southern trench, where southward-flowing meltwater deposited sand and gravel of glacial outwash. In contrast, in the northern trench, the ice lobe frequently dammed the Kootenai River, each time backing up a temporary glacial lake that extended south in the trench to the drainage divide, where the lake's outlet drained south. Most of the fill in the northern trench consists of lake deposits of silt and clay, which are very unstable when saturated, producing landslides that plague the area.

The highway crosses distinct southwest-trending channels that cut the landscape at milepost 499. These formed when the terminus of the Purcell ice lobe was located in the area around Bonners Ferry, just north of here, damming the Kootenai River farther upstream to the east. The river spilled through these channels along the south side of the ice dam and drained south across the mid-trench divide.

Just north of milepost 503 is a pullout on the west with geology road signs describing the glacial lakes that inundated the Naples–Bonners Ferry region. From this viewpoint you can see the very flat lake bottom of the trench and rounded granite knobs, ice-carved basement protrusions that stick out above the trench floor. Higher lake stands created higher lake deposits that were subsequently carved away by both ice advance and glacially fed rivers flowing south, leaving a series of lake terraces. A

View to the northwest along the northern Purcell Trench from geologic signs at the pullout north of milepost 503. Two prominent lake terraces are visible. Bonners Ferry is to the right.

prominent high-elevation terrace can be seen from the pullout. The highway crosses a less prominent lower-elevation terrace between the geologic signs and Bonners Ferry.

At Bonners Ferry, the highway crosses the Kootenai River, which flows west here before swinging north and flowing into Canada to its confluence with the Columbia River. US 95 follows along the east side of the Purcell Trench for 14 miles before diverging from the trench. Roadcuts in the highway north of Bonners Ferry consist of muddy glacial lake sediments that are unstable on the steep hillslope. In 1999 much of this slope collapsed in a large landslide that took out US 95 and the main rail line. Although mostly covered by trees, the slump scarp can still be seen in a large scooped-out area that appears north of milepost 508.

North of milepost 520, cliffs on the east are formed in Copeland granodiorite, a 98-million-year-old pluton of the Kaniksu batholith. North of the ID 1 junction, spectacular roadcuts in the Copeland pluton at milepost 523 expose fresh granodiorite with conspicuously large potassium feldspar crystals in a matrix of quartz (dark-gray clear mineral), plagioclase feldspar (white, blocky mineral), and biotite (black plated mineral). Note the common dark inclusions, called xenoliths (literally "foreign rocks"), within the granodiorite. Most are fine-grained diorite, which may have been injected into the granodiorite as a magma, and others are metasedimentary Belt Supergroup rocks that were plucked from the walls of the magma chamber. Some of these display fine laminations and graded beds that are common in the Belt Supergroup, as well as pyrite-rich rocks of the Prichard Formation. Roadcuts at milepost 525 show rocks of the Prichard Formation and Purcell mafic sills that are near the contact with the Copeland pluton and have been altered by the heat from the pluton. Granodiorite dikes and sills have been injected from the pluton into these wall rocks.

Between the Copeland pluton and the Canadian border, roadcuts consist of more Prichard strata and overlying glacial till and lake deposits. East of milepost 534, the highway enters the south-draining Moyie River valley and follows the river north to the border. The valley is centered on the Moyie thrust fault, a major structure in the Sevier fold-and-thrust belt that thrusts lower Prichard Formation strata to the west over upper Prichard Formation strata to the east.

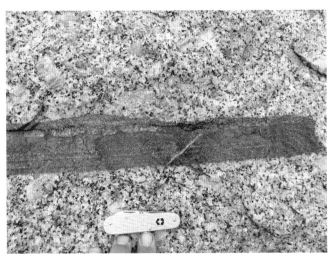

The 98-million-year-old Copeland granodiorite contains abundant xenoliths, some of which, like this one at milepost 523, consist of Belt Supergroup sandstone and siltstone. In the xenolith, note the graded bedding (upward fining sediment sizes) originally formed in Belt sediments. (48.8915, -116.3503)

Just south of the Canadian border, north of the final milepost (537), is the signed turnoff to Copper Falls, 2.3 miles up National Forest Route 2517 and a short, but steep, quarter-mile hike from the parking area. From an overlook you can see a quintessential Northern Idaho waterfall splash down the sloping bedding surface in upper Prichard Formation rocks. These rocks, which are east of the Moyie thrust fault, are in the steeply west-dipping western limb of the north-trending Purcell anticlinorium, a nearly 10-mile-wide anticline that stretches from here eastward into Montana.

ID 3
US 12 Junction—St. Maries—I-90 Junction
117 miles

ID 3, a curvy, slow-paced road in Northern Idaho, follows the eastern margin of the Columbia Plateau where Miocene basaltic lava lapped against the Bitterroot Mountains, partially covering rocks of the Proterozoic Belt Supergroup and Cretaceous Idaho batholith. The margins of volcanic plateaus are geologically dynamic regions of intense interaction between volcanic and sedimentary processes, and outstanding examples of features from these environments keep coming one after another along ID 3.

From its junction with US 12 on the Clearwater River, ID 3 follows the Potlatch River upstream through lava flows of the Imnaha and Grande Ronde Basalts of the Columbia River Basalt Group of Miocene age. Just north of US 12, ID 3 crosses the Lewiston Hill structure, which here forms a large anticline in the basalt due to reactivation of the Mesozoic suture zone between the Wallowa terrane and the North American margin in basement rocks below the basalt. Roadcuts on the west for the first half mile of the route show north-dipping Imnaha Basalt in the northern limb of the anticline. Basalt higher up on the canyon sides consists of Grande Ronde Basalt, including a thick entablature zone that appears straight ahead in the first large right-hand turn north of the junction. Part of this zone is prone to repeated rock fall, ripping out some of the screen and bolts that have been installed to help stabilize it.

The eastern margin of the Columbia Plateau is about 10 miles east of the Potlatch River, and in Miocene time, rivers draining the mountainous highlands to the east deposited sand, silt, and mud across newly erupted basalt on the plateau. The Columbia River Basalt Group exposed in the Potlatch River valley is literally riddled with sedimentary interbeds—lenses and continuous layers—sandwiched between basalt layers. Once a new river channel was established on a freshly solidified basalt surface, a subsequent lava flow would dam it, producing a temporary lake upstream that filled with silt and mud. Eventually new lava flows would fill depressions on the landscape, and the entire process would start anew.

The first interbed along the route appears north of milepost 1, sandwiched between Imnaha Basalt below and Grande Ronde Basalt above. The interbed consists of cross-bedded sand overlain by laminated silt and clay with a buff to gray-black paleosol (a soil horizon) at the top. The cross beds dip northwest, so we know a northwest-flowing river deposited the sand, probably in sandbars along the main river channel. We also know the sand wasn't transported very far from its source because in addition to quartz, it contains lots of feldspar and mica, minerals that weather quickly and eventually disappear. The sand was largely derived from granitic rocks of the Idaho batholith to the east. The laminated siltstone and clay above the sand was deposited

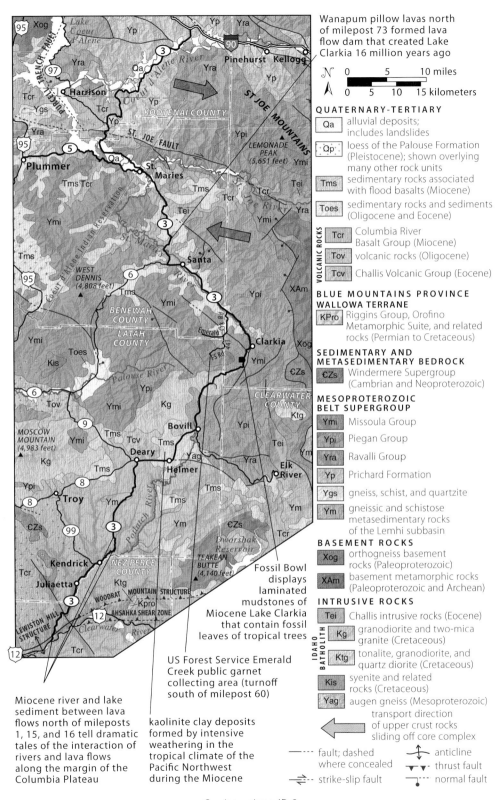

Geology along ID 3.

on a floodplain, indicating that the river channel migrated laterally. The buff to gray-black paleosol above the silt-clay sequence indicates that an ancient soil developed in the floodplain. When one of the lava flows of the Grande Ronde Basalt flowed into the ancient river valley and across the floodplain, it incinerated all vegetation in its path, including along the lush, tropical banks of this ancient river. Bulbous basalt pillows with chilled rinds formed at the top of this interbed where the molten lava solidified instantly against the saturated soil.

The highway continues up the Potlatch River, passing several more roadcuts in the same sedimentary interbed over the course of the next 2 miles before traversing a fairly straight stretch of the valley between Juliaetta and Kendrick. Grande Ronde Basalt is exposed on both sides of the valley here. North of Kendrick in the large roadcut at milepost 14, at the base of the Deary Grade, is a spectacular warbonnet, a section of columnar jointed basalt shaped like a fan. One way that warbonnets form is when basalt fills a river channel, and as the lava solidifies, it creates a fan of columns that are oriented perpendicular to the curved channel sides as seen in cross section.

As the highway climbs up Deary Grade, another spectacular sedimentary interbed appears in a large roadcut on the west just north of milepost 15, with a large pullout on the east side. This interbed, sandwiched between flows in the Grande Ronde Basalt, was deposited in a river channel into the underlying lava flow. Most of the lower part of this interbed consists of rhythmic, thinly bedded layers, each of which fines upward systematically from tan-colored sand at the bottom, to gray silt in the middle, and finally

In the roadcut north of milepost 1, bulbous pillows that developed at the base of the Grande Ronde Basalt (tan rock at top) froze instantly as the fresh lava pushed into the wet sediment. The light-colored mudstone interbed developed a paleosol (dark-colored top) before the lava flowed over it. (46.4919, -116.7629)

A warbonnet developed in a flow of Grande Ronde Basalt at the base of the Deary Grade at milepost 14. (46.6221, -116.6379)

darker-colored mud at the top. These beds were deposited in a lake that formed when the river channel was dammed by basalt lava flows farther downstream from here. There are a couple of options for how the cyclic layers formed. One possibility is that they were deposited seasonally, with sand and silt deposited by faster currents in the lake during wet seasons, while mud was deposited during periods of stagnation during dry seasons. Another option is that this location was close to an inlet to the lake, and during rainfall, increased river flow brought loads of sediment to the lake that settled during deposition, with the larger material falling to the bottom first followed by ever smaller sediment. Interbedded in the lake sediment and overlying the lake sediments are thicker sand beds that were deposited when the lava rock dams were breached, and the river flowed once again. The road traverses through this thick sedimentary interbed over the next half mile. The top of the interbed is exposed in a distinctive yellowish-orange roadcut at a tight westward curve in the road. Here the coarse sand is clearly cross bedded with current directions varying between west and east through the sand bed. The distinctive yellow-orange coloration is due to iron oxidation in groundwater that coursed through the sand layer long after it was deposited.

The highway continues up the Deary Grade through Grande Ronde lava flows, with another poorly exposed interbed at milepost 16, and a final sedimentary interbed exposed on the grade just south of milepost 17. The one near the top consists of mostly silt that was deposited in another lava flow–dammed lake. The distinctive white to reddish-brown to black layering at the top of the interbed shows the development of another ancient soil that pre-dated the eruption of the overlying basalt lava flow.

North of milepost 17, the highway climbs into the overlying Wanapum Basalt lava flows to the top of the grade and then traverses more Wanapum Basalt as the highway continues north on the Palouse at the top of the plateau. Steptoe buttes dotting the landscape consist of metamorphosed rocks of the Mesoproterozoic Belt Supergroup intruded by Cretaceous granites of the Idaho batholith. These buttes were hills that became surrounded by Columbia River Basalt Group lava flows in the Miocene. The radio tower–adorned butte to the north of Deary consists of rhyolite and dacite volcanic rocks that lie above the metamorphic and granitic rocks and are presumed to be part of the Challis Volcanic Group of Eocene age.

Between Deary and Bovill, the highway crosses sediments deposited on the Wanapum Basalt that are notably clay-rich but also contain gravel, sand, and silt. The

Lake sediments at the base of a sedimentary interbed on the Deary Grade at milepost 15. (46.6325, -116.6238)

River sediments at the top of the sedimentary interbed on the Deary Grade north of milepost 15. A lava flow of Grande Ronde Basalt overlies the yellowish-orange sediments. (46.6353, -116.6181)

Miocene Epoch was not kind to the rocks of north-central Idaho. A warm, humid tropical climate accompanied the eruption of Columbia River Basalt Group lava flows, and these conditions produced deep weathering horizons on exposed rock surfaces. In tropical climates, chemical weathering is so intense that most minerals completely break down to clays. Streams transported the clay and deposited it along with the other sediments that occur along ID 3. The region around Bovill produces a prized kaolinite clay that is mined from the sediments and used in the manufacturing of high-temperature ceramics. Kaolinite is a nearly pure aluminum silicate derived from intense chemical leaching of most ions from the soil except for aluminum, silicon, and oxygen. Distinctly white deposits in many of the roadcuts between Helmer and Bovill (between mileposts 30 and 39) consist of these mostly kaolinite clays.

North of Bovill, the highway traverses basement rocks of metamorphosed Belt Supergroup and granitic rocks of the Idaho batholith. The distinctly green metamorphic rocks in roadcuts are mostly calc-silicate quartzite. The green color is from the mineral diopside, a deep-green, calcium-rich pyroxene that formed during the metamorphism of calcite in the original limy sedimentary rocks of the Wallace Formation. Dikes and sills of Cretaceous granite intrude the metamorphic rocks.

ID 3 crosses a low drainage divide north of milepost 47 and descends into the St. Maries River basin. North of milepost 51 and south of Clarkia, the highway enters meadows on the flat floor of Lake Clarkia that existed here in Miocene time. Back then, hills consisting of metamorphic rocks and granite protruded above the lake water in the flooded valley. The high stand for the ancient lake was almost 700 feet above the present valley floor. The thick lake sediments and volcanic ash in this basin were deposited in a period of about 10,000 years between 16.0 and 15.9 million years ago. Lake Clarkia formed as lava flows of the Wanapum Basalt dammed streams in the basin, and water backed up behind the lava dams.

Within the clay-rich Clarkia sediments are spectacular fossil leaves, insects, and fish that provide a snapshot of a tropical Miocene world in the Pacific Northwest.

Calc-silicate quartzite cut by aplite dikes in a roadcut between mileposts 50 and 51. The quartzite is metamorphosed Wallace Formation in the Piegan Group of the Mesoproterozoic Belt Supergroup. Aplite, a fine-grained intrusion, is part of the Cretaceous Idaho batholith. (46.9708, -116.2862)

Trees that grew on the shores of Lake Clarkia include ancestors to such exotic tropical groups as the magnolia, avocado, metasequoia, yew, cypress, cedar, gum (tupelo gum and sweet gum), sycamore, elm, chestnut, oak, and birch tree families that can be found in the southeastern United States today. The Clarkia fossils are in many cases not actually fossils, but rather the preserved original tissue from the remains of these organisms. The clay sediment provided an airtight seal that preserved the original organic material for 16 million years! Prying open clay-rich layers to find plant and insect remains exposes the organic material to our oxygenated atmosphere. In a matter of seconds, many beautiful colors of the original leaves will fade. Fossil Bowl, a family-run business that operates an off-road racecourse, appears on the west, north of milepost 52. They charge a small fee for fossil hunters to collect from the Clarkia lakebeds on their property.

North of the turnoff for the town of Clarkia, roadcuts in pillow basalt appear on the west just north of milepost 55. This Wanapum Basalt advanced across the lake floor, blurping out lobes of lava that quickly quenched upon contact with the lake water to form the pillow structures. The lava had flowed into deeper parts of Lake Clarkia, so it was under enough water pressure to prevent it from fragmenting into tuff.

Miocene-age Lake Clarkia mudstones, accessible at Fossil Bowl north of milepost 52, contain fossil leaves. (46.9915, -116.2766)

The hillsides surrounding Clarkia and the St. Maries River, which the highway follows for about 15 miles north of Clarkia, consist of metamorphosed sedimentary rocks of the Wallace Formation of the Belt Supergroup. Rocks include relatively high-grade schist, gneiss, and quartzite, and some of the spectacular metamorphic minerals that Idaho is known for, including garnet, kyanite, staurolite, muscovite, and biotite. The Idaho state mineral, star garnet, is relatively rare, but this area in Idaho and one location in India are the two main locations for finding them. Idaho has lots of garnets, but only a very few show the highly prized star asterism caused by the diffraction of light by microscopic, aligned crystals of rutile, a titanium oxide mineral, within the garnet. The US Forest Service operates a facility for collecting garnets from river gravels that are quarried and brought to the Emerald Creek site during the summer. The turnoff for this area is on the west between mileposts 59 and 60. Follow signs for the facility on National Forest Route 447. As of this writing, the area has been closed with plans to reopen it when a site for new material is located. Check with the St. Joe National Forest for current status.

A six-sided Idaho star garnet from Clarkia. The polished stone is approximately 1 inch across.
—Photo by Mickey Gunter, University of Idaho

Metamorphic rocks around Clarkia slid west off the top of the Boehls Butte–Clearwater core complex to the east. The metamorphic grade in rocks along the highway decreases to the north toward Santa. High-grade schist and gneiss of the core complex to the east are part of a series of core complexes between the Bitterroot front, well to the east in Montana, and the Priest River core complex in the western side of the Idaho Panhandle to the north.

Between milepost 63 and 83 (south of the town of St. Maries), the highway traverses another extensive area of Clarkia sediments and Wanapum Basalt. North of milepost 73 roadcuts on the west show exquisite pillow lavas in Wanapum Basalt. These lavas formed part of the dam that backed up streams to form Lake Clarkia to the southeast. North of milepost 83, the highway passes a large outcrop on the east with thinly bedded dark and light-gray shale and siltstone of the Snowslip Formation of the lower Missoula Group of the Belt Supergroup.

Wanapum pillow basalts that were part of the lava dam on the ancestral St. Maries River that impounded Lake Clarkia in Miocene time. Bulbous pillows are approximately 2 feet across. (47.1946, -116.4808)

    St. Maries, a classic Northern Idaho lumber town, lies at the confluence of the St. Maries and St. Joe Rivers, less than 5 miles from the inlet of the St. Joe River with Lake Coeur d'Alene. At St. Maries, the highway crosses the St. Joe River. On the north side of the river, St. Joe River Road heads east through mostly rocks of the Wallace Formation of the Piegan Group. ID 3 follows the river downstream to its connection with Lake Coeur d'Alene, passing roadcuts in the lower Missoula Group. During the Pleistocene Epoch, the St. Joe River, along with water from the Coeur d'Alene River, was dammed by gravel left by repeated glacial outburst floods on the Rathdrum Prairie from Glacial Lake Missoula.

    At the south end of Lake Coeur d'Alene, north of milepost 92, ID 3 climbs north out of the St. Joe River valley. Roadcuts along the lower part of the grade are cut into the Grande Ronde and Wanapum Basalts. In the basement rocks hidden below the basalts is the major east-trending St. Joe fault, a right-lateral strike-slip fault that links the Boehls Butte–Clearwater core complex near Clarkia to the southeast with the Priest River core complex to the northwest. At this location, the St. Joe fault juxtaposes lower Missoula Group rocks to the south that slid *west* off the top of the Boehls Butte–Clearwater core complex with Prichard Formation rocks to the north that slid *east* off the top of the Priest River core complex. Thus, the St. Joe fault accommodates the opposite sense of motion for core complexes to the south and north. The first roadcuts of basement rocks appear between mileposts 94 and 95, and these outcrops, which are north of the St. Joe fault, consist of rusty-weathering, laminated quartzite of the Prichard Formation.

    North of the junction with ID 97, the highway passes more exposures of the Prichard Formation and overlying Wanapum Basalt before dropping into the Coeur d'Alene River valley. ID 3 follows the Coeur d'Alene River upstream past a series of lakes and wetlands that have been augmented as part of the heavy-metal

Lakes and wetlands along the Coeur d'Alene River viewed to the west from the viewpoint east of milepost 102. (47.4554, -116.6384)

contamination cleanup efforts on the river. The wetlands trap sediment laden with heavy metals from the Coeur d'Alene mining district on the South Fork of the Coeur d'Alene River, in an effort to limit the contamination of Lake Coeur d'Alene, the Spokane River, and the Rathdrum Prairie aquifer farther downstream. Most outcrops along the road in the Coeur d'Alene River valley consist of biotite quartzite and siltstone of the Prichard Formation.

GLOSSARY

aa. A sluggish lava that forms a blocky, sharp surface as it cools.

alluvial fan. A fan-shaped deposit of sediment that forms where a stream exits its canyon in the mountains onto a valley floor. Deposits typically consist of a mixture of boulders, gravel, and sand.

amphibole. A group of silicate minerals, such as hornblende, common in many igneous and metamorphic rocks.

amphibolite. A metamorphic rock containing a high percentage of amphibole minerals; it is often a metamorphosed basalt or limy mudstone.

andesite. A medium-gray, fine-grained igneous rock with silica, iron, and magnesium content that falls between that of basalt and rhyolite. Crystals of plagioclase feldspar and lesser amounts of amphibole may be visible, surrounded by a finer-grained matrix. When volcanic, it may erupt either as lava, ash, or cinder. It is a fine-grained equivalent of a diorite.

anticline. An upward, arch-like fold in layered rock. The two sides of an anticline, called its limbs, dip outward from the crest or fold hinge.

aplite. A very fine-grained type of granite.

aquifer. Any geologic unit (either sediment or rock) with interconnected pores that store groundwater.

ash. Fine-grained (less than 0.08 inch in diameter), broken fragments of lava violently erupted by volcanoes. Typically, most of these fragments consist of volcanic glass.

asthenosphere. The mechanically weak (ductile or "plastic") part of Earth's upper mantle that lies below the more rigid lithosphere and above the deeper mantle. This weak layer allows the overlying lithosphere to move as tectonic plates.

augen gneiss. A type of gneiss that contains large, eye-shaped grains of feldspar or quartz. In German, *augen* means "eye."

basalt. A dark-gray to black, fine-grained igneous rock with mafic composition. Crystals of plagioclase feldspar, olivine, and pyroxene may be visible, surrounded by a finer-grained matrix. When volcanic, it may erupt either as lava, ash, or cinder. It is the fine-grained equivalent of gabbro.

basement rock. A general term referring to the oldest, often deepest rocks in a region. In Idaho, they often consist of strongly metamorphosed rocks (i.e., schist and gneiss) and intrusive igneous rocks.

basin. A low area where sediment is deposited.

batholith. A large pluton, typically granitic, covering more than 40 square miles. Many batholiths are composed of multiple plutons.

bed. An individual layer in sedimentary rock that is greater than about half an inch thick and has clearly defined upper and lower surfaces.

bedding. The depositional layers, or beds, in sedimentary rocks.

bedrock. The solid rock of the Earth's crust, sometimes underlying loose sediment or soil.

GLOSSARY

biotite. A black to dark-brown, shiny, iron-rich mica mineral that breaks into thin, flexible sheets. It is a common component of granite, gneiss, and schist.

boulder. A rock particle larger than 10 inches in diameter.

breccia. A coarse-grained rock composed of angular rock fragments. Occurs in both sedimentary and volcanic environments, and as the ground-up rock product of faulting.

butte. An isolated hill. In southern Idaho, most buttes are volcanoes.

calcite. A common, light-colored mineral composed of calcium carbonate. It is a major component of limestone.

caldera. A large, collapsed depression that forms after a massive, explosive volcanic eruption.

carbonate. Rocks such as limestone or dolostone composed of calcite (calcium carbonate) or magnesium carbonate.

chert. A hard, fine-grained sedimentary rock composed of microscopic quartz crystals. It forms as a chemical deposit, found primarily as nodules precipitated from water percolating through limy sediments after deposition, but it can also be precipitated directly from water on the seafloor.

chlorite. A platy, green iron- and magnesium-rich mica mineral commonly associated with low-grade metamorphic rocks.

cinder. A marble- to golf ball–sized vesicular basalt fragment thrown out of a volcano.

cinder cone. A small volcanic cone made of cinders that erupt explosively and accumulate around the vent.

cirque. A very steep-walled basin eroded by a glacier near the top of a mountain. Some cirques are now occupied by lakes.

clay. The smallest particle of sediment, less than 0.00016 inch in diameter.

coarse-grained. Term used to describe igneous rocks with relatively large crystals, easy to see with the naked eye.

colonnade. The part of a thick basaltic lava flow that shows near-parallel, typically steep fractures that break into regular-shaped columns.

columnar joints. Columns typically five or six sided in cross section and formed where lava contracts while it cools after crystallizing.

conglomerate. A sedimentary rock consisting of rounded gravel, cobbles, and/or boulders cemented together.

contact. The physical surface separating two adjoining rock bodies.

core complex. See **metamorphic core complex.**

craton. The ancient core of a continent; usually composed of Archean rocks.

cross beds. Inclined beds of sediment oriented at an angle to the primary beds. Cross beds form on the lee side of sand dunes or ripples created by flowing water or wind.

crust. The outer, brittle rock layer of Earth, about 4 miles thick under the oceans and tens of miles thick beneath the continents. Oceanic crust is mafic in composition; most of the continental crust is felsic to intermediate composition.

dacite. A dark, fine-grained igneous rock with a composition that falls between that of rhyolite and andesite. It is the fine-grained equivalent of granodiorite.

deformation. The physical changes to rock caused by applied stress, such as extension, compression, or shearing. Includes faulting and folding, and mostly results from tectonic motion.

deposition. The process by which sediment settles from moving water, wind, or ice. The opposite of erosion.

detachment fault. A nearly horizontal or gently dipping normal fault associated with large-scale extension, often with large displacement (tens of miles) that juxtaposes low-grade metamorphosed rocks above from higher-grade metamorphic rocks below.

detrital. A term pertaining to broken rock and mineral fragments.

diamictite. A poorly sorted sedimentary rock consisting of a muddy groundmass and a wide range of rock fragments.

dike. A sheetlike intrusion of igneous rock that cuts across layers of another rock.

diorite. A plutonic igneous rock with a composition intermediate between granite and gabbro. Typically contains the minerals plagioclase feldspar and amphibole. Coarse-grained equivalent to an andesite.

dip. The steepness or tilt of inclined rock layers.

dolostone. A sedimentary rock composed of the mineral dolomite, a calcium-magnesium carbonate. Most dolostone forms when calcium in limestone is replaced by magnesium.

earthquake. A sudden release of energy caused by movement along a fault.

entablature. The central to upper part of a basaltic lava flow that displays numerous irregularly jointed, small columns of random orientation.

erosion. The physical or chemical (or both) removal of rock, mineral, sediment, and soil by moving water, wind, and ice. The opposite of deposition.

fault. A fracture along which the rock to either side has shifted. This movement commonly produces earthquakes. See also **normal fault** and **thrust fault**.

fault plane. The plane along which faulting has occurred.

fault scarp. A steep slope or small cliff produced by vertical movement along a fault that offsets the ground surface.

fault zone. A zone consisting of multiple faults within which blocks of rock are broken or rotated. Fault zones can be inches to miles wide.

feldspar. The most abundant rock-forming mineral group. Potassium feldspar is typically white, beige, or pink. Plagioclase feldspar, the sodium- and calcium-bearing variety, is typically white or gray. Most feldspars occur as rectangular or square crystals with shiny faces.

felsic. A term used to describe both igneous rocks and silicate minerals that are rich in silica, sodium, potassium, and aluminum; poor in iron and magnesium; and also tend to be lighter in color. Granite is a felsic igneous rock; quartz, potassium feldspar, and muscovite are felsic minerals.

fine-grained. Term used to describe igneous rocks with small to tiny crystals, usually sand size or smaller.

fissure. An elongated fracture where basaltic lava erupts at the surface.

foliation. A sheeted rock fabric consisting of platy minerals that are mostly aligned with one another in a metamorphic rock. It is caused by pressure that is not equal in all directions. The largest pressure direction is oriented perpendicular to the foliation. For example, if the dominant pressure is oriented vertically, foliation will form horizontally.

gabbro. A dark-colored, mafic, plutonic igneous rock composed mainly of plagioclase feldspar and pyroxene. The coarse-grained equivalent of basalt.

garnet. A hard, usually dark-red silicate mineral commonly found in high-grade metamorphic rocks.

glass. Lava that cooled very quickly and has no crystalline structure.

gneiss. A coarse-grained metamorphic rock having a banded or striped appearance. Gneisses form at higher metamorphic grades.

graben. A down-dropped block of Earth's crust bounded on both sides by normal faults. Grabens are produced by extension.

granite. A light-colored, felsic, plutonic igneous rock consisting of approximately equal proportions of potassium feldspar, plagioclase feldspar, and quartz. It is the coarse-grained equivalent of rhyolite.

granitic rock. An igneous rock whose texture and composition is broadly similar to that of granite.

granodiorite. A coarse-grained, light-colored igneous rock that looks very similar to granite, but contains more plagioclase feldspar and less potassium feldspar. It is the intrusive equivalent of dacite.

gravel. Sediment with a diameter between 0.08 and 2.5 inches.

groundmass. See **matrix**.

groundwater. Water occupying open spaces of rocks or sediment in the subsurface, mostly derived from rainfall and snowmelt.

horn. A pointy, pyramidal mountain peak sculpted by alpine glaciers on three or more sides.

hornblende. A common mafic silicate mineral of the amphibole mineral group. It crystallizes into glossy black needles and is commonly found in many igneous and metamorphic rocks.

hot spot. A chain of volcanic centers of progressive ages fed by a plume of hot mantle. Movement of the lithosphere above the relatively stationary mantle plume produces the progressive series of volcanic centers. In Idaho, Yellowstone represents the present volcanic center above the hot spot. A series of progressively older volcanic centers extend southwest along the Snake River Plain.

hyaloclastite. A deposit of angular volcanic material formed by the explosive fragmentation of lava in water. It usually consists of glassy fragments along with ash and other materials.

hydrothermal. Related to hot water below Earth's surface.

ignimbrite. A rock formed by the deposition and consolidation of hot volcanic ash. Often ignimbrite beds have vitrophyre, or glassy welded ash, at the base.

intrusion. A mass of igneous rock formed by the cooling and crystallization of magma injected into existing rock below Earth's surface.

intrusive. Types of igneous rocks that form when magma cools and crystallizes within the Earth. Opposite is *extrusive*.

isotope. A variety of a particular chemical element having a different number of neutrons. Radioactive isotopes, for example uranium-238 (^{238}U), are used to determine the age of rocks and other material.

joint. A fracture.

knickpoint. A change or break in stream gradient or slope. Waterfalls are an example.

lava. Molten rock erupted onto the Earth's surface.

lava dome. A steep-sided volcano formed by the slow extrusion of rhyolitic lava. Also called a **rhyolite dome**.

lava field. A large area of basaltic lava flows.

lava tube. A subsurface conduit, or tunnel, through which lava flowed, eventually draining out and leaving the hollow tube behind. Large lava tubes form caves.

limestone. A sedimentary rock composed entirely of calcium carbonate. Limestone often contains microscopic and/or visible shells of marine organisms as fossils. Many limestones form in marine settings, but they form in freshwater environments as well.

lineation. A linear rock fabric consisting of elongate minerals that are mostly aligned with one another to resemble parallel rods. In shear zones, lineation commonly forms parallel to the direction of shear.

lithosphere. The rigid, low-density, outer layer of the Earth, approximately 60 miles thick, including the crust and uppermost mantle. It breaks into tectonic plates that float on underlying asthenosphere.

mafic. A term used to describe both igneous rocks and silicate minerals that are relatively poor in silica and rich in iron and/or magnesium. They tend to be dark in color. Basalt and gabbro are mafic igneous rocks, and olivine, pyroxene, amphibole, and biotite are mafic minerals.

magma. Molten rock beneath Earth's surface; magma becomes lava if it pours out on the surface.

magma chamber. A large underground mass of molten magma that may feed a volcano.

magmatic arc. See **volcanic arc**.

magmatism. Processes related to the formation and movement of magma.

mantle. The ultramafic—magnesium-rich, silica-poor—rock layer within the Earth below the crust. It includes the asthenosphere.

marble. A crystalline metamorphic rock made of calcite. It is metamorphosed limestone or dolostone.

matrix. A term used to describe the visually dominant portion of a rock. Distinguished from less common, generally larger crystals or rock fragments in the rock.

megaflood. A rare, catastrophic flood.

metamorphic core complex. The high-grade metamorphic rocks exposed at the surface after the overlying low-grade rocks slid off them along a shallow detachment fault.

metamorphism. The recrystallization of a rock at high temperature and/or pressure to produce a new rock that commonly differs considerably in texture from the original but largely retains the original composition. Low-grade metamorphic rocks are those that metamorphosed at relatively low pressures and temperatures; high-grade rocks have experienced relatively high pressures and temperatures. **Meta**, as a prefix, means a particular rock type is metamorphic, such as **metavolcanic** and **metasedimentary**.

mica. A group of shiny silicate minerals, including both biotite and muscovite, among others, that break into thin, flexible sheets.

moraine. A ridge of loose till deposited along the margins of a glacier.

mud. A sediment size that includes a mixture of silt and clay.

mudstone. A fine-grained sedimentary rock composed chiefly of silt and clay.

muscovite. A colorless to silver-colored, shiny mica mineral that breaks into thin, flexible sheets.

mylonite. A generally fine-grained, intensely sheared rock that forms in a shear zone. The minerals in a mylonite are typically drawn out into ribbons during shear by metamorphic processes deep within the Earth.

normal fault. A fault in which the block of rock above the fault slides down relative to the block of rock below the fault. It forms in response to extension of Earth's crust.

olivine. An olive-green iron and magnesium silicate mineral common in mafic igneous rocks.

orogeny. An episode of compressional tectonic deformation during which mountains form.

pahoehoe. A fluid lava that forms smooth or ropy surfaces as it cools.

pegmatite. A very coarse-grained granitic rock with large, interlocking crystals.

permeable. Said of a material that allows water to pass through it easily.

pillow basalt (pillow lava). Rounded and lobate masses of basalt with a glassy exterior and a fine-grained interior that formed where basaltic lava flowed underwater.

placer gold. Small flecks and nuggets of gold found in stream deposits.

plagioclase feldspar. A common white mineral in granitic rocks. It is a silicate mineral containing calcium and/or sodium, and typically forms square or rectangular crystals.

plate. A tectonic plate consisting of the Earth's lithosphere, the brittle outer portion of the Earth that includes crust and upper mantle.

pluton. A body of intrusive igneous rock. Adjective *plutonic*.

potassium feldspar. A white to pinkish silicate mineral containing potassium and aluminum. It forms square or rectangular crystals and is common in granitic rocks.

pumice. A light-colored, lightweight, spongelike volcanic rock explosively erupted from gas-rich magma.

pyroclastic flow. A dense mixture of volcanic ash, fragments, and gas expelled by an explosive volcanic eruption. These flows, which travel up to 100 miles per hour, often form where tall ash columns collapse during an eruption.

pyroxene. A black to dark-green iron and magnesium silicate mineral commonly found in mafic igneous and some metamorphic rocks. Crystals are often square or rectangular.

quartz. A common colorless to white or gray mineral composed of silica. Quartz is the most common constituent of sand.

quartz diorite. A coarse-grained igneous rock that contains mainly plagioclase feldspar and quartz and common hornblende and biotite. It has the appearance of a quartz-bearing diorite.

quartzite. A metamorphic rock consisting of tightly interlocked quartz crystals. Originates as a quartz-rich sandstone.

reverse fault. A steeply dipping fault in which the overlying block moved up relative to the block underlying the fault. Reverse faults commonly form during compression of Earth's crust. It is the shallowly dipping equivalent of a thrust fault.

rhyolite. A light-colored, fine-grained igneous rock with felsic composition. Crystals of potassium feldspar and quartz may be visible, surrounded by a finer-grained matrix. Occurs as lava flows and domes when volcanic; rhyolite magma typically erupts explosively to produce tuffs. It is the fine-grained equivalent of granite.

rift. An elongate zone of extension in the Earth's crust typically hosting normal faults, basins, and fissure eruptions. The crust is thinned in these regions.

ripples. Wavy structures on the surface of a sedimentary bed, formed by either moving water or wind. Collectively, these features are known as **ripple marks** in rocks.

sand. Sediment with a diameter ranging between 0.0025 and 0.08 inch.

sandstone. A sedimentary rock composed of compacted and cemented sand.

scarp. A linear cliff produced by faulting or erosion. The term is an abbreviation of *escarpment*.

schist. A coarse-grained, foliated metamorphic rock typically composed mainly of mica.

shale. A sedimentary rock, consisting of compacted and cemented silt and clay, that breaks into thin sheets along bedding planes.

shear zone. A broad zone, typically tens of yards to miles across, that has accommodated shifting of rocks on either side. Used in this book to describe ductile—as opposed to brittle—deformed fault zones. Mylonite is a common shear zone rock.

shield volcano. A volcano that is much broader than it is tall, resembling a warrior's shield in profile, and composed of overlapping flows of fluid basaltic lava.

silica. Silicon dioxide, the compound that makes up quartz in all its varieties, including chert. Also constitutes the compositional framework of all **silicate minerals**.

sill. An igneous sheet injected parallel to sedimentary or volcanic bedding.

silt. Sediment ranging between 0.00016 and 0.0025 inch in diameter.

siltstone. A sedimentary rock composed of compacted and cemented silt.

slate. A fine-grained, foliated metamorphic rock formed from the low-grade metamorphism of shale. It is tougher than shale because it contains microscopic mica in place of clay minerals.

soil. The loose surface material composed of minerals, organic material, water, and air.

sorting. A term used to describe the uniformity of grain size in a sedimentary rock. A well-sorted rock has a uniform grain size. A poorly sorted rock contains grains of many sizes.

spring. Groundwater that reaches Earth's surface.

stock. A pluton smaller than 40 square miles in area. A cluster of stocks may be large enough to form a batholith.

striation. A linear scratch on a rock surface that results from rocks grinding against one another along a fault or by abrasion caused by rocks in glacial ice moving over bedrock.

strike-slip fault. A near-vertical fault in which the two sides slip laterally with respect to one another. Described as *right-lateral* when one side of the fault moves to the right when viewed from the other side of the fault and *left-lateral* when the opposite is true.

stromatolite. An internally layered, mound-shaped fossil formed by cyanobacteria mats that trap silt suspended in water. The bacterial mats keep growing as they are covered in sediment to produce layers. Among the oldest definitive fossils on Earth.

subduction. The process of sliding an oceanic plate beneath either a continental plate or another oceanic plate at a convergent tectonic boundary. Subduction occurs because oceanic plates are denser than continental ones. In the case of ocean-ocean convergent zones, the older, and thus denser, oceanic plate subducts beneath the younger one.

subduction zone. The plate boundary where subduction occurs. It is marked by a trench where the two plates meet at the surface, a chain of volcanoes on the overriding plate, and frequent earthquakes.

subsidence. The settling, sinking, or downward motion of Earth's surface.

suture zone. A fault zone where two or more terranes have joined.

syncline. A downward trough-shaped fold, resembling a smile, in layered rocks. The two sides of a syncline, called its limbs, dip inward toward the center or fold hinge.

tailings. Waste rock produced during mining.

talus. Angular boulders from rockfall lying on a slope.

terrace. A level bench of older stream deposits at a higher level than the active stream deposits.

terrane. A fault-bounded crustal fragment that has been attached, or moved along, a continental margin. Exotic terranes have moved vast distances before docking with a continent.

texture. The physical characteristics of a rock, including the size, shape, and arrangement of grains or crystals, as well as other features such as layering, foliation, and vesicles.

thrust fault. A shallowly dipping reverse fault.

till. A sedimentary deposit produced by a glacier and characterized by a wide assortment of sediment types and sizes.

tonalite. A coarse-grained granitic igneous rock that contains mainly plagioclase feldspar and quartz with some biotite. Hornblende is absent or uncommon.

travertine. A sedimentary rock produced by the precipitation of calcite at springs or in caves.

tributary. A river or stream that flows into a larger river or stream system.

trondhjemite. A light-colored, coarse grained granitic igneous rock that contains mostly plagioclase feldspar and quartz with little to no mafic minerals. Commonly contains muscovite.

tuff. A volcanic rock composed chiefly of volcanic ash.

tuff cone. A steep-sided volcano formed by the explosive interaction between magma and water.

turbidite. Sands and muds that settle on the seafloor from clouds of sandy, muddy water that flow as underwater density currents. They form alternating layers of sandstone overlain by shale because the coarser, denser sediment settles out first, followed by finer sediment. The sandstone layers grade upward from coarse sand to fine sand.

ultramafic. An igneous rock composed mainly of mafic minerals such as olivine.

unconformity. A contact between two rock bodies representing a period of erosion or lack of deposition during a significant length of geologic time.

uplift. Upward movement of the Earth's lithosphere caused by thickening of the crust or removal of dense material from the base of the lithosphere.

vein. A fracture filled with mineral material precipitated from water, usually quartz or calcite. Some veins contain valuable minerals.

vent. An opening in Earth's surface where lava or ash erupts.

vesicles. Holes or cavities left by gas bubbles in volcanic rocks. Adjective is *vesicular*.

viscosity. A measure of a fluid's resistance to flow. A high-viscosity material flows slowly, whereas a low-viscosity material flows quickly and easily.

vitrophyre. A felsic volcanic rock composed of visible crystals (usually quartz and feldspar) surrounded by a black, glassy matrix. Essentially, it is a hybrid of obsidian and rhyolite.

volcanic arc. A chain of volcanoes that form on the overriding plate adjacent to a subduction zone.

volcanism. Processes involved with the eruption of lava and ash at Earth's surface.

volcano. An opening in Earth's crust where magma reaches the surface.

weathering. The breakdown of rock at Earth's surface by physical or chemical processes.

welded tuff. A rock composed of ash particles fused together by heat.

zircon. A zirconium-bearing silicate mineral formed in igneous and metamorphic rocks. It is durable enough to survive erosion and deposition in sedimentary environments. Zircon contains trace amounts of uranium, which is radioactive and allows the crystallization age of the mineral to be determined.

REFERENCES

State Geologic Map

Lewis, R. S., P. K. Link, L. R. Stanford, and S. P. Long. 2012. *Geologic Map of Idaho*. Idaho Geological Survey Map M-9, 1:750,000.

General Reading

Alt, D. D., and D. W. Hyndman. 1995. *Northwest Exposures: A Geologic Story of the Northwest*. Missoula, MT: Mountain Press Publishing.

Lewis, R. S., M. McFaddan, J. Burch, and C. M. Feeney. 2020. *Idaho Rocks! A Guide to Geologic Sites in the Gem State*. Missoula, MT: Mountain Press Publishing.

Link, P. K., and E. C. Phoenix. 1996. *Rocks, Rails, and Trails*. 2nd ed. Pocatello, ID: Idaho Museum of Natural History.

Maley, T. S. 2018. *Exploring Idaho Geology*. 3rd ed. Boise, ID: Mineral Land Publications.

Miller, M. B. 2014. *Roadside Geology of Oregon*. Missoula, MT: Mountain Press Publishing.

Willsey, S. 2017. *Geology Underfoot in Southern Idaho*. Missoula, MT: Mountain Press Publishing.

Websites

Digital Atlas of Idaho https://digitalatlas.cose.isu.edu

Digital Geology of Idaho http://digitalgeology.aws.cose.isu.edu/Digital_Geology_Idaho

Ice Age Floods Institute http://iafi.org

Idaho Geological Survey https://www.idahogeology.org

US Geological Survey National Geologic Map Database. https://ngmdb.usgs.gov/ngmdb/ngmdb_home.html

Field Guides and Collections of Geological Papers

Bonnichsen, B., C. M. White, and M. McCurry, eds. 2002. *Tectonic and Magmatic Evolution of the Snake River Plain Volcanic Province*. Idaho Geological Survey Memoir 30.

Chamberlain, V. E., R. M. Breckenridge, and B. Bonnichsen, eds. 1989. *Guidebook to the Geology of Northern and Western Idaho and Surrounding Area*. Idaho Geological Survey Bulletin 28.

Hughes, S. S., and G. D. Thackray, eds. 1999. *Guidebook to the Geology of Eastern Idaho*. Pocatello: Idaho Museum of Natural History.

Lee, J., and J. P. Evans, eds. 2013. *Geologic Field Trips to the Basin and Range, Rocky Mountains, Snake River Plain, and Terranes of the US Cordillera*. GSA Field Guide 21.

Lewis, R. L., and K. L. Schmidt, eds. 2016. *Exploring the Geology of the Inland Northwest*. GSA Field Guide 41.

Link, P. K., and W. R. Hackett, eds. 1988. *Guidebook to the Geology of Central and Southern Idaho*. Idaho Geological Survey Bulletin 27.

Link, P. K., and R. S. Lewis, eds. 2007. *Proterozoic Geology of Western North America and Siberia*. SEPM Special Publication 86.

Link, P. K., M. A. Kuntz, and L. B. Platt, eds. 1992. *Regional Geology of Eastern Idaho and Western Wyoming*. GSA Memoir 179.

MacLean, J. S., and J. W. Sears, eds. 2016. *Belt Supergroup: Window to Mesoproterozoic Earth.* GSA Special Paper 522.

Roberts, S. M., ed. 1986. *Belt Supergroup: A Guide to Proterozoic Rocks of Western Montana and Adjacent Areas.* Montana Bureau of Mines and Geology Special Publication 94.

Shaw, C. A., and B. Tikoff, B. 2016. *Exploring the Northern Rocky Mountains.* GSA Field Guide 37.

Worl, R. G., P. K. Link, G. R. Winkler, and K. M. Johnson, eds. 1995. *Geology and Mineral Resources of the Hailey 1° x 2° Quadrangle and the Western Part of the Idaho Falls 1° x 2° Quadrangle, Idaho.* USGS Bulletin 2064.

Snake River Plain

Anders, M. H., J. Saltzman, and S. R. Hemming. 2009. Neogene tephra correlations in eastern Idaho and Wyoming: Implications for Yellowstone hotspot-related volcanism and tectonic activity. *GSA Bulletin* 121: 837–856.

Andrews, G. D. M., and M. J. Branney. 2005. Folds, fabrics, and kinematic criteria in rheomorphic ignimbrites of the Snake River Plain, Idaho: Insights into emplacement and flow. In *Interior Western United States,* GSA Field Guide 6, eds. J. Pederson and C.M. Dehler, pages 311–27.

Beranek, L. P., P. K. Link, and C. M. Fanning. 2006. Miocene to Holocene landscape evolution of the western Snake River Plain region, Idaho: Using the SHRIMP detrital zircon provenance record to track eastward migration of the Yellowstone hotspot. *GSA Bulletin* 118: 1027–50.

Bonnichsen, B., and M. M. Godchaux. 2006. *Geologic Map of the Murphy 30 X 60 Minute Quadrangle, Ada, Canyon, Elmore, and Owyhee Counties, Idaho.* Idaho Geological Survey Digital Web Map 80, 1:100,000.

Cluer, J. K., and B. L. Cluer. 1986. The late Cenozoic Camas Prairie Rift, south-central Idaho. *Contributions to Geology,* University of Wyoming 24: 91–101.

Knott, T. R., M. J. Branney, M. K. Reichow, and others. 2016. Mid-Miocene record of large-scale Snake River–type explosive volcanism and associated subsidence on the Yellowstone hotspot track: The Cassia Formation of Idaho, USA. *GSA Bulletin* 128: 1121–1146.

Lamb, M. P., B. H. Mackey, and K. A. Farley. 2014. Amphitheater-headed canyons formed by megaflooding at Malad Gorge, Idaho. *Proceedings of the National Academy of Sciences of the USA* 111: 57–62.

Malde, H. E. 1968. *The Catastrophic Late Pleistocene Bonneville Flood in the Snake River Plain, Idaho.* US Geological Survey Professional Paper 596.

O'Connor, J. E. 1993. *Hydrology, Hydraulics, and Geomorphology of the Bonneville Flood.* GSA Special Paper 274.

Struhsacker, D. W., P. W. Jewell, J. Zeisloft, and S. H. Evans Jr. 1982. The Geology and Geothermal Setting of the Magic Reservoir Area, Blaine and Camas Counties, Idaho. In *Cenozoic Geology of Idaho,* Idaho Bureau of Mines and Geology Bulletin 26: 377–93.

Wells, R., D. Bukry, R. Friedman, and others. 2014. Geologic history of Siletzia, a large igneous province in the Oregon and Washington Coast Range: Correlation to the geomagnetic polarity time scale and implications for a long-lived Yellowstone hotspot. *Geosphere* 10: 692–719.

Wood, S. H., and D. M. Clemens. 2002. Geologic and Tectonic History of the Western Snake River Plain, Idaho and Oregon. In *Tectonic and Magmatic Evolution of the Snake River Plain Volcanic Province,* Idaho Geological Survey Bulletin 30, eds. B. Bonnichsen, C. M. White, and M. O. McCurry, p. 69–103.

Basin and Range

Amidon, W. H., and A. C. Clark. 2015. Interaction of outburst floods with basaltic aquifers on the Snake River Plain: Implications for Martian canyons. *GSA Bulletin* 127: 688–701.

Brennan, D. T., D. M. Pearson, P. K. Link, and K. R. Chamberlain. 2020. Neoproterozoic Windermere Supergroup near Bayhorse, Idaho: Late-stage Rodinian rifting was deflected west around the Belt basin. *Tectonics* 39: e2020TC006145.

Carney, S. M., and S. U. Janecke. 2005. Excision and the original low dip of the Miocene-Pliocene Bannock detachment system, SE Idaho: Northern cousin of the Sevier Desert detachment? *GSA Bulletin* 117: 334–53.

Coogan, J. C. 1992. *Thrust Systems and Displacement Transfer in the Wyoming-Idaho-Utah Thrust Belt*. PhD dissertation, University of Wyoming.

Coogan, J. C., D. M. Feeney, and R. S. Lewis. 2021. *Geological Maps of the Pegram and Montpelier Canyon Quadrangles, Bear Lake County, Idaho*. Idaho Geological Survey Digital Web Maps, 1:24.000.

Evans, K. V., and G. N. Green, compilers. 2003. *Geologic Map of the Salmon National Forest and Vicinity, East-central Idaho*. US Geological Survey Geologic Investigations Series I-2765, 1:100,000.

Hildebrand, R. S. 2009. *Did Westward Subduction Cause Cretaceous–Tertiary Orogeny in the North American Cordillera?* GSA Special Paper 457.

Hodges, M. K. V., P. K. Link, and C. M. Fanning. 2009. The Pliocene Lost River found to west: Detrital zircon evidence of drainage disruption along a subsiding hotspot track. *Journal of Volcanology and Geothermal Research* 188: 237–49.

Janecke, S. U., B. F. Hammond, L. Snee, and J. Geissman. 1997. Rapid extension in an Eocene volcanic arc: Structure and paleogeography of an intra-arc half graben in central Idaho. *GSA Bulletin* 109: 253–67.

Janecke, S. U., and R. Q. Oaks, Jr. 2011. New insights into the outlet conditions of late Pleistocene Lake Bonneville, southeastern Idaho, USA. *Geosphere* 7: 1–23.

Keeley, J. A., P. K. Link, C. M. Fanning, and M. D. Schmitz. 2013. Pre- to synglacial rift related volcanism in the Neoproterozoic (Cryogenian) Pocatello Formation, SE Idaho: New SHRIMP and CA-ID-TIMS constraints. *Lithosphere* 5: 128–50.

Konstantinou, A., A. Strickland, E. Miller, and J. P. Wooden. 2012. Multistage Cenozoic extension of the Albion–Raft River–Grouse Creek metamorphic core complex: Geochronologic and stratigraphic constraints. *Geosphere* 8: 1429–66.

Link, P. K., C. M. Fanning, and L. P. Beranek. 2005. Reliability and longitudinal change of detrital-zircon age spectra in the Snake River system, Idaho and Wyoming: An example of reproducing the bumpy barcode. *Sedimentary Geology* 182: 101–42.

Pearson, D. P., and P. K. Link. 2017. Field guide to the Lemhi Arch and Mesozoic-Early Cenozoic faults and folds in east-central Idaho; Beaverhead Mountains. *Northwest Geology* 46: 101–12.

Pierce, K. L., and L. A. Morgan. 2009. Is the track of the Yellowstone hotspot driven by a deep mantle plume? Review of volcanism, faulting, and uplift in light of new data. *Journal of Volcanology and Geothermal Research* 188: 1–25.

Rodgers, D. W., and K. L. Othberg. 1999. *Geologic Map of the Pocatello South Quadrangle, Bannock and Power Counties, Idaho*. Idaho Geological Survey Geologic Map GM-26, 1:24,000.

Rodgers, D. W., S. P. Long, N. N. McQuarrie, and others. 2006. *Geologic Map of the Inkom Quadrangle, Bannock County, Idaho*. Idaho Geological Survey Technical Report T-06-2, 1:24,000.

Yonkee, W. A., C. D. Dehler, P. K. Link, and others. 2014. Tectono-stratigraphic framework of Neoproterozoic to Cambrian strata, west-central US: Protracted rifting, glaciation, and evolution of the North American Cordilleran margin. *Earth-Science Reviews* 136: 59–95.

Yonkee, W. A., and A. B. Weil. 2015. Tectonic evolution of the Sevier and Laramide belts within the North American Cordillera orogenic system. *Earth-Science Reviews* 150: 531–93.

Central Mountains

Beranek, L. P., P. K. Link, and C. M. Fanning. 2016. Detrital zircon record of mid-Paleozoic convergent margin activity in the northern US Rocky Mountains: Implications for the Antler orogeny and early evolution of the North American Cordillera. *Lithosphere* 8.

Fayon, A. K., B. Tikoff, M. Kahn, and R. M. Gaschnig. 2017. Cooling and exhumation of the southern Idaho batholith. *Lithosphere* 9: 299–314.

Gaschnig, R. M., J. D. Vervoort, and R. S. Lewis. 2019. Migrating magmatism in the northern US Cordillera: In situ U-Pb geochronology of the Idaho batholith. *Contributions to Mineralogy and Petrology* 159: 863–83.

Gaschnig, R. M., J. D. Vervoort, R. S. Lewis, and B. Tikoff. 2011. Isotopic evolution of the Idaho batholith and Challis intrusive province, northern US Cordillera. *Journal of Petrology* 52: 2397–2429.

Link, P. K., K. M. Autenrieth-Durk, A. Cameron, and others. 2017. U-Pb zircon ages of the Wildhorse Gneiss Complex, Pioneer Mountains, south-central Idaho and tectonic implications. *Geosphere* 13: 681–698.

Link, P. K., M. K. Todt, D. M. Pearson, and R. C. Thomas. 2017. 500-490 Ma detrital zircons in Upper Cambrian Worm Creek and correlative sandstones, Idaho, Montana, and Wyoming: Magmatism and tectonism within the passive margin. *Lithosphere* 9: 910–28.

Vallier, T. L. and H. C. Brooks, eds. 1987. *Geology of the Blue Mountains Region of Oregon, Idaho, and Washington: The Idaho Batholith and Its Border Zone*. USGS Professional Paper 1436.

Vogl, J. J., D. A. Foster. C. M. Fanning, and others. 2012. Timing of extension in the Pioneer metamorphic core complex with implications for the spatiotemporal pattern of Cenozoic extension and exhumation in the northern US Cordillera. *Tectonics* 31: TC1008.

Western Margin

Armstrong, R. L., W. H. Taubeneck, and P. O. Hales. 1977. Rb-Sr and K-Ar geochronometry of Mesozoic granitic rocks and their Sr isotopic composition, Oregon, Washington, and Idaho. *GSA Bulletin* 88: 397–411.

Blake, D. E., K. Gray, S. Giorgis, and B. Tikoff. 2009. A tectonic transect through the Salmon River suture zone along the Salmon River Canyon in the Riggins region of west-central Idaho. In *Volcanoes to Vineyards: Geologic Field Trips through the Dynamic Landscape of the Pacific Northwest*. GSA Field Guide 15, eds. J. E. O'Connor, R. J. Dorsey, and I. P. Madin, p. 345–72.

Dorsey, R. J., and T. A. LaMaskin. 2008. Mesozoic collision and accretion of oceanic terranes in the Blue Mountains province of northeastern Oregon: New insights from the stratigraphic record. *Arizona Geological Society Digest* 22: 325–32.

Garwood, D.L, K. L. Schmidt, J. D. Kauffman, and others. 2008. *Geologic Map of the White Bird Quadrangle, Idaho County, Idaho*. Idaho Geological Survey Digital Web Map Series DWM-101, 1:24,000 scale.

Giorgis, S., W. McClelland, A. Fayon, and others. 2008. Timing of deformation and exhumation in the western Idaho shear zone, McCall, Idaho. *GSA Bulletin* 70: 1119–1133.

Gray, K. D. 2013. *Structure of the Arc-Continent Transition in the Riggins Region of West-central Idaho: Strip Maps and Structural Sections.* Idaho Geological Survey Technical Report 13-1, 1:24,000.

Hamilton, W. 1969. *Reconnaissance Geologic Map of the Riggins Quadrangle, West-Central Idaho.* USGS Miscellaneous Geologic Investigations Map I-579, 1:250,000.

Kasbohm, J., and B. Schoene. 2018. Rapid eruption of the Columbia River flood basalt and correlation with the mid-Miocene climate optimum. *Science Advances* 4: eaat8223.

Kauffman, J. D., K. L. Schmidt, R. S. Lewis, and others. 2014. *Geologic Map of the Idaho Part of the Grangeville 30 x 60 Minute Quadrangle, and Adjoining Areas of Washington and Oregon.* Idaho Geological Survey Geologic Map 50, 1:100,000.

Kauffman, J. D., D. L. Garwood, K. L. Schmidt, and others. 2009. *Geologic Map of the Idaho Parts of the Orofino and Clarkston 30 x 60 Minute Quadrangles.* Idaho. Idaho Geological Survey Geologic Map 48, 1:100,000.

Kurz, G. A., M. D. Schmitz, C. J. Northrup, and T. L. Vallier. 2017. Isotopic compositions of intrusive rocks from the Wallowa and Olds Ferry arc terranes of northeastern Oregon and western Idaho: Implications for Cordilleran evolution, lithospheric structure, and Miocene magmatism. *Lithosphere* 9: 235–64.

Lewis, R. S., R. F. Burmester, E. H. Bennett, and D. L. White. 1990. *Preliminary Geologic Map of the Elk City Region, Idaho.* Idaho Geological Survey Technical Report 90-2, 1:100,000.

Lewis, R. S., R. F. Burmester, R. W. Reynolds, and others. 1992. *Geologic Map of the Lochsa River Area, Northern Idaho.* Idaho Geological Survey Geologic Map GM-19, 1:100,000.

Lewis, R. S., R. F. Burmester, J. D. Kauffman, and others. 2007. *Geologic Map of the Kooskia 30 x 60 Minute Quadrangle, Idaho*: Idaho Geological Survey Digital Web Map 93, 1:100,000.

Lund, K. 2004. *Geology of the Payette National Forest, Valley, Idaho, Washington, and Adams Counties, West-Central Idaho.* USGS Professional Paper 1666.

Lund, K., and L. W. Snee. 1988. Metamorphism, structural development, and age of the continent-island arc juncture in west-central Idaho. In *Metamorphism and Crustal Evolution of the Western United States*, ed. W. G. Ernst, p. 296–331. Englewood Cliffs, New Jersey: Prentice-Hall.

Lund, K. I., J. N. Aleinikoff, K. V Evans, and C. M. Fanning. 2003. SHRIMP U-Pb geochronology of Neoproterozoic Windermere Supergroup, central Idaho: Implications for rifting of western Laurentia and synchroneity of Sturtian glacial deposits. *GSA Bulletin* 115: 349–72.

Lund, K. I., J. N. Aleinikoff, E. Y. Yacob, and others. 2008. Coolwater culmination: Sensitive high-resolution ion microprobe (SHRIMP) U-Pb and isotopic evidence for continental delamination in the Syringa embayment, Salmon River suture, Idaho. *Tectonics* 27: TC2009.

Manduca, C. A., M. A. Kuntz, and L. T. Silver. 1993. Emplacement and deformation history of the western margin of the Idaho batholith near McCall, Idaho: Influence of a major terrane boundary. *GSA Bulletin* 105: 749–65.

McKay, M. P., H. H. Stowell, K. D. Gray, and others. 2017. Metamorphism records thrust faulting during prolonged terrane accretion: Sm-Nd garnet and U-Pb zircon geochronology and P-T paths from the Salmon River suture zone, west-central Idaho. *Lithosphere* 9: 683–701.

Reidel, S. P., V. E. Camp, T. L. Tolan, and others. 2013. Tectonic evolution of the Columbia River flood basalt province. In *The Columbia River Flood Basalt Province*, GSA Special Paper 497, eds. S. P. Reidel, V. E. Camp, M. E. Ross, and others, p. 293–324.

Selverstone, J., B. Wernicke, and E. A. Aliberti. 1992. Intracontinental subduction and hinged uplift along Salmon River suture zone in west-central Idaho. *Tectonics* 11: 124–44.

Vallier, T. L. 1998. *Islands and Rapids: A Geologic Story of Hells Canyon*. Lewiston, Idaho. Confluence Press.

Vallier, T. L., and H. C. Brooks. 1987. *The Idaho Batholith and Its Border Zone*: USGS Professional Paper 1436.

Vallier, T. L., and H. C. Brooks, eds. 1995. *Geology of the Blue Mountains Region of Oregon, Idaho, and Washington: Petrology and Tectonic Evolution of Pre-Tertiary Rocks of the Blue Mountains Region*. USGS Professional Paper 1438.

Vallier, T. L., K. L. Schmidt, and T. A. LaMaskin. 2016. Geology of the Wallowa terrane, Blue Mountains province, in the northern part of Hells Canyon, Idaho, Washington, and Oregon. In *Exploring the Geology of the Inland Northwest:* Geological Society of America Field Guide 41, eds. R. S. Lewis and K. L. Schmidt, p. 211–49.

Northern Idaho

Breckenridge, R. M., R. F. Burmester, R. S. Lewis, and M. D. McFaddan. 2014. *Geologic Map of the East Half of the Bonners Ferry 30 x 60 Minute Quadrangle, Idaho and Montana*. Idaho Geological Survey Digital Web Map 173, 1:75,000.

Buddington, A. M., Da Wang, and P. T. Doughty. 2016. Pre-Belt basement tour: Late Archean-Early Proterozoic rocks of the Cougar Gulch area, southern Priest River complex, Idaho. In *Exploring the Geology of the Inland Northwest*, Geological Society of America Field Guide 41, eds. R. S. Lewis and K. L. Schmidt, p. 265–84.

Doughty, P. T., and K. R. Chamberlain. 2007. Age of Paleoproterozoic basement and related rocks in the Clearwater complex, northern Idaho, USA. In *Proterozoic Geology of Western North America and Siberia*, Society for Sedimentary Geology Special Publication 86, eds. P. K. Link and R. S. Lewis, p. 9–35.

Doughty, P. T., and K. R. Chamberlain. 2008. Protolith age and timing of Precambrian magmatic and metamorphic events in the Priest River complex, northern Rockies. *Canadian Journal of Earth Sciences* 45: 99–116.

Doughty, P. T., and R. A. Price. 1999. Tectonic evolution of the Priest River complex, northern Idaho and Washington: A reappraisal of the Newport fault with new insights on metamorphic core complex formation. *Tectonics* 18: 375–93.

Doughty, P. T., R. A. Price, and R. R. Parrish. 1998. Geology and U-Pb geochronology of Archean basement and Proterozoic cover in the Priest River complex, northwestern United States, and their implications for Cordilleran structure and Precambrian continent reconstructions. *Canadian Journal of Earth Sciences* 35: 39–54.

Gaschnig, R. M., J. D. Vervoort, R. S. Lewis, and B. Tikoff. 2013. Probing for Proterozoic and Archean crust in the northern US Cordillera with inherited zircon from the Idaho batholith. *GSA Bulletin* 125: 73–88.

Hobbs, S. W., A. B. Griggs, R. E. Wallace, and A. B. Campbell. 1965. *Geology of the Coeur d'Alene District, Shoshone County, Idaho*. USGS Professional Paper 478.

Lewis, R. S., R. F. Burmester, R. M., Breckenridge, and others. 2002. *Geologic Map of the Coeur d'Alene 30 x 60 Minute Quadrangle, Idaho*. Idaho Geological Survey Geologic Map 33, 1:100,000.

Lewis, R. S., R. F. Burmester, R. M. Breckenridge, and others. 2020. *Geologic Map of the Sandpoint 30 x 60 Minute Quadrangle, Idaho and Montana, and the Idaho Part of the Chewelah 30 x 60 Minute Quadrangle*. Idaho Geological Survey Digital Web Map 189, 1:100,000.

Lewis, R. S., R. F. Burmester, J. D. Kauffman, and T. P. Frost. 2000. *Geologic Map of the St. Maries 30 x 60 Minute Quadrangle, Idaho*. Idaho Geological Survey Geologic Map 28, 1:100,000.

Lewis, R. S., R. F. Burmester, M. D. McFaddan, and others. 1999. *Digital Geologic Map of the Wallace 1:100,000 Quadrangle, Idaho.* USGS Open-File Report 99-390, 1:100,000.

Lewis, R. S., J. H. Bush, R. F. Burmester, and others. 2005. *Geologic Map of the Potlatch 30 x 60 Minute Quadrangle, Idaho.* Idaho Geological Survey Geological Map 41, 1:100,000.

MacInnis, J. D., Jr., B. B. Lackaff, R. M. Boese, and others. 2009. The Spokane Valley–Rathdrum Prairie Aquifer Atlas 2009 Update. Spokane, WA: City of Spokane. Available at www.spokaneaquifer.org.

Smiley, C. J., and W. C. Rember. 1979. *Guidebook and Road Log to the St. Maries River (Clarkia) Fossil Area of Northern Idaho.* Idaho Bureau of Mines and Geology Information Circular 33.

Smyers, N. B., and R. M. Breckenridge. 2003. Glacial Lake Missoula, Clark Fork ice dam, and the floods outburst area: Northern Idaho and western Montana. In *Western Cordillera and Adjacent Areas*, GSA Field Guide 4, ed. T. W. Swanson, p. 1–15.

Stevens, L. M., J. A. Baldwin, J. L. Crowley, and others. 2016. Magmatism as a response to exhumation of the Priest River complex, northern Idaho. Constraints from zircon U-Pb geochronology and Hf isotopes. *Lithos* 262: 285–97.

Waitt, R. B., Jr. 1985. Case for periodic colossal jokulhlaups from Pleistocene Glacial Lake Missoula. *GSA Bulletin* 96: 1271–86.

Wintzer, N. E., and R. S. Lewis. 2016. Eocene deformation at Tubbs Hill of Coeur d'Alene, Idaho, southeast Priest River Complex. *Geological Society of America Abstracts with Programs* 48: Abstract no. 22-1.

INDEX

Page numbers in bold face include photographs.

aa, **20**, 72, 95
Absaroka thrust fault, 125, 156, 157
accretion, xiv, 9, 184, 187, 249
actinolite, 217, 258
Ahsahka shear zone, 188, 199, **200**, 257
Albion, 180
Albion core complex, 40, 176, 178
Albion detachment, 41, 123, 176, 177, 178
Albion Mountains, 6, 10, 40, 114, 123, 176, 178, 180
alluvial fans, 4, 24, 82, 120, 126, 141, 145, 154, 157, 219, **220**, 225, 284
Almo pluton, 178
Alturas Lake, 244, 287
American Falls, 21, 42, 108, 172, 173
American Falls Lake, 42
ammonites, 224
amphibole, 215
amphibolite, 186, 199, 214, **215**, 246, 293, 297, **321**, **322**, 330, **331**; garnet, **301**, 316, **320**, 323
amygdules, **103**
Ancestral Rocky Mountains, 8
andesite, 2, 252; Challis dikes of, 224, 251, 260 270, 274, 300; of island arc, 3, 207, 215, 218, 225; Miocene sill of, 126, **128**; Oligocene, 227; of Weiser volcanics, 192
Ankareh Formation, 134, 137, **138**
Antelope Creek, 142
Antelope Flats, 146
anticlines, 83, **85**, **144**, 148, 156, **158**, 198, 230–32, **291**, 316, 342
antigorite, 217
Antler orogeny, xiv, 8, 76, 145, 248, 278, 2 85, 290
aplite, **259**, 260, **347**
Apple Creek Formation, 151, 155, 245, 262, **263**
aquifers, 22, 49, 58, 59, 63, 66, 85, 107, 132, 304, 305, 310, 332, 336, 351
Arbon Valley Tuff, 173, 174, **175**
Arco, 47, 49, 78, 142
argentite, 281

ash, **69**; altered, 78; from Challis magmatic event, 10, 261; dating of, 22, 108, 347; from hot spot calderas, 12, 17–18, 43, **45**, 47, 55, 66, 68, **69**, **80**, 92, 93, 99, **124**, 134, 168, **175**, 202, 271; from Mazama, xiv, 207. *See also* tuff
Ashton, 50, 51, 53, 82
Ashton Grade, 50, 51
Aspen Range, 134
asthenosphere, 5, 6, 20
Athol, 304, 334, 336
Atlanta lobe, 3, 239, 249, 250, 267, 287, 288, 328
Atomic City, 56
augen gneiss, 238, **242**, 246, **247**, 249, 259, 297, 320, 321, 322, 338
axial volcanic zone, 49, 56, 58

Baker Creek, 282
Baker terrane, 182, 183, 203
Bald Mountain, 281, 284
Bancroft, 130, 163
Banks, 58, 213, 274, 275, 311, 344
Banner Summit, 270
Bannock detachment fault, 114, 116, 138, 140, 159, 165
Bannock Pass, 149
Bannock Range, 42, 117, 121, 165, 167, 168
barite, 329
Barney Hot Springs, 154
basalt, 2, 3, 12, 15, 17, 19–22, **20**, **21**, 188–92, 302; dikes of, 19, 20, 266, **268**, 269; diktytaxitic, **57**; gas bubbles in, 57, 76, 103, **190**; Pleistocene, 12, **14**, 21, 30, 33, **35**, 40, **42**, 49, **53**, **64**, **70**, **71**, 72, 74–75, **77**, 87, 125, **127**, 131, 162, **267**. *See also* Columbia River Basalt Group; columnar jointing; flood basalt; pillow basalt
Basalt of Lucky Peak, 267
basalt of Portneuf Valley, 118, **119**, 120, 121
basement, 4, 5, 10
Basin and Range, 11, 105–14

368 INDEX

batholiths, 2, 3, 249–50. *See also* Idaho batholith; Kaniksu batholith
Bayhorse, 247, 251, 291
Bayview, 323, 339
Bear Canyon thrust fault, 120
Bear Lake, 133, **135**, 136
Bear Lake fault (East), 132, 135, 136, 137
Bear Lake graben, 136, 165, 172
Bear Lake Plateau, **135**
Bear Lake Valley, 112, 165, 172
Bear River, 129, 133, 134, 135, 136, 139, 140, 159, **162**, 167, 170, 172; ancestral, 120, 121; landslide complex along, 139, 140, 167, 168; lava diversion of, 24, 25, 119, 131, 161, 168
Bear River Massacre, 139, 140, 141
Bear River Range, 131, 136, 161, 165, 167, 168, 170, 172
Bear River Range syncline, 170
Bear River valley, 161, **168**
Beaver Creek Summit, 270
Beaverhead fault, 148, 149, **153**, 265
Beaverhead impact structure, 147
Beaverhead Mountains, 30, 86, 108, 110, 111, 148, 155, 156, 264, **265**; Lemhi Group in, 155, 246; limestones in, 111, **150**, 152; thrust faults in, 153, 154, 249
Beaverhead pluton, xiv, 110, 113, 151, 153
Bellevue, 88, 94, 251, 278, 279
Bell Mountain, 151
Belt basin, xiv, 4, 7, 239, 245–46, 261, 294–99, 305, 308, 325. *See also* Lemhi subbasin
Belt Supergroup, 7, 108, **240**, 245–46, 261, 294–99, **329**; metamorphosed, 242, **247**, 258, **259**, 260, 316, **320**, 328, 330, **331**, 345, **347**, 349; ore in, 264, 305, 311; xenoliths of, **341**; in upper plate, 301, 308, 309, 323, 324–25, **326**, 337. *See also* Lemhi Group; Missoula Group; Piegan Group; Ravalli Group; Pritchard Formation
Berg Creek amphibolite, 214, 215
Big Bar, **207**
Big Cinder Butte, 76
Big Craters, 74
Big Creek, 313
Big Creek Formation, 245
Big Creek suite, 110
Big Grassy Butte, 29
Big Hole Mountains, 50

Bighorn Crags, 264
Big Lost River, 47, 49, 64, 85, 142, 145, 284
Big Southern Butte, 26, **28**, 42, 49, 56, 77
Big White arc, 264
Big Wood River, 43, 46, 94, 95, 96, 97, 278
Billingsley Creek, 58
biotite: in Challis dikes, **251**, 252; granodiorite, **46**, 239, 249, **250**, 259, 269, 323, 334, **341**; in granite, 260, 328; in metamorphic rock, 215, 216, 239, **240**, 242, 297, 330 349; in tonalite, **187**, 216, **237**, 255, **258**, 274; in two-mica granite, 249, **250**, 259, 275, 328; in ultramafic rock, 217
Birch Creek, 85, 86, 148, 152
Bitterroot complex, 301
Bitterroot lobe, xiv, 3, 188, 249, 250, 259, 260
Bitterroot Mountains, 194, 242, 255, 293, 305, 342
black and tan couplets, **298**, 314, 325
Black Butte Crater, 72, 73, 94, 95, 96, 97
Black Canyon (Bear River), 131, 160
Black Canyon (Payette River), 273
Blackfoot, 27, 58
Blackfoot lava field, 131, 162, 163
Blackfoot Mountains, 30
Blackfoot River, 27, 30, 162, 163
Black Magic Canyon, 94, **96**
Black Pine Mountains, 40, 123, 124
Blackrock Canyon Limestone, 120
Blacksmith Limestone, 165, **167**, **170**, 172
Blacktail Creek Tuff, 86
Blacktail pluton, 187, **237**
Bliss, 58, 59, 61, 62, 88
Bliss landslide, **61**, 62
Bloomington Formation, 109, 170, **171**, 172
Blue Mountains Province, 8, 182, 183, 208, 233, 249
Blue Mountains terrane, xiv, 188, 208, 217, 249, 271
Boehls Butte–Clearwater complex, 6, 10, 301, 349, 350
Bogus Basin, 251
Boise, 16, 23, 31, 33, 197, 251, 267, 271
Boise Basin, 197, 265, 268
Boise front, 271
Boise Mountains, 16, 22
Boise River, 31, 33, 80, 81, 267, 269, 270
Bonner Formation, 297, 325, 329
Bonners Ferry, 300, 303, 340, 341
Bonneville, Lake, 13, 24–25, 114, 161, 167; dam of, 141; deposits of, 138, 140, 159,

161, 165, **168**; shorelines of, 114, 116, 125, 138, **140**, 141, 165, 167, 168, **169**, 173, **175**, 176
Bonneville Flood, xiv, 13, 24–25, 38–39, 141, 193; channels of, 36, 39, 66, 71, 72; deposits of, 79, **81**, 93, 97, 120, 194, **196**, 225, 228; erosion by, 67, 72, 99, 118, 119, **207**, 225; Melon Gravel of, **26**, 35, 42, 59, **60**, 66, **101**; megaripples from, **226**
Borah Peak, 105, 143, 145, **146**, 254
Borah Peak earthquake, 11, 106, 107, 142, 143, 145
Borah Peak syncline, **146**
Boulder batholith, 249
Boulder Mountains, 10, 46, 252, 282, **283**
Box Canyon Springs Nature Preserve, 63–65, **65**
brachiopods, 136, 162, 170, 207
breccia, 18, 40, 42, 123, 131, 137, 147, 190, 205, 211, 231
Brigham Group, 109, 117, 120, 161, 162, 168, 169, 172, 176
Brownlee Dam, 205, 208
Browns Bench, 66
Bruce's Eddy, 199, 200
Bruneau, 93, 104
Bruneau Dunes State Park, **102**, 103
Bruneau-Jarbidge volcanic field, 16, 66, 91, 92, 93
Bruneau River, 24, 90, 91, 93, 104
Buck Peak, 160
Buffalo Cave, 75
Buhl, 58, 63, 66, 88
Bunker Hill Mine, 310, 312, 313
Bunker Hill Superfund Site, 312
Burke, 314, **315**
Burke Formation, 297, 310
Burley, 38, 39, 114
Buttermilk Slough, 81

Cabinet Gorge Dam, 325
Cabinet Mountains, 293, 301, 335, 338
Cache Valley, 114, 117, 138, 140, 141, 165, 167
Caddy Canyon Quartzite, 109
calcite, 111, 199, 217, 218, 286, 295, 325, 347
calcium carbonate, 57, 132, 170, 199, 295
calc-silicate, 199, 239, **347**
calderas, 12, 16–19, 55. *See also* Heise volcanic field; Henrys Fork caldera; Island Park caldera; Picabo volcanic field; Twin Falls volcanic field; Twin Peaks caldera; Yellowstone caldera
caliche, 57
Camas Creek, 45, 46, 88
Camas Prairie (Grangeville), 200, 227, **229**, **230**, 233
Camas Prairie (Fairfield), 43, 44, 45, 89, 90
Cambridge, 182, 204, 208, 272
Camelback Mountain Quartzite, 109
Canfield Butte, 332
Canyon Creek, 270, 314, 315
Cape Horn Creek, 270
carbonate bank, 108, 111, 112, 151, 207, 248
Carbonate Mountain, 280
carbonates, 8, 174, 176, 245, 295, 297, 298, 299. *See also* calcium carbonate; dolostone; iron carbonate; limestone
Carey, 43, 46, 72, 74, 279
Carey Lake, 72
Caribou mining district, 162
Caribou Mountain, 162, 163
Cascade, 254, 272, 274, 275
Cascade Range, 11, 17
Cascade Reservoir, 275
Cascade Valley, 272, 275
Cassia Mountains, 39, 66, 70
Castle Rocks State Park, 176, 178
Cat Creek Summit, 43, 45
caves, **20**, 21, 74, 75, 94, 95, 136, 218
Cedar Butte, 56
Celebration Park, **101**
Centennial Mountains, 31, 55, 56
chalcedony, **69**
chalcopyrite, 311
Chalk Hills Formation, 22, 23, 92, **93**, 100
Challis, 145, 147, 152, 261, 289, 291
Challis Hot Springs, 147, 261
Challis magmatic event, xiv, 3, 10, 242, 251–52, 258, 270, 274, 282, 300; dikes of, 224, 251, 260, 270, 274, 300; granite of, 281, **283**, 300. *See also* Challis Volcanic Group
Challis Volcanic Group, 144, 145, 146, 155, 251–53, **261**, 263, **264**, 281, 282, **283**, **287**, 290, 346
Channeled Scablands, 13
charcoal kilns, **152**
chert, 4, 96, 109, 183, 235
Chesterfield, 130, 131
Chilly Buttes, 145
China Hat, 162, **164**

chlorite, 213, 215, 217, 235, 236, 258, 316, 335
cinder cones, 20, 33, 72, 74, 75, 76, 77, 130, 161
cinders, 33, 75, 76, **77**, 161
cirques, 152, 287, **292**, 318, 338
City of Rocks National Reserve, 178, **179**, 180
clams, 207, 224
Clark Fork River, 13, 316, 304, 323, 324, 325
Clark Hill rest area, 125, 127
Clarkia, 300, 302, 347, 348, 349, 350
Clarkia, Lake, 347–48, 349, 350
Clayton, 290
Clayton Mine Quartzite, **291**
Clearwater Canyon, 199, 230
Clearwater core complex, 6, 10, 300, 349, 350
Clearwater embayment, 194, 231
Clearwater River, 194, 197, **198**, 199, **201**, 202, 230, 231, 232, 257; North Fork, 199; shear zone along, 185, 188; South Fork of, 187, 233, 235, 238, 241
Clifton, 141, 166
coal, 3, 111
Cobalt mining district, 264
Cocolalla Lake, 334, 337
Coeur d'Alene, 294, 304, 305, 307, 332
Coeur d'Alene, Lake, 304, 305, 307, **308**, 309, 330, 350
Coeur d'Alene mining district, 299, 305, 307, 309, 310, 311, 312, 313, 351
Coeur d'Alene Mountains, 200, 305
Coeur d'Alene River, 307, 309, 350, **351**; North Fork of, 310; South Fork of, 293, 310, **311**, 351
Cold Canyon, 286
colonnade, **190**, 231
Columbia (supercontinent), 245, 247, 293, 294
Columbia, Glacial Lake, 304
Columbian mammoth, 193, 227, **229**
Columbia Plateau, 1, 3, 12, 181, 185, 189, 193, 194, 202, 209, 302, 326, 328, 342, 343
Columbia River Basalt Group, xiv, 12, 17, 188–92, 302, 346, 347; breccia of, 211; canyons cut in, **181**, 194, 201, **220**, **222**, 273; canyons filled by, 202, 219, 255; folds in, 198, 231; plateau of, 227, 233; sedimentary units of, 271, 342, **344**, **346**. *See also* Grande Ronde Basalt; Imnaha Basalt; Payette Formation; Saddle Mountains Basalt; Wanapum Basalt
Columbia River basin, 189
columnar joints, **20**, 21, 52, **119**, 178, **180**, **190**, **267**
Conda Mine, 162
conglomerate, 3, 121, 128, 146, 165, 174, 184, 207, 218, 224, 225, 248, 278, **280**
Connor Summit, 178, 180
Connor's Columns, 178, **180**
Continental Divide, 15, 19, 31, 55, 86
Coolwater culmination, 257–58, 259
Coolwater Ridge gneiss, **258**
Coon Hollow Formation, 184, 186, 209, 222, 224, **225**
Copeland granodiorite, **3**, **341**
copper, 144, 152, 184, 205, 206, 207, 236, 311
Copper basin, 145, 280
Copper Basin fault, 249, 286, 290, **291**
Copper Basin Group, xiv, 8, 72, 76, 145, 248
Copper Falls, 339, 342
corals, 136, 142, 162, 207, **224**
Cordilleran ice sheet, 196, 293, 303, 334
core complexes, 10, 113–14, 300–301. *See also* Albion core complex; Boehls Butte–Clearwater core complex; Pioneer core complex; Priest River core complex
Cotterel Mountains, 40, 123, 176, 178, 180
Cottonwood Butte, 227, 228, **229**, 230
Cougar Creek complex, 222, 225
Cougar Gulch gneiss, 327
Council, 23, 208
Council Mountain, 208, 276
Crater Rings, 33, 34
Craters of the Moon National Monument and Preserve, 21, 72, 74–75, 77
cratons, 6, 31, 113, 116, 186, 281
crinoids, 136, 162
Croesus granodiorite, 278
Crooked River, **241**
cross bedding, 35, 104, 169, 174, 246, 297, 299, 328, 342
Cryogenian, 108, 121
Cuddy Mountains, 182, 204, 208
Cuprum mining district, 184, 206
Curlew National Grassland, 174
Custer mining district, 291
Custer Motorway, 261, 291

dacite, 251, 252, 260, 274, 281, 282, 284, **287**, 300, 346; dikes of, 252, 260
Dayton, 167
Deadman's Hole, **253**
Deary Grade, 344, 345, 346
Deer Trail Group, 299, 319
deformation, 10, 112, 185, 187, 238, 257, 299, 320, 321, 323
desert varnish, **101**
detachment faults, 107, 113–14, 252, 253; Albion, 40, 41, 123, 176, 177, 178; Bannock, 114, 116, 138, 159, 165; Newport, 316, 318; Priest River, 297, 300, 301, 308, 323, 335, 338; Salmon Basin, 156, 264, 265; Wildhorse, 279, 281, 282
detrital zircon, 113, 151, 155
Devil Creek Reservoir, 117
Devils Corral, 39
Devils Washbowl, **37**
Dewey Mine, 236
diamictite, **7**, 8, 108, 109, 117, 121, 121, 141, 167
Diamond Peak, **86**, 148, 149
diatomite, 202, **204**
dikes: of basalt, 19, 20, 266, **268**, 269; of dacite, **251**, 252, 260, 300; of andesite, 224, 251, 260 270, 274, 300; granitic, 240, 259, 260, 308, 322, 323, 332, 341, 347; trondhjemite, **237**. See also aplite; pegmatite
Dinosaur Ridge, 97
dinosaurs, 97, 162, 163
diopside, 347
diorite, 2, **46**, 184, 199, 220, 227, 233, **235**, 252, 260, 300, 341. See also quartz diorite
Dinwoody Formation, 109
Dollarhide Formation, 278, 281, 282, 287
dolomite, 109, 151, 282
dolostone, 8, 77, 83, 108, 111, 146, 147, 170, **171**, 313, 314
domes, rhyolite, 26, **28**, 40, 42, 45, 49, 50, 56, **57**, 77, **97**, 162, 164
Donkey Hills, 114, 143
Donnelly, 275
Downey, 120, 142, 173
Doyle Creek Formation, 183, 184, 207, **224**, 225, 233
Driggs, 82
Drummond, 82
Dubois, 86

Duck Valley, 90, 92
dunes, 4, 50, 85, **102**, 103
Dworshak Reservoir, 199

Eagle, 271
Eagle Creek Member, 285, 286
earthquakes, 5, 11, 24, 105, 106, 107, 114, 141, 142, 145, 254, 271, 289
East Butte, **14**, 26, 42, 56, 87
eastern Snake River Plain aquifer, 22, 49, 58, 63, 85
East Fork of the Salmon River, 146
East Fork of the Wood River, 281
East Newport fault, 300, 316, 318, 319
Eden, 38, 39, 59, 71, 72, 135, 177
Eden Channel, 38, 39, 71, 72
Edwardsburg, 8, 110, 247, 251, 281
Eimers-Soltman Park, 227, 229
Elk City, 238, 242, 243, 251, 254
Elkhorn alluvial fan, 145
Elkhorn Peak, 116
Ella marble, 282
Ellis, 162, 261, 263, 264
Emerald Creek, 349
Emmett, 23, 273
enclaves, 235
entablature, **190**, 231, 328, 329, 342
epidote, **187**, 201, 235, 236
Equus simplicidens, **63**
erosion, 65: dissection, **220**; glacial, 12, 13, 55, 271; headward, 23, 192; by megafloods, 13, 61,67, 71, 72, 99, 118, 119, **207**, 225; period of, 135, 183, 188, 224; in Proterozoic time, 263; uplift and, 146, 192, 239, 254. See also landslides
exfoliation, 239

Fairfield, 88
Falls Park, **307**
Farragut State Park, 323, 334
faults, 6. See also detachment faults; normal faults; reverse faults; strike-slip faults; thrust faults
fault scarp, **107**, **132**, 145, 146, 161
feldspar, 3, 205, 206, 208, **250**, 322, 335; in Belt rock, 239, 325, 330; in sand, 342. See also plagioclase feldspar; potassium feldspar
Felt, 145, 271
Ferdinand, 227, 229, 234
ferns, 162, 224, 225

Ferry Butte, 26
Fiddle Creek Schist, 185, 186, 214, 215, 218
fish (fossils), 22, 63, 110, 189, 302, 347
fish, tectonic, **285**, 286
Fish Creek, 72, 130, 131, 260
Fish Creek Summit, 131
Fish Haven Dolomite, 109, 151
fissures, 3, 12, **20**, 21, 40, 74, 75, 108, 181, 188, 302
Flat Top Butte, 59, 72
flood basalt, 11, 12, 188, 189, 302, 343
Florence mining district, 219
foliation, 4, 186, 215, 216, 241, 257, 277, 278, 308, 335
Fort Hall, 26, 28, 110, 130
Fossil Bowl, 348
Fossil Butte, 104
fossils, 4, 23, 35, 42, 183, 189, 207, 225, 302, 347–48. *See also* ammonites; brachiopods; clams; corals; crinoids; dinosaurs; ferns; fish; invertebrates; *Helicoprion*; leaves; mammals; mollusks; oncolites; plants; stromatolites; vertebrates
Fourth of July Pass, 305, 309
Frank Church River of No Return Wilderness, 242, **243**, 244
Franklin, 138, 140, 165, 166
Freeman Peak, 264
French John Hill, 78
Frontier Formation, 109, 157, **158**

gabbro, 2, 184, 199, 227, **235**, 260, 293, 295, **301**, 320, 323
galena, 152, 244, 281, 282, 287, 310, 311
Galena Summit, 244, 282, 287
Gannett Group, 162, **164**
Garden Creek Gap, **118**, 119
garnets, 214, 215, 216, 237, 238, 299, **301**, 316, **320**, 322, 323, **349**
Gay Mine, 28
Gem Valley, 119, 130, 131, 161, 168
Geneva Summit, 137
geodes, 67, 68, **69**, 143, 144
Georgetown, 132, 134
Georgetown Summit, 134
geysers, 132, 163
Gibbonsville, 246, 265
Gilmore, 152
Givens Hot Springs, 98, 100
glacial outwash, 227, 271, 323, 326, 334, 337, 340

glaciation, 12, 117, 121, 167, 254, **286**, **326**
glaciers, xiv, 12–13, 24, 86, 123, 145, 170, 192, 254, 271, 318, 338; moraines left by, 82, 277, 281, 284, 287; scour by, 334. *See also* cirques; moraines; Purcell lobe
Glade Creek fault, 256, 259
glass, 2, 17, 202
Glenns Ferry, 35
Glenns Ferry Formation, 22, **35**, 58, 63, **99**, 100, 104
gneiss, 211, **215**, **216**, 297, **307**, **320**, **321**, **322**, **331**, 349; Archean, 83, 282, 293, 330; augen gneiss, 238, **242**, 246, **247**, 249, 259, 297, 320, 321, 322, 338, 343; biotite-bearing, 239, 240, 297; calc silicate, 199, 259; hornblende-bearing, 186, 199, 211, 236, 237, 255. *See also* Coolwater Ridge gneiss; Hauser Lake gneiss; Laclede gneiss; Pend Oreille gneiss
Goff Bridge, 216
gold, 10, 77, 124, 184, 197, 205, 207, 236, 242, 245, 265, 299, 311, 329; first in Idaho, 195, 201, 257; in placer deposits, 148, 162, 201, 218, 219, 234, 238, 242, 243, 254, 268, 289, 290, 310
Goldbug Hot Springs, 262, 263
Gold Cup Quartzite, 294, 297, 316, **320**, 330
Gold Hill stock, **229**, 327, 329
Gooding, 37, 58, 94
Gooding Butte, 38, 88
grabens, 16, 33, 66, 92, 106, 107, 117, 125, 135, 145, 165, 172, 202, 209, 211, 275, 278; half, 107, 148, 172
Grace, 121, 130, 131, 161, 163
graded bedding, 341
Grande Ronde Basalt, 191, 192, 199, 208, 219, 220, 221, 227, 231, 233, 302, 342, **344**, **345**, **346**
Grand Prize Formation, 282, 287
Grand Valley fault, 125
Grand View Canyon, **147**
Grand View Member, 146, 147
Grangeville, 188, 209, 227, 229, 233, 236, 250
granite, 2, 18, 249, **250**, 299, 328; Archean, 176, 245, 281, 293; of Beaverhead pluton, 151–52, 153; dikes of, 240, 259, 260, 308, 322, 323, 332, 341, 347; Eocene, **13**, 252, 260, 264, 282, **283**, 287, 330; exfoliation of, 239; ice-rafted, **194**; of Little Goose Creek Complex, 275, 277; metamorphosed, 242, 246, 320, 321; mylonitized,

335; Oligocene, 114, 178, **179**, 180; in Owyhees, 24, **100**; two-mica, 88, 94, 186, 214, 249, 339. *See also* Idaho batholith; Kaniksu batholith
granite gneiss, 293, 330
Granite Mountain, 211
granodiorite, 2, **3**, 274, 288; biotite in, 46, 239, 249, **250**, **259**, 269, 323, 334, **341**; dikes of, 260; Eocene, 265, 300, 316, 318, 322; exfoliation of, **239**; of Idaho batholith, **43**, **46**, 216, **243**, 249, **250**, 267, **268**, 269, 270, **290**; of Kaniksu batholith, 299, 332, **341**; weathering of, 270. *See also* Copeland granodiorite; Croesus granodiorite; Hailey granodiorite; Silver Point pluton; Wrencoe pluton
graphite, 211, 218, 297
Grasmere escarpment, 92
Grassy Cone, 76, 77
gravels: of alluvial fans, 129, 141, 318; glacial, 270, 304, 323, 340; of lakeshores, 100, 138; of megafloods, 101, 193, 194, **196**, 207, 225, **226**, 304, 305, 332, 334, 336, **337**, 350; of rivers, 33, 50, 78, 87, 92, **93**, 96, 104, 111, 208, 218, 231, 267, 268, **269**, 271, 277, 346, 349. *See also* conglomerates; Melon Gravel; placer deposits
Grays Lake National Wildlife Refuge, 162
Great Falls tectonic zone, xiv, 244, 252
Great Rift, **21**, 40, 75, 108
Great Salt Lake, 24, 25, 114, 124, 131, 193; basin of, 121, 122, 124
Green Monarch Mountain, 323, **324**
Greer, 201
Griffin Butte, 281
groundwater, 45, 63, 81, 145, 168; in aquifers, 22, 49, 85, 93, 107, 304–5, 332; and hot springs, 16, 66, 260, 270; mineral deposition by, 31, 68, 125, 290; magma encountering, 40, 42, 87, 99, 103; oxidation by, 269, 345. *See also* eastern Snake River Plain aquifer
grus, **270**
Guffey Butte, 103
Gunsight Formation, 155, 245

Hagerman Fossil Beds National Monument, 35, 58, 63
Hagerman horse, **63**
Hailey, 250, 252, 278, 280

Hailey Conglomerate, 278, **280**
Hailey stock, 252, 278, 279
half grabens, 107, 172
Hammer Creek, 222
Hammer Creek assemblage, 183, 184, 186, 221, 224, 225
Hammett, 35, 97, 98
Harpster, 233
Harpster pluton, 233, **235**
Harriman State Park, 50, 56
Harrison Lake, **292**
Hat Butte, 100, 103
Hatwai Canyon, 232
Hauser Lake gneiss, 301, 309, 316, 320, **322**, 323, 330, **331**, 332
Hawkins Basin volcanic center, 119
Hawley Creek thrust fault, 153, 249
Hayden Lake, 304, 305, 332
Hazard Creek Complex, 187, 208, 211, 215, **216**, 278
Hazelton, 38, 39, 177
Heavens Gate Lookout, 213
Heavens Gate plate, 212, 218
Heavens Gate thrust fault, 187, 218
heavy metals, 110, 309, 310, 313, 351
He Devil, 193
Heise Cliffs, 82, 125, **127**
Heise Group, 86
Heise Hot Springs, 125
Heise volcanic field, 16, 31, 82, 85, 86, 125, 129, 148, 152
Helena Formation, 297, 325
Helicoprion, 110
Hellhole thrust, 137
Hells Canyon, xiv, 15, 23, 24, **181**, 184, 192–93, **193**, 194, 202, 220, 222–26, 230, 231
Hells Canyon Reservoir, **205**, 206
Hells Canyon Road, 205–7
Hells Gate State Park, 197
Hells Half Acre, 28, 30, 50, 58
hematite, 330
Henrys Fork (Snake River), 30, 50, 53, **54**, 84, 85, 87
Henrys Fork caldera, 16, 50, 54, 55, **56**
Hollister, 59, 70
Home Canyon thrust, 137
Homestead mining district, 184
Hoodoo Channel, 334
Hoodoo Quartzite, 262
Hope fault, 300, 324

hornblende, 201, 235, 320, 329; in andesite, 252, 270; in dacite; **251**, 252; in gneiss, 186, 199, 211, 236, 237; in granodiorite, 216, 249, **250**, 274, 323; in tonalite, 216, 238, 299
Horseshoe Bend, 271, 273, 274
horsts, 106, 107
hot spots, 6, 12, 16–19, 45, 86, 188, 254. *See also* Yellowstone hot spot
hot springs, 16, 66, 161, 270. *See also specific hot spring names*
Howe Point, 85
Huckleberry Ridge Tuff, 50, 52, 55, 81, 83, **128**, 157, **159**
Hudspeth Cutoff, 131, 174
Hughes Creek, 265
hummocky topography, **273**
Hunsaker Creek Formation, 184, 205, 218, 235
Hurwal Formation, 183, 184, 185
hyaloclastite, 59, **60**, 125, 126, **127**
hydrothermal, 86, 103, 144, 148, 152, 235, 261, 290
hydrovolcanic vents, 23, 42, 59, 97, 98, 99, 103, 104
Hyndman shear zone, 279, 282

ice age, xiv, 12–13, 24, 136, 192–93, 227, 302–4, 323–24. *See also* Bonneville, Lake; Bonneville Flood; glaciers; Missoula floods; Terreton, Lake
ichthyosaurs, 207
Idaho: state capitol, 197; state fossil, **63**; state mineral, **349**
Idaho, Lake, xiv, 12, 22–24, 31, **35**, 63, 92, **93**, **99**, 100, 103, 192, 202
Idaho, Mt., 188, 227, 236, 237
Idaho batholith, xiv, 3, 9, 10, 187, 188, 249–51. *See also* granite; granodiorite; tonalite
Idaho City, 251, 268, 272
Idaho Falls, 30, 50, 52, 125
Idaho fold-and-thrust belt, 113, 129
Idaho National Laboratory, 47, 56, 148
igneous rocks, 2–3
ignimbrite, 123, 125, 129, 172, 173, 175, 261
Imnaha Basalt, 191, 192, 199, 208, 211, **212**, 219, 220, 230, 302, 342
impact structure, 147
inclusions, **46**, **341**
Inkom, 118, 119, 120, 130, 142

Inkom Formation, 109
insects, 189, 302, 347
interbeds, 17, 189, **190**, 198, 199, 233, 342, **344**, 345, **346**
invertebrates, 83, 136, 170
inverted topography, 119
iron carbonate, 278, 311
Irondyke Creek, 207
Iron Lake thrust fault, 261, 263
iron oxides, 298, 330
iron sulfides, 311, 298, 330
Irwin, 126, 128, 157
Island Park, 50, 53
Island Park caldera, 16, 54, 55, 81, 128, 159
isotopes, 182, 185

Jackpot, 66, 67
Jefferson Formation, 146, 147, 152
Jerry Johnson Hot Springs, 256, 260
Jim Sage Mountains, 176
John Day Creek, 218
Johns Creek, 239
joints, 20, 21, 52, 178, 180, 282, **283**
J. R. Simplot Company, 28
Juan de Fuca Plate, 10, 11
Juliaetta, 344
Jump Creek Rhyolite, 78

Kamiah, **201**, 202, 227, 229, 303
Kamiak Butte, **229**, 328
Kaniksu batholith, 3, 10, **292**, 299, 318, 338, 339, 341
Kasiska Quartzite, 168, **169**
Kellogg, 310, 312–13
Kellogg Mountain, 312, 313
Kelly Forks fault, 300
Kelso Lake pluton, 334
Kendrick, 344
Ketchum, 8, 113, 114, 278, 281, 284
Kilgore Tuff, 86
Kings Bowl, **21**, 74
Kingston, 310, 311
Kinnikinic Quartzite, 148, 151, 152, 154, 245, 282, 290, **291**
Kinport Peak, **28**
Kinzie Butte, 94, 95
kipukas, 76, **77**, 78
Kirkham Hot Springs, 266, 270
Klopton Creek thrust, 222, 224
knickpoints, 65, 211, 241, 274, 275
Kooskia, 182, 202, 233, 255

Kootenai River, 339, 340, 341
Kootenay, Glacial Lake, 340
Kuna–Mountain Home rift zone, 33
kyanite, 257, 349

Laclede gneiss, 293, 297, 316, 317, 320, **321**, 322, 338
lagerstätte, 302
lakebeds, 141, 156, 165, 264, **265**, 348
Laketown Dolomite, 109, 151
laminations, 341
Land of the Yankee Fork State Park, 148, 291
landslides, **61**, **104**, 140, 144, 167, **198**, 202, **207**, 213, 219, 221, 222, 225, 226, **240**, 241, **273**, 275, 340, 341
Lapwai, 229, 230, 327
Lapwai Creek, 229
Laramide orogeny, 10, 185, 249
Latah Formation, 189
lava, 2, 12, 15, 18, 20–21. *See also* andesite; basalt; dacite; lava domes; lava fields; lava tubes; rhyolite
Lava Creek Tuff, 50
lava domes, 18, 94
lava fields, 1, 21, 28, 40, 49, 50, 58, 72, 74, 75, 76, 94, 95, 96, 131, 162
Lava Hot Springs, 129, 130, 131
lava trees, 75
lava tubes, **20**, 59, 75, **95**, 189
Lawson Creek Formation, 151, 245, 261, 265
Lawyer Creek, 229
lead, 10, 94, 148, 151, 152, 278, 281, 284, 287, 290, 293, 305, 309–13
Lead Bell Formation, 109
Leadore, 110, 148, 152, 153, 154, 155
Leatherman Peak, **112**
leaves (fossil), 33, 78, 156, 174, 202, 225, 267, 336, 337, 347, 348
Lemhi arch, xiv, 85, 110, 111, 148, 152, 153
Lemhi Group, 151, 245, 246, 263, 264, 265
Lemhi Pass, 155, 262
Lemhi Pass fault, **155**
Lemhi Range, 30, 83, 85, 86, 108, 109, 111, 112, 148, 152, 153, 154, 245, 246, 254, 261, 264
Lemhi subbasin, 148, 155, 209, 238, 239, 240, 245, 246, 252, 264, 280
Lemhi Valley, 142, 155, 156
Lewis and Clark zone, 305, 309, 324
Lewiston, 181, 193, 194, 197–98, 199

Lewiston basin, 194, 198, 229, 230, 231, **232**, 233, 326, 327
Lewiston-Clarkston Valley, 196
Lewiston Hill, 198, 227, 230–33, 326, 342
Lewiston Hill anticline, 198, 230–31, 232
Lewiston syncline, 229, 230, 233
Libby Formation, 297, 299, 325, 326, 329
Liberty Butte, 97, 98
Lidy Hot Springs, 84, 86
Lightning Creek Schist, 185, 186, 214, 215, 217, 218
limestone, 8, **47**, 83, 108, 109, 111, 136, **150**, 151; Cambrian, 117, **118**, 119, 128, 138, **141**, 165, **167**, 168, **170**, 207, 224; caves in, 136; lake deposited, 86; Mesozoic, 134, 135, 137, **164**, 183, 184, **205**, 207, 224; metamorphosed, 4, 185, 214, 218, 259; Mississippian, **47**, 85, 129, 136, **137**, 142, 144, 146, 148, 162, 290; ore in, 144, 148, 151, 152; Pennsylvanian, **123**, 124, 153, 281, 282, 285; Proterozoic, 120, 295, **296**, 297, 298, 325. *See also* carbonates; Blackrock Canyon Limestone; Blacksmith Limestone; Lodgepole Limestone; Mission Canyon Limestone; Scott Peak Formation; Twin Creek Limestone; White Knob Limestone
lithosphere, 5, 12
Little Blacktail Mountain, 334
Little City of Rocks, **89**
Little Goose Creek, 277
Little Goose Creek Complex, 212, 272, 275, **277**, 278
Little Lost River Valley, 85, 144
Little Mountain, 138, **140**
Little Salmon Falls, 210, 211
Little Salmon River, 187, 208, 209, 211, 213, 216, 278
Little Wood River, 37, 58, 64, 72, 94, 143
Lizard Butte, **99**
Lochsa River, 255, 257, 258, 259, 260
Lockman Butte, 33
Lodgepole Limestone, 136
loess, 120, 122, 130, 156, 173, 174, 193, 227, 326, 328
Lolo Hot Springs batholith, 260
Lolo Pass, 260
Lone Pine, 148, 150, 153
Lone Pine basalt, **150**
Long Valley, 208, 254, 274, 275, 277
Lookout Pass, 305, 308, 315, 343

Lost River fault, 47, 105, 142, 144, 145, 146, 280
Lost River Playa, 85
Lost River Range, **28**, 47, 78, 83, **85**, **105**, 106, **112**, 142, 144, 145, **146**, 151
Lost River Valley, 78, 85, 107, 142, 144, 145, 286
Lost Trail Pass, 261, 265
Lowell, 257, 258, 272
Lucile Caves, 218
Lucile slate, 185, 186, 209, 214, 218
Lucky Friday Mine, 315
Lucky Peak Dam, 267

Mackay, 11, 105, 107, 142, 144, 145
Mackay mining district, 144
Mackay Reservoir, 144
Mackay stock, 143, 144
mafic sills, 295, 297, 298, 318, 323, 330, 331, 337, 339, 341, 347; metamorphosed, **301**, 316, **320**, 331
Magic Reservoir, 45, 46, 96
magma, 2, 4, 5, 6, 9, 10, 17, 18–20; chambers of, 2, 10, 17, 18, 68, 100, 183, 184, 222, 235, 259, 260, 341; and groundwater, 40, 42, 87, 99, 103; rhyolitic, 17, 18, 49, 56. *See also* Challis magmatic event
magmatic arcs, 155, 245, 249
magmatism, 10, 187, 207, 246, 249, 251, 253, 260, 278, 281, 290, 293, 300
magnesite, 217
Malad City, 114, 116, 117, 165, 172, 173, 176
Malad Gorge, **37**, **38**
Malad Range, 114, 116, 140, 176
Malad River, **37**, 38, 46, 58
Malad Summit, 117
Malad Valley, 114, 116, 117, 165, 176
Malm Gulch, 290
Malta, 122, 176, 177
mammals, 42, 92, 129, 227
mammoths, wooly, 85, 193, 227, 229
Mammoth Cave, 95
Mann Creek, 202
mantle, 5, 6, 12, 17, 19, 183, 188, 192, 217, 246, 249, 295
marble, 4, 185, 186, 214, 218, 236, 238, 282
Market Lake, 30
Marsh Creek, 117, 118, 119, 141, 180, 271
Marsh Valley, 117, 118, 120, 141
Marsing, 97, 99

Martin Bridge Formation, 183, 184, 185, 186, **205**, 207
Massacre Rocks State Park, 42
Massacre volcanic complex, 40, **42**, 173
Mazama ash, xiv, 207
McCall, 81, 254, 271, 274, 277
McCammon, 118, 119, 129, 130, 173
McGowan Creek Formation, 151
McKinney, Lake, 61, 62
McKinney Basalt, **26**
McKinney Butte, 35, 38, 58, 59, 61, 62
Meade thrust, 110, 132, 134, 136, 172
Meadow Lake Creek, 152
Meadows Valley, 208, 209, 211, 277, 278
Medicine Lodge Creek, 85, 86
Medicine Lodge Formation, 86
megafloods. *See* Bonneville Flood; Missoula floods
Melon Gravels, 25, **26**, 35, 42, 59, **60**, 66, **101**
Melon Valley, 66
megaripples, **226**, 336
Menan Buttes, 50, **87**
Mesa Falls, **54**
Mesa Falls Scenic Byway, 50, 53–54
Mesa Falls Tuff, 50, **54**, 55
metamorphic core complexes, 10, 113–14, 300–301. *See also* Albion core complex; Boehls Butte–Clearwater core complex; Pioneer core complex; Priest River core complex
metamorphic rocks, 4–5
metamorphism, 4–5, 188–86. *See also* amphibolite; gneiss; greenstone; marble; phyllite; quartzite; schist
metasedimentary, 103, 114, 178, 199, 239, 242, 259, 341
metavolcanic, 82, 218, 244
meteor impacts, 17, 147
miarolitic cavities, 252
mica, 178, 211, 247, 342. *See also* biotite; chlorite; granite, two-mica; muscovite
Mica Summit, 330
Michaud Flats, 119
Middle Butte, **14**, 26, 42, 49, 56, **57**
mid-ocean ridges, 5
Midvale Canyon, **204**
Midvale Hill, 202, 203
migmatite, **259**, 272, **276**, 321
Milligen Formation, 278, 280, 281, 282, 284
Mine Hill, 144
mineralization, 144, 148, 236, 287, 299, 305

mining, 110, 144, 148, 151, 152, 201, 242–43, 264, 268–69, 278, 291, 305, 309–15. *See also* gold; lead; phosphate; silver
Mink Creek, 168, 169
Minnetonka Cave, 136, 166
Minnie Moore Mine, 278, 279
Miracle Hot Springs, 59
Mission Canyon Limestone, 108, 156
Missoula, Glacial Lake, 193, 303, 304, 323, 324, 325, 332, 334, 336, 350
Missoula floods, 13, 194, 196–97, 302–4, 307, 336, 337; deposits of, **196**, **197**, 305, 307, 326
Missoula Group, 245, 246, 295, 297, 299, 325, **329**, 338, 349, 350
molar tooth structures, 295, **296**, 297, 298, 325
mollusks, 165
Mollys Nipple, **174**
molybdenum, 146, 289, 290
Monida Pass, 26, 31, 149
Montpelier, 110, 132, 134, 136, 137
Montpelier Canyon, **137**
Montpelier Reservoir thrust, 137
Moonstone Mountain, 45
moraines, 244, 271, 277, 281, 282, 284, 287, 304, 337
Mores Creek, 268, 269
Morning Mill, 311
Moscow, 326, 328, 343
Mountain Home, 33, 35, 43, 58, 90, 93
Moyie River, 341
Moyie thrust fault, 339, 341, 342
Mt. Bennett Hills, 35, 43, 45, 88, 89
Mt. Idaho fault, 227, 236
Mt. Idaho shear zone, 188, 227, 236, 237
Mt. Shields Formation, 297, 325
mud chips, 295, **296**, 325, 329
mud cracks, 263, 295, **296**, 298, 325
Mud Lake, 85, 148, 149
mudstones, 111, 125, **225**, 298, 299, 314, 323, **329**, **344**, **348**; diamictites, 8, 108, 117, 167; metamorphosed, 4, 216
Mullan, 309, 310, **311**, 343
Murphy, 103
muscovite, 215, 237, 249, 250, 255, 259, 275, 299, 328, 349
Mutual Formation, 109, 161
mylonite, 113, 199, 233, **236**, **277**, 301, 307, 308, 309, 330, **335**, 338

Naples, 339, 340
Nat-Soo-Pah Hot Springs, 67, 70
Newdale, 82
New Meadows, 202, 209, 236, 271, 272
Newport, 302, 316
Newport plate, 300, 318, 319; detachment of, 316, 318, 319
Nez Perce, 201, 202, 221, 242
Nicholia mining district, 151
Niter, 161
nodules, 170
normal faults, 6, 10, 11, 92, **93**, 106, 107, 131, 132, 155, 300; Eocene, 252, 260, 261, 263; lack of in Belt basin, 246, 247; bounding Snake River Plain, 16, 19, 31, 43, 66, 89, 267; of Basin and Range, 75, 81, 105, 106, 107, 113, 117, 141, 168, 172, 176, 208, 211, 220–21, 254, 274–75, 282; scarps of, **107**, **132**. *See also* Bear Lake fault; Beaverhead fault; Lemhi Pass fault; Lost River fault; Purcell Trench fault; Rexburg fault; Salmon River fault; Sawtooth fault; Snake River fault; Wasatch fault
North American Plate, 8, 9, 11, 12, 16, 17, 19, 45
Northern Idaho, 293–305
North Fork, 265
North Fork Pine Creek, 156
North Hansel Mountains, 175
North Menan Butte, 87
Notch Butte, 72, 88
Nounan Formation, 109, 170, 171, 172
Nugget Sandstone, 108, **134**, 135, 137
Nuna supercontinent, 247, 294

Oakley, 178
oil, 23, 110, 135, 136
Okanogan region, 299
Olds Ferry terrane, 182, 183, 202, 208, 210
olivine, 58, 217, 258
Onaway Volcanics, 329
oncolites, **170**
Oneida Narrows, 159, 161, **162**, 168
opal, **30**, 31
Oquirrh basin, 174, 176
Oquirrh Group, 109, 111, **123**, 124
ore, 10, 110, 146, 148, 152, 162, 184, 201, 284, 305, 309, 310, 311, 312. *See also* gold; mining; silver

Oregon Trail, 26, 35, 40, 42, 129, 130, 131, 132, 134, 174
Orofino, 199, 200
Orofino Bridge, 199
Orofino Metamorphic Suite, 186, 199, 200, 255, 257, 258
orogenies, 8, 9, 10. *See also* Antler orogeny; Laramide orogeny; Sevier orogeny
Osborne Bridge, 50
Osburn fault, 300, 308, 309, 310, 311, 313, 314, 315
Ovid, 136, 165, 172
Owyhee Highway, 97
Owyhee Mountains, 14, 16, 24, 33, 78, 97, 100, 103, 104
Owyhee River, 90
Oxbow, 183, 184, 205, 206
Oxford Peak, **117**, 141, 142, 168

Pacific Ocean (paleo), 7, 108, 238, 247
pahoehoe, **20**, 95, 103
Pahsimeroi Mountains, 146, 147
paleosols, 68, **69**, **124**, 227, 231, **232**, 342, 344
Palisades Reservoir, 128, 129
Palouse, 232, 326, 328, 329, 345
Palouse Formation, 195, 228, 306, 327. *See also* loess
Palouse Hills, 326, **328**
Palouse Range, 328
Panhandle, 13, 293, 299, 300, 301, 302, 309, 324, 330, 349
Paradise Ridge, 328
Paris, 135, 136
Paris Canyon, 133, 136
Paris-Putnam thrust, 132, 163
Paris thrust, 112, 116, 129, 130, 132, 133, 136, 172
Pass Creek, 144
passive margin, xiv, 108, 111, 281
Payette, 23, 81
Payette Formation, 189, 202, **204**, 208, 271, 273
Payette Lakes, 272, 277
Payette River, 78, 81, 212, 265, 266, 273; Middle Fork of, 251; North Fork of, 272, 274, 275, 277, 278; South Fork of, 270
Payette River Complex, 212
Peck, 199
pegmatite, 239, 259, 260
Pend Oreille, Lake, 13, 302, 316, 323, **324**, 325, 332, **334**, 336, 337, 339

Pend Oreille gneiss, 293, 297, 316, **320**, 338
peneplains, 253, 254
peridotite, 217, 258
Perrine Bridge, 67, 70, **71**
Petit Lake, 287
petroglyphs, **101**
Phi Kappa Formation, 282, 286
Phosphoria Formation, 8, 28, 109, 110, 134, 136, 151, 156, **158**, 161
phyllite, 218, 313
Picabo Hills, 47
Picabo Tuff, 97
Picabo volcanic field, 16, 28, 85, 97, 174
Piegan Group, 246, 260, 295, 2978, 298, 308, 313, 325, 338, 339, 347, 350
Pierce, 201, 254, 256
pillow basalt, 23, 59, 61, 62, **70**, 71, 126, 189, **190**, 221, 222, 227, 268, 269, **344**, 348, 359, 350
Pine Creek, 156, 157
Pine Creek Pass, 156, 159
Pinehurst, 311, 343
Pink House Recreation Site, 199
Pioneer core complex, 280, 281, 282, 286
Pioneer Intrusive Suite, 281
Pioneer Mountains, 6, 10, 40, 72, 74, 95, 114, 145, 244, 248, 249, 280, 281, 284
Pioneer thrust, 248, 249, 286
Pittsburg Landing, 193, 209, 220, 222, **224**, 225, 226, 236
placer deposits, 148, 162, 201, 218, 219, 238, 241, 242, 243, 253, 254, 265, 268, 269, 290, 310
plagioclase feldspar, 3, 57, 58, 78, 242, **251**, 252, 259, 329, 341
plant (fossils), 76, 161, 189, 263, 295, 302. *See also* leaves
Pleistocene, xiv, 12–13, 24, 136, 192–93, 227, 302–4, 323–24. *See also* basalt, Pleistocene; Bonneville, Lake; Bonneville Flood; glaciers; Missoula floods; Terreton, Lake
Plummer, 310, 329, 330
plutons, 2–3, 8, 182, 184, 185, 187; Eocene, 144, 148, 242, 260, 265, 270, 281, **282**, 300, 302, 316; mafic, 19, 199; tonalite, 211, 249. *See also* Almo pluton; Beaverhead pluton; Blacktail pluton; Copeland pluton; Harpster pluton; Kelso Lake pluton; Klopton pluton; Silver Point

pluton; Six Mile Creek pluton; Wrencoe pluton
Pocatello, 14, **28**, 118, 119, 120, 121, 131
Pocatello Formation, **7**, 8, 108, 109, 115, 117, **118**, 119, 120, 141, 167, 281
Poison Creek thrust fault, 261, 263
Pollock Mountain amphibolite, 214, 215
Pollock Mountain plate, 211, 212, 214; thrust of, 211, 215
Portneuf Gap, 7, 108, 118, 120, **121**, 131
Portneuf Range, 14, 42, 117, **118**, 119, **121**, 141
Portneuf River, 25, 119, 120, 129, 130
Post Falls, 304, 305, 307, 332, 336
potassium feldspar, **3**, **242**, **247**, **251**, 259, 260, 270, 288, 320, 323, 329, 341
Potlatch, 327, 329
Potlatch River, 342, 344
Powell, 258, 259
Preston, 25, 108, 114, 138, 140, 159, 165, 167
Preuss Formation, 109, 137
Preuss Range, 134, 136, 137
Prichard Formation, 246, 295, 297, 301, **314**, 324, 330, 337, 338, 341, 342; metamorphosed, 301, 302, 316, 330; ore in, 312; rusty weathering of, 297, 298, 308, 309, 318, 350; turbidites in, 295
Priest Lake, 13, 316, **318**
Priest River core complex, 293, 297, 300–302, 307–9, 318–19, 320–23, 330, 331, 335, 338, 350
Priest River valley, 318
Provo, Lake, 116, 138, 142, 161, 167; level of, **140**, **168**, **175**
pumice, 46, 116
Purcell anticlinorium, 342
Purcell lobe, 302, 304, 305, 325, 332, 334
Purcell Mountains, 248
Purcell sills, 295, 297, 298, 318, 323, 330, 331, 337, 339, 341, 347; metamorphosed, **301**, 302, 316, **320**, 331
Purcell Supergroup, 245, 294, 309
Purcell Trench, 293, 301, 323, 332, 338, **340**, 341; drainage divide in, 339–40; ice lobe in, 193, 302, 303, 316, 323, 332
Purcell Trench fault, 300, 301, 307, 308, 309, 323, 330, 332, 337, 338
pyrite, 235, 298, 311, 330, 341
pyroclastic, 17, 18, 68, 89, 124, 180, 261
pyroxene, 329, 347
pyrrhotite, 298, 311, 330

quartz, 2; in granitic rock, 2, 3, 201, 242, **250**, 260, 341; in gneiss, 322, 330; microcrystalline, 68, **69**; in rhyolite, 205, 235, **251**; in sand, **102**, 103, 111, 316, 330, 342; in schist, 211; shearing of, 255, 335; smoky, 260
quartz diorite, 184, 195, 199, 201, 229, 233, **235**, 255, 278
quartzite, 4, 76, 111, 178, 186, 214, 237, 261; in Belt Supergroup, 155, 238, 239, 245, 260, 297, 310, 313, 329, 337, 338, **347**, 350; breccia, **131**, 147; in Brigham Group, 117, 160, 162, 168, 172; Cambrian, 118, 261; cross-bedded, **246**, 328; in conglomerate, 248; in gravel, 87, 96; Ordovician, **77**, **131**; in till, 170; in Swauger Formation, **246**; in Windermere Supergroup, 255, **257**, 328. *See also* Clayton Mine Quartzite; Gold Cup Quartzite; Kasiska Quartzite; Kinnikinic Quartzite; Razorback quartzite
Quartzite Mountain, 264
Queens Crown, 47

Rabbit Spring, 68
Raft River Valley, 40, 123, 176
Railroad Ranch, 50
rapakivi, 259
Rapid River plate, 211, 212, 214, 215, 218; thrust of, 218
Rathdrum Prairie, 13, 302–5, 307, 324, 332, 334, 336
Rathdrum Prairie aquifer, 304–5, 310, 332, 336, 351
Ravalli Group, 260, 295, 297, 298, 308, 310, 313, 315, 324, 325, 329, 332, 338, 339
Razorback quartzite, 294, 297, 330
Redfish Canyon, 13
Redfish Lake, 271, 287, 288
Red Rock Pass, 24, 25, 117, 141, 166
Register Rock, 42
reverse faults, 6, 10, 231, **232**
Revett Formation, 297, 313, 325
Rexburg, 50, 52, 87
Rexburg fault, 50, 87
rhyolite, 2, 3, 12, 15, 16–19, **45**, **71**, **80**, **100**, 251; columns in, **180**; dikes of, **251**, 260, 274, 300; domes of, 26, **28**, 40, 42, 45, 49, 50, 56, **57**, 77, **97**, 162, **164**; flow folds, **46**, 89, **90**; geodes in, 68; oceanic, 184, 205, 218, 235, 236; opal in, 30, 31; quarry

in, 116. *See also* Jump Creek Rhyolite; ignimbrite; tuff, rhyolitic; Yellowstone hot spot, rhyolite of
rhythmites, 196, **197**
rifting, 6, 7, 8, 108, 182, 185, 216, 217, 238, 247, 255, 281, 299, 321, 328. *See also* Great Rift; Kuna–Mountain Home rift zone
Riggins, 184, 212, 214, 216
Riggins Group, 185, 186, 211, 215, 237, 258
Riggins Hot Springs, 214, 215
Riggins syncline, 216
ripples: in Belt rock, **263**, 294, **296**, 297, 298, 325; mega, **226**, 304, 336; in silt, 80, **81**, 159, **161**, **197**
rip-up clasts, 295, **296**, 325
Ririe Reservoir, 125
River of No Return, 216, 242, 243, 244, 264, 265
Ritter Island State Park, 59, 63, 64
Rock Creek (Twin Falls), 66, 70
Rock Creek (Rockland), 172, 173, 174
Rockland Valley, 172, 173
Rock Roll Creek, 286
Rocky Butte, 33, 72
Rodinia, xiv, 7, 8, 108, 185, 216, 238, 247, 299, 321, 328
Rogerson, 66, 69, 70
Rogerson graben, 66
Ross Fork Creek, 26
Ross Park, 119, 121
Round Mountain, 301, 304, 335
Round Valley, 275
rutile, 349

Saddle Mountains Basalt, 191, 192, 198, 202, 221, 233
Salmon, 156, 264, 265
Salmon Basin, 156, 264, 265
Salmon Basin detachment fault, 156, 264, 265
Salmon Butte, 67, 69, 70
Salmon Falls Creek, 66
Salmon Falls Dam, 59, 70
Salmon River: ancestral, 23, 155, 192; canyons of, 216, 219, 220, 227, **230**, 253, 254; dams along, 288; East Fork of, 146; along faults, 261, 263, 264; headwaters of, 148, 287; Middle Fork of, 264; rocks along, 212, 214–16, **220**, **253**, **261**, **265**, 287, **290**; terraces of, 218, 219; time zone along, 216; Yankee Fork of, 161, 288, 289

Salmon River assemblage, 290, 291
Salmon River fault, 263, 264
Salmon River Mountains, 146, 244, 245, 263, 275, 288
Salmon River Road, 214–16
Salmon River suture zone, xiv, 182, 184–88, 202, 209, 211, 212, 214, 216, 217, 233, 234, 237, 275, 276, 278
salt casts, 299, 325, **329**
Salt Lake Formation, 117, 129, 134, 159, 165, 167, 168, **169**, 173
Samaria Mountains, 116, 176
sand dunes, 4, 50, 85, **102**, 103
Sandpoint, 245, 294, 300, 302, 316, 323, 332, 334, 337, 339
sandstone, xiv, 3, 7, 111; in Belt basin, 298, 299, 308, 313, 323, 325, 329; Cambrian, 116, 168, **169**; Cretaceous, 157, 159; Jurassic, **134**, 135, 137, 184, 214, 224, 225; metamorphosed, 4, 176, 216, 235, 313, 330; Miocene, 32; Mississippian, 76; Ordovician, 108; Pennsylvanian, 124, 137; Pliocene, **35**; Proterozoic (not Belt), 119, 210, 161, 255, 294, 316, 330; Triassic, 207, 224. *See also* cross bedding; quartzite
Sawtell Peak, 55, **56**
Sawtooth fault, 271
Sawtooth granite, **13**, 287
Sawtooth pluton, 270
Sawtooth Range, 3, 10, 40, 106, 148, **244**, 252, 254, 255, 265, 270, 271, 287, **288**
Sawtooth Valley, 271, 287
scabland, 13, 38, 72, 303, 304
scarps, 107, 131, 132, 145, 146, 161, 218, 341
schist, 4, 178, 180, 185, 216, 236, 237, 239, 240, 241, 257, 260, 297, 307, 316, 328, 338, 349. *See also* Fiddle Creek Schist; Lightning Creek Schist; Squaw Creek Schist
Scott Peak Formation, 47, 78, **112**, **150**, 151
Scout Mountain Member, **7**, 109, **118**, 119, 120, 121
sedimentary interbeds, 17, 189, **190**, 198, 199, 233, 342, **344**, 345, **346**
sedimentary rocks, 3–4
Selkirk Mountains, **292**, 299, 301, **318**, 323, 337, 338, 339
Selway River, 257
serpentine, 217
Seven Devils Group, 184, 200, 205, 209, 233, 234

Seven Devils Mountains, 181, 184, 192, 193, 213, 230
Sevier fold-and-thrust belt, 10, 111–13, 125, 153, 156, 165, 299, 341
Sevier orogeny, xiv, 10, 83, 110, 111–13, 136, 146, 147, 185, 249, 252, 281, 293, 299, 302, 322
shale, 3, 7, 111; of Belt basin, 298, 299, 308, 313, 315, 325, **326**, 329, 349; black carbonaceous, 183, 224, 278, 280, 281, 325; Cambrian, 170, **171**, 172; Cretaceous, 157, 159, 164; Jurassic, 183, 184, 222, 224, **225**; metamorphosed, 4, 235, 330; Miocene, 121; Mississippian, 76, 290, **291**; phosphatic, 134, 156; Triassic, 156. See also Woodside Shale
shear zones, 218, 219; Ahsahka, 188, 199, 200, 257, 343; Mt. Idaho, 188, 227, 236, 237; Western Idaho, 188, 214, 216, 237, 238, 271, 275, 276, 277
She Devil, 181, 210
Sheep Rock, 130, 131, **132**, 161, 206
Shepard Formation, 297, 325, **326**
shield volcanoes, 20, **33**, 35, 39, 40, 49, **57**, 59, 61, 69, 72, 86, 88, 94, 95, 100, 103
shorelines, 114, 116, 122, 125, 138, 173, 196, 304
Shoshone, 72, 94, 96
Shoshone County Mining and Smelting Museum, 312
Shoshone Falls, 13, 39, 70
Shoshone Ice Caves, 94, 95
Shoshone lava field, 94, **95**
Shoshone tribe, 85, 155
siderite, 278, 311
Silent Cone, 74, 76
silica, 2, 4, 15, 17, 31, 68, 77, 97, 184, 199, 202
silicate, 2, 199, 211, 215, 217, 239, 334, 347
sillimanite, 216, 257
sills: andesite, 126, **128**; granitic, **240**, **259**, 260, 330, 341, 347; mafic, 295, 297, 298, 318, 323, 330, 331, 337, 339, 341, 347; metamorphosed, **301**, 316, **320**, 331
siltite, 151, 245, 297
silts, **71**, **81**, **93**, **161**, **197**
siltstone: of Belt basin, 245, 297, 298, 299, 308, 310, 315, 325, 329, 330, 337, **341**, 349, 351; Miocene, 342; Paleozoic, 284, 286, 290; Triassic, 134, 137; of Wallowa terrane, 183, 214, 224, 235

silver, 10, 94, 148, 151, 152, 205, 207, 264, 278, 281, 284, 287, 290, 293, 305, 311–13
Silver City, 98
Silver Creek, 44, 94, 278
Silver Point pluton, 302, 316, 318
Silverton, 313
Silver Valley, 311, 313
Sinker Creek, 103
Six Mile Creek pluton, 187, 201
Skeleton Butte, 39
Skull Canyon, 148, **150**
slate, 4, 185, 186, 209, 214, 218, 219, 282
Slate Creek, 219
slickensides, 157, **159**, 281, 286
Smiley Creek Conglomerate, 146
Smith Prairie, 267
Smiths Ferry, 274, 275
Smoky Mountains, 252, 281
Snake River, **26**, 30, 31, 40, 59, 81, 127, **181**, 202, **205**, **207**, **224**; Bonneville Flood in, 24–25, 38–39, 64, 65, 66, 71, 72, 193, **194**, 225, **226**; canyons of, 13, 35, 39, 58, 59, 66, 70–71, **71**, 85, 125; drainage development of, xiv, 12, 24, 192, 254; irrigation with, 56, 58, 93; landslides along, 61–62, **61**, **104**; lava dams in, 35, 39, 58; Missoula floods in, **194**, 196, 197; sediments of, 50, 87; South Fork of, 125. See also Hells Canyon; Henrys Fork
Snake River fault, 126
Snake River Plain, xiv, 1, 3, 12, **14**, 15–22, 106; aquifer of, 22, 49, 58, 63, 85; axis of, 30, 49, 56; Basin and Range extension in, 107; subsidence of, 19, 28, 45, 89; western, 22–24, 33, 35, 99, 100
Snake River Range, 125, 128, 129, 156
Snake River transfer, 248
Snowball Earth, xiv, 7, 108, 121, 167
Snowbank Mountain, 275, 276
Snowslip Formation, 297, 325, 349
Soda Point, 130, 131
Soda Springs, 8, 110, 129, 132, 162, 174
Soda Springs geyser, **132**
soils, 26, 31, 81, 148, 229, 318, 329. See also paleosols
Soldier Mountains, 45, 90
South Fork of the Clearwater River, 187, 233, 235, 238, 241
South Fork of the Payette River, 270
Spalding, 231
spatter cones, 74, 75

Spencer, 30, 31, 51
sphalerite, 152, 311
spinel, 217
Spirit Lake, 304, 305, 336
Spokane dome mylonite zone, 301, 307, 308, 309, 319, 330, 335, 338
Spokane River, 305, 307, 309, 351
Spokane Valley, 304, 336
Spring Valley Summit, 271, 273
Squaw Creek Schist, 185, 186, 211, 212, 213, 214, 216, **217**
St. Anthony dunes, 50, 82
St. Charles, 135, 136
St. Charles Canyon, 136
St. Charles Formation, 109
St. Joe fault, 301, 330, 350
St. Joe River, 350
St. Mary–Moyie transfer, 248
St. Maries, 349, 350
St. Maries River, 347, 349, 350
St. Regis Formation, 297, 313, 325
Stanley, 11, 251, 271, 287, 288
Stanley Lake, 271, 287
star garnets, 349
Starlight Formation, 121, 172, **174**
Star Valley fault, 125
steptoe buttes, 189, 328, 345
Stibnite, 251
Stites, 256
stocks, 144, 152, 205, 207, 224, 278, 329
Strawberry Summit, 170
striations, **13**, 325
strike-slip faults, 6, 247, 248, 301, 330, 350
Striped Peak Formation, 329
stromatolites, 170, 295, **296**, 297, 298, 325, **326**
stumps, 290
Sturgill Peak, 208
subduction, xiv, 5, 6, 8, 9, 10, 11, 17, 100, 105, 111, 182, 183, 187, 208, 249, 299
Sublette Mountains (WY), 135, 137
Sublett Range (ID), 40, 123, 172, 174
Succor Creek Formation, 78, **80**
sulfide mineral, 295, 309
Summit Creek, 281
Sunbeam Hot Springs, 288
Sunset Cone, 72, 74, 77
Sunshine Mine disaster, 313, 314
Sun Valley, 174, 281, 284, 286
Sun Valley Group, 8, 47, 278, **280**, 282

supercontinents, xiv, 7, 8, 108, 185, 216, 238, 245, 247, 293, 294, 299, 321, 328
suture zone, xiv, 8, 9, 182, 184–88, 214–16, 249–50; high-grade rocks in, 211, 237, 276; below Lewiston Hill, 231, 233, 342; plutons in, 201, 208, 216, 234, 275, 278; ultramafic rocks in, **217**, 258; thrusts of, 212, 220; Woodrat Mountain, 255
Swan Peak Formation, 109, **131**, 170
Swan Valley, 50, 125, 128, 156, 157
Swauger Formation, 151, 245, **246**, 261, 265
Sweetwater Creek interbed, 198
synclines, 83, **85**, 116, 146, 170, 198, 216, 229, 230, 284, 285, 286, 327
Syringa, 237
Syringa fault, 257

Table Legs Butte, 56, **57**
Table Rock, 32
Table Mountain, 41
tailings, 309, 310, 311, 313, 314, 315
talc, 217, 258
talus, 284, 286
tectonic plate, 1, 5, 6, 16, 245, 248, 299
Teewinot Formation, 129
Tempskya, 162, 163
Tendoy, 155
Tensed, 329, 343
terraces, 33, 124, 132, 155, 218, 267, 326, **340**, 341
terranes, 10, 181, 182–84. *See also* Baker terrane; Olds Ferry terrane; Wallowa terrane
Terreton, Lake, 28, 30, 50, 85, 86, 148
Teton Dam, **52**
Tetonia, 55, 81, 82, 83
Teton Range, 31, 50, 81, 82, **83**, 159
Teton Valley, 82, 156, 159
tetrahedrite, 311
Thatcher, Lake, 161
Thaynes Formation, 109, 134, 137
Thompson Creek Mine, 289, 290
Thousand Springs, 59, 63, **64**
Thousand Springs Scenic Byway, 58
Thousand Springs State Park, 37, 58, 63, 64
Thousand Springs Valley, 145
thrust faults, xiv, 6, 10, 111–13, 156, **219**, 248, 249, **291**, 299. *See also* Absaroka thrust; Bear Canyon thrust; Cabin thrust; Copper Basin thrust; Hawley Creek thrust; Heavens Gate thrust;

Hellhole thrust; Home Canyon thrust; Iron Lake thrust; Klopton Creek thrust; Meade thrust; Medicine Lodge thrust; Moyie thrust; Mt. Idaho thrust; Paris thrust; Pioneer; Poison Creek thrust; Pollard Mountain thrust; Putnam thrust; Rapid River thrust; Willard thrust
Thurmon Ridge, 50, **56**
till, **7**, 82, 167, 270, 318, 341
Time Zone Bridge, 212, 216, 217
Timmerman Hills, 46, 97
Tincup Creek, 162
titanite, 334
Tolo Lake, 227, 229
tonalite, 2; in Idaho batholith, 238, 249, **258**, 274; in Kaniksu batholith, 299; in suture zone, **187**, 211, **216**, **237**, 255, **258**, 278; in Wallowa terrane, 184, **219**, **236**
Trail Creek, 281, 286
Trail Creek Road, 145, 284–86
Trail Creek Formation, 282, 286
Trail of the Coeur d'Alenes, 310, **313**
Trans-Challis fault zone, 252, 261, 270
travertine, 86, **132**
Treasureton Summit, 159, 161
Treasure Valley, 31, 78
tree mold, **75**, 191
Trestle Creek, 323
Triumph Mine, 281
trondhjemite, 186, 187, 200, 201, 209, 211, 214, 216, 237, 238
Tubbs Hill, 308, 309, 332
tuff, 18, 19, **53**, 85, 55, 99, 121, 123, 152; basaltic, 42, 99, 189, **190**, 192, 202, 211; Eocene, 260, 261, 263; Jurassic, 224; Proterozoic, 108, 120; rhyolitic, 40, 47, 66, 86, **89**, 125, 168, 172, 205; in Salt Lake Formation, 134, 165, 168; welded, **45**, 46, 124, 174, **175**, 263. *See also* Huckleberry Ridge Tuff; Kilgore Tuff; Lava Creek Tuff; Mesa Falls Tuff; Picabo Tuff; Walcott Tuff
tuff cones, 50, **87**
tumuli, 57, 72
turbidite, **76**, 145, 248, 295
Tuttle, 37
Twin Creek Limestone, 109, 135, 137
Twin Falls, 24, 38, 39, 66, 70, 113
Twin Falls volcanic field, 16, 43, 66, 69, 70, 89, 93, 124, 180

Twin Peaks, 146, **147**, 261
Twin Peaks caldera, 146, 147, 261

Uinta Mountains, 131
ultramafic, 217, 258
unconformity, 109, 151, 245, 280, 282
Union Pacific Railroad, 31, 58, 134, 138
Upper Reynolds Creek, 100
uranium-lead dating, 249, 301

veins, 148, 152, 187, 201, 205, 236, 242, 278, 281, 286, 305, 310, 311, 329
vertebrate fossils, 22, 129
vesicles, 57, 76, 103, 180, **190**, 204, 211
Victor, 82, 156, 157
Vienna Mine, 279, 287
Vinegar Creek, 214, 216
vitrophyre, **45**, **68**, **69**, 78, **80**, **124**, 180
volcanic arcs, 182, 202, 207, 224
volcanism, 11, 12, 14, 15, 19, 40, 66, 69, 107, 108, 147, 183, 192, 281
volcanoes, 2, 6, 12, 17, 18, 20, 182, 184; subaqueous, 59, 60. *See also* calderas; cinder cones; domes; shield volcanoes; tuff cones

Waha escarpment, 198, 229, 231, 232, 233
Walcott Tuff, **86**
Wallace, 314
Wallace District Mining Museum, 314
Wallace Formation, 297, **298**, 313, 314, **315**, 325, **347**, 349, 350
Wallowa island arc, 184, 207, 222, 231, 235, 255
Wallowa terrane, 182, 183–86, 188, 192, 205–7, 215, 218, **219**, 222–26, 258; boundary of, 182, 185, 238, 255, 278; ore in, 236; plutons intruding, 199, 201, 211, 227, 229. *See also* Coon Hollow Formation; Hunsaker Creek Formation; Orofino Metamorphic Suite; Riggins Group; Seven Devils Group
Wallula Gap, 193, 196, 197
Walters Bar, 101
Wanapum Basalt, 191, 192, 198, 231, **232**, 302, 307, 329, 330, 332, 345, 347, 348, 349, **350**
Wapi lava field, 40, 74
warbonnet, **345**
Warm River, **53**, 54, 50

Wasatch fault, 114, 116, 140, 141, 165, 172
Wasatch Formation, 109, 135
Wasatch Range, 114
Wayan Formation, 109, 162, 163
Weatherby Formation, 182, 183, 185, 200
weathering, 57; along fractures, 178, 179, 211; resistance to, 224, 264; of feldspar, 239, 251; rusty, 297, 308, 316, 318, 322, 323, 350; during tropical climate, 347. *See also* exfoliation; grus
Wedge Butte, 94, **97**
Weir Hot Springs, 256, 260
Weiser, 22, 81, 202
Weiser embayment, 189, 191, 202, 208
Weiser River, 78, 81, 202, 204, 208, 272
Weiser volcanic sequence, 191, 192, 204, 271; basalt of, 202, **204**, 208
welded tuff, **45**, 46, 124, 174, **175**, 263
Wells Formation, 109, 137
Wendell, 59, 64, 88
Western Idaho shear zone, 188, 214, 216, 237, 238, 271, 275, 276, 277
West Mountains, 202, 208, 275, 276
Weston, 165, 167
Weston Canyon, 165, **167**
West Yellowstone, 50, 55
White Bird, 217, 218
White Bird Battlefield, 210, 220, 221
White Bird fault, 220, 221, 227
White Bird Hill, 221, 227
White Cloud Mountains, 146, 287
White Knob Limestone, 144, 145, 151
White Knob Mountains, 142, 144
Wilbert Formation, 261
Wildhorse detachment fault, 281, 282
Wildhorse gneiss complex, 282
Wild Sheep Creek Formation, 184, 205, 207, **213**, 218, 222, **224**, 233
Willard thrust fault, 116
Williams Lake, 264

Wilson Creek Member, 286
Winchester Grade, 229
Windermere Supergroup, xiv, 7, 108, 109, 130, 216, 238, 247, 255, **257**, 258, 281, 299, 328
wind gaps, 272, 275, **276**, 277
Windy Ridge Formation, 183, 184, 205
Woodrat Mountain suture, 200, 255, 257
Wood River Formation, 278, 281, 282, 284, 285, 286
Wood River mining district, 278
Wood River Valley, 46, 97, 251, 280, 281, 282
Worley, 329, 330, 343
Worm Creek, 140
Wrencoe pluton, 302, 322, 323
Wyoming Province, 31

xenolith, **341**

Yahoo Clay, 61, **62**
Yankee Fork, 261, 288, 289
Yellowjacket mining district, 264
Yellowstone caldera, 11, 54, 55
Yellowstone hot spot, xiv, 12, 16–19, 50, 55, 107, 188, 254; rhyolite of, 24, 35, 43, 66, 72, 123. *See also* Bruneau-Jarbidge volcanic field; Heise volcanic field; Henrys Fork caldera; Island Park caldera; Picabo volcanic field; Twin Falls volcanic field
Yellowstone National Park, 12, 15, 19, 55, 63, 83

zeolites, **103**, 211, **212**
Zims Hot Springs, 210, 211
zinc, 10, 94, 278, 293, 305, 309, 311, 312
zircons, 113, 151, 155, 247, 249, 264, 301
zoisite, 215

Paul K. Link began teaching geology at Idaho State University (ISU) in 1980. He has a BS from Yale, a BSc (Hons) from the University of Adelaide, South Australia, and a PhD from the University of California, Santa Barbara. Before coming to ISU, he was a mountaineering instructor at the National Outdoor Leadership School in Lander, Wyoming, and he returned to the Wind River Range in the 1980s to conduct geological research with students. For 15 years, starting in 2003, he directed the ISU Geology Field Camp at the Lost River Field Station north of Mackay, Idaho. He is coauthor of the *Geologic Map of Idaho*, published in 2012 by the Idaho Geological Survey, and *Rocks, Rails, and Trails*, published by the Idaho Museum of Natural History. Link's career in sedimentary geology was inspired by Lehi Hintze, the careful stratigrapher from BYU; Steve Oriel of the US Geological Survey, a demanding thrust belt mapper; and Betty Skipp, also of the US Geological Survey, whose zest for geology and life inspired Link and countless ISU students.

Shawn Willsey, a geology professor at the College of Southern Idaho, also teaches whitewater rafting and rock climbing courses. Shawn lived all over the western United States during his childhood as an Air Force brat. He earned a BS in geology from Weber State University and a MS in geology from Northern Arizona University. Shawn has traveled and led geologic field trips to a variety of locations including Scotland, Iceland, Hawaii, and Costa Rica. He is the author of *Geology Underfoot in Southern Idaho*, now in its second printing. Shawn lives in Twin Falls with his wife, Erika, and three children.

Keegan Schmidt is a professor of earth science at Lewis-Clark State College in Lewiston, Idaho. He earned a BA from the University of Colorado, MS from Idaho State University, and PhD from the University of Southern California. He has worked for the Idaho Geological Survey numerous summers mapping Idaho's spectacular geology. Highlights of his career include working in the Peninsular Ranges batholith of Baja California on a suture zone analogous to the one in western Idaho. One of Keegan's passions is leading field trips for students and community members to showcase the geology of Idaho and the Pacific Northwest. He lives in Lewiston—with his wife, Karen, and son, Liam—where he enjoys recreational pursuits on the rivers and in the mountains.